SYSTEMS BIOLOGY

Applications in Cancer-Related Research

SYSTEMS BIOLOGY

Applications in Cancer-Related Research

Edited by

Hsueh-Fen Juan

National Taiwan University, Taiwan

Hsuan-Cheng Huang

National Yang-Ming University, Taiwan

NEW JERSEY · LONDON · SINGAPORE · BEIJING · SHANGHAI · HONG KONG · TAIPEI · CHENNAI

Published by

World Scientific Publishing Co. Pte. Ltd.

5 Toh Tuck Link, Singapore 596224

USA office: 27 Warren Street, Suite 401-402, Hackensack, NJ 07601

UK office: 57 Shelton Street, Covent Garden, London WC2H 9HE

British Library Cataloguing-in-Publication Data
A catalogue record for this book is available from the British Library.

SYSTEMS BIOLOGY
Applications in Cancer-Related Research

ISBN-13 978-981-4324-45-8
ISBN-10 981-4324-45-0

Typeset by Stallion Press
Email: enquiries@stallionpress.com

Printed by FuIsland Offset Printing (S) Pte Ltd Singapore

CONTENTS

CONTRIBUTORS

Stuart Brown: New York University Langone Medical Center, New York, NY 10016, USA

Hsin-Yi Chang: Department of Life Science, Institute of Molecular and Cellular Biology and Graduate Institute of Biomedical Electronics and Bioinformatics, National Taiwan University, Taipei 106, Taiwan

Jer-Wei Chang: Department of Pharmacology and Institute of Basic Medical Science, National Cheng Kung University, Tainan 701, Taiwan

Chien-Yu Chen: Department of Bio-Industrial Mechatronics Engineering, National Taiwan University, Taipei 106, Taiwan

I-Hsuan Chen: Department of Chemistry, National Taiwan Normal University, Taipei 106, Taiwan

Pei-Mien Chen: Agricultural Biotechnology Research Center, Academia Sinica, Taipei 115, Taiwan

Yu-Ju Chen: Institute of Chemistry, Academia Sinica, Taipei 115, Taiwan

Lihua Cheng: Systems Biology Division, Zhejiang-California International NanoSystems Institute, Zhejiang University, Hangzhou, Zhejiang 310029, China

Chih-Wei Chien: Department of Chemistry, National Tsing Hua University, Hsinchu 300, Taiwan

Chia-Li Han: Institute of Chemistry, Academia Sinica, Taipei 115, Taiwan

Chuan-Chih Hsu: Institute of Chemistry, Academia Sinica, Taipei 115, Taiwan

Chun-Hua Hsu: Department of Agricultural Chemistry, and Program of Genome and Systems Biology, National Taiwan University, Taipei 106, Taiwan

D. Frank Hsu: Department of Computer and Information Science, Fordham University, New York, NY 10023, USA

Hsien-Da Huang: Department of Biological Science and Technology, and Institute of Bioinformatics and Systems Biology, National Chiao Tung University, Hsinchu 300, Taiwan

Hsuan-Cheng Huang: Institute of Biomedical Informatics, Center for Systems and Synthetic Biology, National Yang-Ming University, Taipei 112, Taiwan

Tsui-Chin Huang: Department of Life Science, Institute of Molecular and Cellular Biology, and Graduate Institute of Biomedical Electronics and Bioinformatics, National Taiwan University, Taipei 106, Taiwan

Hsueh-Fen Juan: Department of Life Science, Institute of Molecular and Cellular Biology, and Graduate Institute of Biomedical Electronics and Bioinformatics, National Taiwan University, Taipei 106, Taiwan

Biaoyang Lin: Systems Biology Division, Zhejiang-California International NanoSystems Institute, Zhejiang University, Hangzhou, Zhejiang 310029, China; Swedish Neuroscience Institute, Swedish Medical Center, Seattle, WA 98122, USA; Department of Urology, University of Washington, Seattle, WA 98195, USA

Tsunglin Liu: Institute of Bioinformatics and Biosignal Transduction, National Cheng Kung University, Tainan 701, Taiwan

Christina Schweikert: Department of Computer and Information Science, Fordham University, New York, NY 10023, USA

Arthur Chun-Chieh Shih: Institute of Information Science and Research Center for Information Technology Innovation, Academia Sinica, Taipei 115, Taiwan

Yu-Ni Sun: Institute of Chemistry, Academia Sinica, Taipei 115, Taiwan, Institute of Bioscience and Biotechnology, College of Life Science, National Taiwan Ocean University, Taiwan

Zuojian Tang: New York University Langone Medical Center, New York, NY 10016, USA

Chia-Feng Tsai: Institute of Chemistry, Academia Sinica, Taipei 115, Taiwan

Feng-Sheng Wang: Department of Chemical Engineering, National Chung Cheng University, Chiayi County 62102, Taiwan

Yi-Ching Wang: Department of Pharmacology and Institute of Basic Medical Science, National Cheng Kung University, Tainan 701, Taiwan

Yi-Ting Wang: Institute of Chemistry, Academia Sinica, Taipei 115, Taiwan, Institute of Biochemical science, National Taiwan University and chemical Biology and Molecular Biophysics, Taiwan International Graduate Program, Taiwan; Institute of Bioscience and Biotechnology, College of Life Science, National Taiwan Ocean University, Taiwan

Hsin-Yi Wu: Institute of Chemistry, Academia Sinica, Taipei 115, Taiwan

Wu-Hsiung Wu: Department of Computer Science and Information Engineering, National Chung Cheng University, Chiayi County 62102, Taiwan

Hong Xu: Systems Biology Division, Zhejiang-California International NanoSystems Institute, Zhejiang University, Hangzhou, Zhejiang 310029, China

Part I

Gene/Protein Networks and Pathways

Chapter 1

Introduction to Systems Biology

Hsueh-Fen Juan[*]

1. What is Systems Biology?

Systems biology, which combines computational and experimental approaches to analyzing complex biological systems (Fig. 1), focuses on understanding functional activities from a systems-wide perspective and can be applied to many organisms from bacteria to man. This holistic approach enables rapid advances in elucidating biochemical pathways, regulatory networks, and disease therapies. With the advent of high-throughput global gene expression, proteomics, and metabolomic technologies, systems biology has become a viable approach to improve our knowledge of health and disease.[1,2] To integrate the data generated from these high-throughput technologies, there is much academic focus on developing fundamental, computational, and informatics tools into models of regulatory networks and cell behavior. Classic bench-top experiments are used hand-in-hand to provide high-throughput data and validate simulated predictions. Through an iterative process of experimental testing, computational simulation, and data analyzing, the model is refined to more closely reflect the biological reality (Fig. 1).

2. Network Biology and Pathway Modeling

Systems biology simultaneously explores the complex interaction of many levels of biological information to understand how they work together.

[*]Department of Life Science, Institute of Molecular and Cellular Biology, and Graduate Institute of Biomedical Electronics and Bioinformatics, National Taiwan University, Taipei 106, Taiwan. yukijuan@ntu.edu.tw

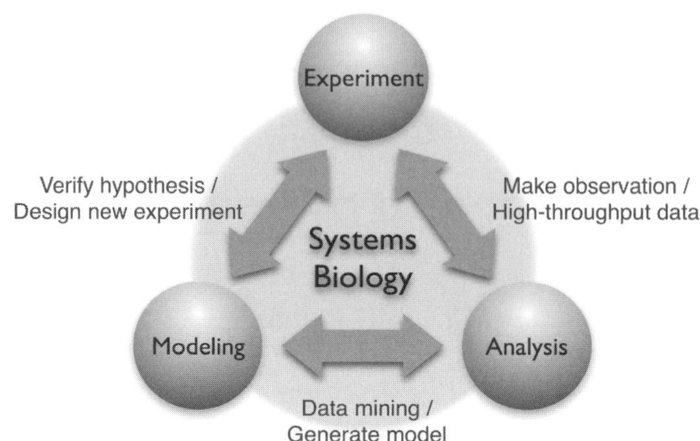

Fig. 1. Diagram of systems biology. Systems biology is a combined computational and experimental approach for analyzing complex biological systems. It provides an iterative process of experimental testing, computational simulation, and data analyzing to model biological behavior.

Networks of interacting proteins can provide researchers with a basic understanding of the molecular mechanisms of cell and tissue function. Network analysis has been performed on large-scale medical data, capturing the global properties of drugs, targets, and disease relationships.[3,4] Here, we organize the subject of network biology and pathway modeling into four categories: (1) gene network construction for molecular regulation, (2) microRNA-regulated cellular networks in cancer, (3) disease-related modules in protein interaction networks, and (4) biomolecular pathway modeling.

2.1. *Gene network construction for molecular regulation*

Most biochemical relationships among genes, proteins, and other organic substrates are described as being many-to-many, meaning that one component can have many functions and that one function can be influenced by many components.[5] To understand these complex relationships, we first need to identify the structure of a biological system, such as the regulatory relationships between a set of genes. Construction of gene networks from time-course data, generated from high-throughput experimental technologies, such as genomics, transcriptomics, proteomics, and metabolomics, can help to understand molecular regulatory processes in living organisms (Fig. 2). In Chapter 2, Juan and Huang introduce the gene network models

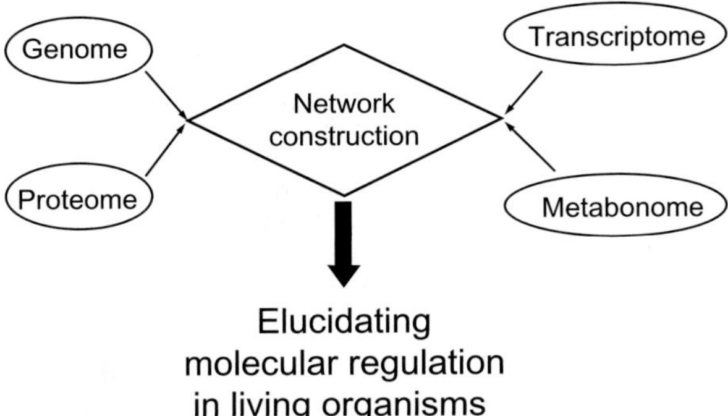

Fig. 2. The importance of omics data in network construction. High-throughput experimental data generated from genomics, transcriptomics, proteomics, and metabolomics are used to construct networks that enable us to elucidate molecular regulation in living organisms.

and describe how to construct a gene network for elucidating molecular regulatory mechanisms.

2.2. *MicroRNA-regulated cellular networks in cancer*

MicroRNAs (miRNAs) are short ($\sim$22 nt) single-stranded noncoding RNAs that play a key role in post-transcriptional regulation of mRNAs. By virtue of base complementarities, they bind to the 3′UTR of their target mRNAs and interfere with translation or accelerate the degradation of the target mRNAs (Fig. 3). The discovery of miRNAs changed the traditional view that gene expression regulators are proteins and revealed an additional

A. Translation repression

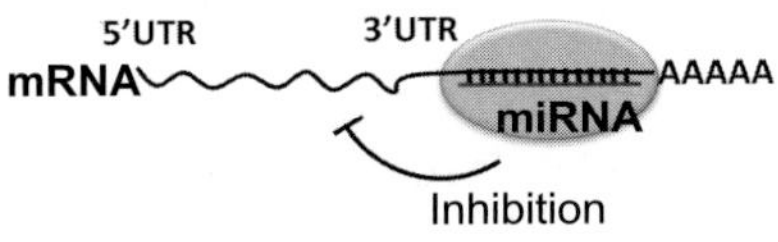

B. mRNA degradation

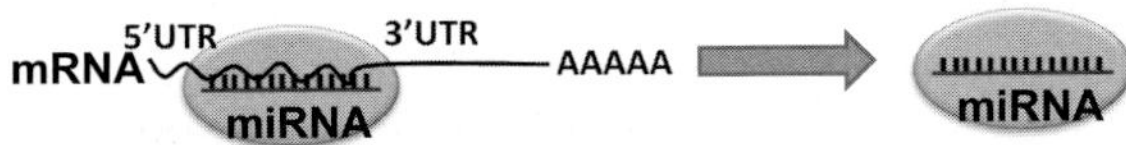

Fig. 3. The interaction between miRNA and its target gene. miRNA binds to the 3′UTR of its target mRNA and interfere with translation (A) or accelerate the degradation of the target mRNA (B).

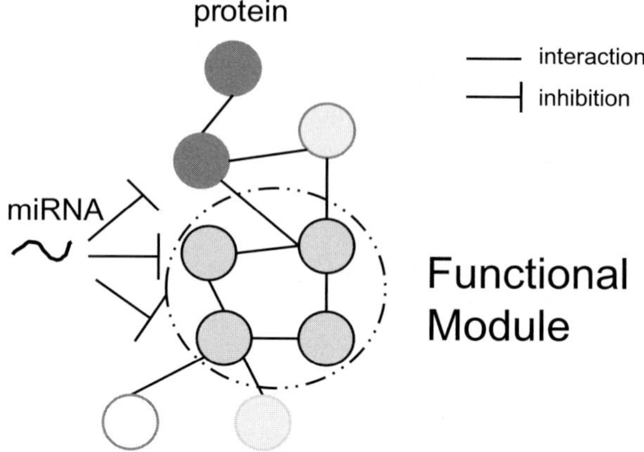

Fig. 4. An miRNA-regulated network. An miRNA regulates its target genes/proteins and their interacting partners that form a functional module.

layer of gene regulation. Interest in miRNAs was further increased by the finding that miRNAs play important roles in many biological processes, including development, differentiation, proliferation, and apoptosis, implicating their involvement in various diseases such as cancer.[6,7] Topological analysis to elucidate the global correlation between miRNA regulation and a protein–protein interaction network in humans revealed that target genes of individual miRNAs tend to be hubs and bottlenecks in the network.[8] Although proteins directly regulated by miRNAs might not form a network module, the target genes and their interacting partners might form functional modules (Fig. 4). In this way, miRNAs could influence specific biological functions through regulating a smaller number of selected genes. In Chapter 3, Juan and Huang introduce the microRNA-regulated cellular networks in cancer.

2.3. *Disease-related modules in protein interaction networks*

Protein interaction networks are sets of interactions formed by two physically interacting proteins and are fundamental to most biological processes in organisms.[9] With the accumulation of protein–protein interaction data, understanding the architecture and function of the cellular network becomes possible.[10] Moreover, computational approaches have allowed the discovery of the functional modules involved in these networks.[11]

Lin *et al.* developed a network-based comparative analysis approach that integrates protein–protein interactions with gene expression profiles and biological function annotations to reveal dynamic functional modules under different biological states.[4] Lin *et al.* found that hub proteins (proteins with more interacting partners), in condition-specific co-expressed protein interaction networks tend to be differentially expressed between biological states. Applying this method to a cohort of patients with heart failure, they identified two functional modules that significantly emerged from the interaction networks. The dynamics of these modules between normal and disease states further suggest a potential molecular model of dilated cardiomyopathy.[4] Juan and Huang introduce the disease-related modules in protein interaction networks in Chapter 4.

2.4. *Biomolecular pathway modeling*

Biomolecular pathways are building blocks of cellular biochemical function. Biomolecular pathway modeling is useful to clarify the relationship between structure, function, and regulation in complex networks that can be reconstructed from genomic or biochemical data. A mathematical model can be designed based on biological knowledge derived from experimental data. Determining the parameters for a mathematical model from quantitative measurements is the main bottleneck in modeling biological systems. In conventional parameter estimation, numerical integration is required to yield dynamic profiles for the model, but this approach is time-consuming. Numerical integration failure constitutes the main problem in the optimization search. Fortunately, experimental verification can help to improve the model. In general, mechanistic mathematical modeling could be used to find key mechanisms in complex biomolecular pathways/networks; therefore, these mechanisms could be potentially exploited for therapeutic benefit in cancer and other diseases. In Chapter 5, Wang and Wu introduce how to model the biomolecular pathway and describe some examples.

3. High-Throughput Omics Data and Analysis

Omics is an umbrella term referring to related fields of study in biology ending in -omics, such as genomics, transcriptomics, proteomics, metabolomics; the related suffix -ome is used to address the objects of study of such fields, such as genome, transcriptome, proteome, and metabolome, respectively. The first draft sequence of the human genome was announced by the International Human Genome Sequencing Consortium and Celera Genomics in February 2001.[12] This and subsequent sequences have been instrumental for the systematic analysis of various human disease genomes, including the cancer genome.[13]

Here, we introduce the high-throughput omics data and analysis from six topics: (1) methods and systems for chromatin immuno precipitation sequence (ChIP-seq) data analysis, (2) discovery of transcription factor-binding sites and its applications in cancer study, (3) cancer epigenomics, (4) tools and emerging applications for mining the phosphoproteome in cancer biology, (5) predicting microRNA, and (6) tools and databases for microRNA regulation.

3.1. *Methods and systems for ChIP-seq data analysis*

Chromatin immunoprecipitation (ChIP) was first described by Varshavsky and colleagues as a method to study protein–DNA interactions.[14] There are two approaches, ChIP-chip and ChIP-seq, which can be used for identifying DNA–protein-binding sites. ChIP-chip is a method of hybridization of fluorescently labeled DNA fragments to an appropriate microarray after ChIP and crosslinking reversal. Alternatively, ChIP-seq generates an adaptor-ligated library of the released immunoprecipitates fragments, which can then be sequenced *en masse.*[15] ChIP-seq offers important advantages over ChIP-chip, including lower cost, minimal hands-on processing, a need for fewer replicate experiments, and requires less starting material.[16] Follow-on bioinformatics analysis enables the genome-wide identification of binding sites of the protein with high specificity[15] and saves time over more traditional methods. The methods and systems for ChIP-seq data analysis will be described in detail in Chapter 6.

3.2. *Discovery of transcription factor-binding sites and its applications in cancer study*

Transcription factors (TFs) are essential for the regulation of gene expression. TFs bind to either enhancer or promoter regions of DNA at specific nucleotide sequences called transcription factor-binding sites (TFBSs), to activate or repress gene expression. Many differential genes discovered in cancer-related microarray studies are caused by abnormal activity of TFs, so identification of TF-binding motifs and their genomic locations is essential to understand the regulatory circuits that control cellular processes such as cell division and differentiation[17] and the mechanism of tumorigenesis. In Chapter 7, Chen introduces the method to discover TFBSs and its applications in cancer study.

3.3. *Cancer epigenomics*

Epigenomics is the study of epigenetic elements that determine stably inherited changes in gene expression without changes in the genomic DNA sequence. Epigenetic modifications include two main categories: DNA methylation and histone modifications. Abnormal DNA methylation may result in silencing of tumor suppressor genes and is common in a variety of human cancer cells.[18] Increasing evidence demonstrates the multiple connections between DNA methylation and histone modifications. Genes that are methylated are usually related to deacetylated and inactive chromatin, whereas unmethylated promoters and active genes are associated with an open euchromatin.[19] Chang and Wang discuss how DNA methylation and histone modification are involved in uncontrolled cell proliferation and, thus, in the development of cancer in Chapter 8.

3.4. *Tools and emerging applications for mining the phosphoproteome in cancer biology*

Phosphoproteomics is a branch of proteomics that identifies, catalogs, and characterizes proteins containing a phosphate group as a post-translational modification (PTM). Major cellular processes, such as cell division, growth, and differentiation, are regulated by PTM through highly dynamic and complex signaling pathways.[20] Quantitative phosphoproteomic profiling allows researchers to investigate aberrantly activated signaling pathways and therapeutic targets in cancers.[21] It can also reveal downstream effectors of kinases, thereby providing an opportunity to understand the molecular mechanisms that lead to oncogenesis.[21] Moreover, phosphoproteomics has become a powerful tool to map signal transduction pathways[22–24] and deliver the functional information that will promote insights into cell biology and systems biology. In Chapter 9, Wu *et al.* will introduce many tools and emerging applications for cancer phosphoproteomics.

3.5. *Predicting microRNAs*

In humans, some 1000 distinct miRNAs have been identified.[25] The miRNAs negatively regulate gene expression by post-transcriptional gene silencing, causing mRNA degradation or translation suppression of their target genes. Previous studies have shown that miRNAs play crucial roles in tumorigenesis by targeting the mRNAs of oncogenes or tumor suppressors[6,7] (see also Sec. 2.2 in this chapter). Many reports indicate that miRNA expression is altered in tumor tissues, suggesting that miRNAs could be potential markers for detection and prognosis in human cancers.[26,27] Because of the critical role of miRNAs as regulators of cell fate, analyzing and manipulating miRNAs within cancer cells may provide powerful new avenues for diagnosis, prognosis, drug discovery, and therapeutics.[25,28,29] Shih and Liu introduce several major approaches for predicting miRNAs in Chapter 10.

3.6. *MicroRNA regulation: databases and tools*

Currently there are many miRNA-related database systems to elucidate miRNAs and their target genes, including miRBase,[30] miRTarBase,[31] TarBase,[32] miRecords,[33] and miR2Disease.[34] Additionally, miRGen,[35] miRGator,[36] miRDB,[37] microRNA.org,[38] and miRNAMap[39,40] also provide miRNA-target interactions based on combinations of extensively adopted target prediction programs. Several computational methods and web servers for identifying target genes of miRNAs are very useful for cancer research. These databases will be introduced in Chapter 11.

4. Applications for Biomarkers and Drug Discovery

New combinations of genomic, proteomic, and bioinformatic research will provide deeper insights into disease mechanisms, novel markers for diagnostics, and new molecular targets for therapeutic intervention.[41] Cancer is a class of life-threatening diseases that constitutes one of the leading causes of death worldwide. Cancer cells are robust due to their characteristics of heterogeneity, redundancy, and feedback.[42,43] Current best practice in cancer treatment involves combined multidisciplinary treatment (also known as multimodality treatment). This treatment model may consist of surgical resection, radiation therapy with systemic chemotherapy, targeted therapy, or other various adjuvant medications. Combinations of these modalities can maximize therapeutic effects, improve patient survival, and give patients a higher quality of life.

Here, we present the applications of systems biology for biomarkers and drug discovery into four categories: (1) next-generation sequencing technologies and their applications in cancer research, (2) membrane proteomics for the opportunity of cancer biomarker and drug target discovery, (3) structure-based systems biology and drug design in cancer research, and (4) discovering drug targets for cancer therapy.

4.1. *Next-generation sequencing technologies and their applications in cancer research*

The first implementation of the next-generation sequencing is the massively parallel signature sequencing (MPSS) technology, which was developed by Sydney Brenner.[44] MPSS is a bead-based method that uses a complex approach of adapter ligation followed by adapter decoding and reading the sequence in increments of four nucleotides. Many diseases are caused by mutations in noncoding sequences but it has been difficult to study this in detail since the function of most noncoding sequences is poorly understood. With the rapidly increasing efficiency and cost-savings of the next-generation sequencing technologies (such as MPSS, 454 pyrosequencing, Illumina (Solexa) sequencing, and SOLiD sequencing), a deluge of data from individual human disease genomes will be useful for purposes including disease detection and personalized medicine. This method is particularly suited for cancer research because it allows the frequency of somatic mutations within a population of cancer cells to be determined.[45] Moreover, next-generation sequencing technology combined with ChIP-seq can provide a genome-wide look at transcription-factor binding (see also Sec. 3.1 in this chapter).[16] Detailed information about next-generation sequencing technologies and their applications in cancer research will be introduced in Chapter 12.

4.2. *Membrane proteomics for the opportunity of cancer biomarker and drug target discovery*

Defining and characterizing the dynamic plasma membrane (PM) proteome is crucial for understanding fundamental biological processes and disease mechanisms and for finding disease markers and drug targets.[46,47] Due to the generally low abundance of PM proteins and their insolubility in water, the study of PM proteins still remains difficult. Recent technological developments in sample preparation, such as a combination of differential centrifugation and sucrose density gradient centrifugation, detergent/polymer two-phase partitioning system, and biotinylation and glycoslyation affinity purification, together with important improvements in mass spectrometric analysis have facilitated analyses of these proteins.[47] In Chapter 13, Han *et al.* introduce methods of sample preparation as well as methods of quantitative membrane proteomics and their applications in the discovery of cancer biomarkers and therapeutic target candidates.

4.3. *Structure-based systems biology and drug design in cancer research*

The main focus of systems biology is to predict the behavior of biological systems based on the molecules involved and understand the interactions between these molecules.[48] A better understanding of these interactions can come from identifying the three-dimensional (3D) structures of these molecules. There are many techniques such as nuclear magnetic resonance (NMR), X-ray crystallography, and homology modeling to determine or predict protein 3D structures. Knowing the 3D structure of target proteins facilitates the discovery of potential drug candidates using protein–ligand docking simulation (Fig. 5).[49] Thus, structure-based systems biology provides a new approach to designing novel drugs for disease therapies including cancer therapy. In Chapter 14, Hsu introduces structure-based systems biology and drug design in cancer research.

4.4. *Discovering drug targets for cancer therapy*

Combinations of these modalities can maximize therapeutic effects, improve patient survival, and give patients a higher quality of life. With the establishment of gene regulatory networks and protein–protein interaction

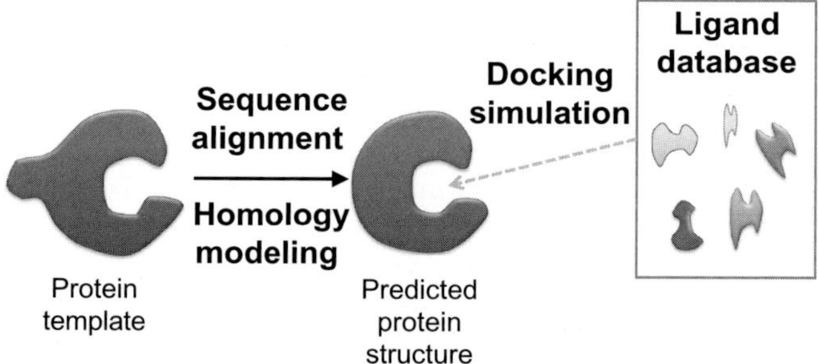

Fig. 5. A diagram to explain virtual screening. Virtual screening based on homology modeling and docking simulation is a useful approach for drug discovery.

networks in cancer cells, the development of cancer therapy has been moved into the era of a systems biology-based strategy, which provides a global perspective for understanding dynamic cancer behavior. There have been many drug targets discovered in recent decades, and many other targets are under investigation at the present time. It is vital to undertake more clinical trials to verify the feasibility of various treatment combinations. Since pharmacologic strategies now exist that effectively prevent cancer development in several types of tumors, a concerted effort to eliminate all malignant cancers in humans will become a real possibility in the near future. In Chapter 15, Huang *et al.* introduce how systems biology can accelerate drug discovery.

5. The Future

Systems biology is a revolutionary approach for analyzing biological complexity and understanding how biological systems function. Yet the field of systems biology has many remaining challenges. The high-throughput experimental techniques are still in an exponential development phase. Undoubtedly, systems biology will break the divide between classical and high-throughput methods. Its role is not to replace any of the classical techniques from biochemistry or genetics but to provide a set of organizing principles that integrate these methods.[50] Systems biology can help to shorten the time for drug discovery in several ways, such as biomarker and target identification, as well as improve predictions of drug mechanism and toxicity, which will ultimately increase our longevity and improve our lives in the near future.

Acknowledgments

This work was supported by National Research Program for Genomic Medicine, Department of Health (DOH), National Science Council of Taiwan, and the National Taiwan University Frontier and Innovative Research Program.

References

1. Whichard ZL, Sarkar CA, Kimmel M *et al.* (2010) Hematopoiesis and its disorders: A systems biology approach. *Blood* **115**: 2339–2347.
2. Rudy Y, Ackerman MJ, Bers DM *et al.* (2008) Systems approach to understanding electromechanical activity in the human heart: A national

14   *H.-F. Juan*

heart, lung, and blood institute workshop summary. *Circulation* **118**: 1202–1211.

3. Ertugrul D, Xuewei W, Neil W *et al.* (2010) Cancer-drug associations: A complex system. *PLoS ONE* **5**: e10031.

4. Chen-Ching L, Jen-Tsung H, Chia-Yi W *et al.* (2010) Dynamic functional modules in co-expressed protein interaction networks of dilated cardiomyopathy. *BMC Systems Biology* **4**: 138.

5. Wu CC. (2004) GeneNetwork: An interactive tool for reconstruction of genetic networks using microarray data. *Bioinformatics* **20**: 3691–3693.

6. Aurora E-K, J. SF. (2006) Oncomirs — microRNAs with a role in cancer. *Nature Reviews Cancer* **6**: 259–269.

7. Ventura A, Jacks T. (2009) MicroRNAs and cancer: Short RNAs go a long way. *Cell* **136**: 586–591.

8. Hsu C-W, Juan H-F, Huang H-C. (2008) Characterization of microRNA-regulated protein–protein interaction network. *Proteomics* **8**: 1975–1979.

9. Cusick ME, Klitgord N, Vidal M *et al.* (2005) Interactome: Gateway into systems biology. *Hum Mol Genet* **14**(Spec No. 2): R171–R181.

10. Rual JF, Venkatesan K, Hao T *et al.* (2005) Towards a proteome-scale map of the human protein–protein interaction network. *Nature* **437**: 1173–1178.

11. Sharan R, Ulitsky I, Shamir R. (2007) Network-based prediction of protein function. *Molecular systems biology* **3**: 88.

12. Lander ES, Linton LM, Birren B *et al.* (2001) Initial sequencing and analysis of the human genome. *Nature* **409**: 860–921.

13. Geurts Van Kessel A. (2010) The "omics" of cancer. *Cancer Genetics and Cytogenetics* **203**: 37–42.

14. Solomon MJ, Larsen PL, Varshavsky A. (1988) Mapping protein–DNA interactions *in vivo* with formaldehyde: Evidence that histone H4 is retained on a highly transcribed gene. *Cell* **53**: 937–947.

15. Mardis ER. (2008) The impact of next-generation sequencing technology on genetics. *Trends in Genetics* **24**: 133–141.

16. Mardis ER. (2007) ChIP-seq: Welcome to the new frontier. *Nature Methods* **4**: 613–614.

17. Euskirchen GM, Rozowsky JS, Wei CL *et al.* (2007) Mapping of transcription factor binding regions in mammalian cells by ChIP: Comparison of array- and sequencing-based technologies. *Genome Research* **17**: 898–909.

18. Yu-Ching F, Hsuan-Cheng H, Hsueh-Fen J. (2008) MeInfoText: Associated gene methylation and cancer information from text mining. *BMC Bioinformatics* **9**: 22.

19. Razin A. (1998) CpG methylation, chromatin structure and gene silencing — a three-way connection. *The EMBO Journal* **17**: 4905–4908.

20. Trost M, Bridon G, Desjardins M *et al.* (2010) Subcellular phosphoproteomics. *Mass Spectrometry Rev* **29**: 962–990.

21. Harsha HC, Pandey A. (2010) Phosphoproteomics in cancer. *Mol Oncol* **4**(6): 482–495.

22. Michaelevski I, Segal-Ruder Y, Rozenbaum M *et al.* (2010) Signaling to transcription networks in the neuronal retrograde injury response. *Sci Signal* **3**: ra53.

23. Oberprieler NG, Lemeer S, Kalland ME *et al.* (2010) High-resolution mapping of prostaglandin E2-dependent signaling networks identifies a constitutively active PKA signaling node in CD8+CD45RO+ T cells. *Blood* **116**: 2253–2265.

24. Rinschen MM, Yu MJ, Wang G *et al.* (2010) Quantitative phosphoproteomic analysis reveals vasopressin V2-receptor-dependent signaling pathways in renal collecting duct cells. *Proc Natl Acad Sci USA* **107**: 3882–3887.

25. Lieberman J. (2009) Micromanaging cancer. *New Engl J Med* **365**: 1500–1501.

26. Mattie MD, Benz CC, Bowers J *et al.* (2006) Optimized high-throughput microRNA expression profiling provides novel biomarker assessment of clinical prostate and breast cancer biopsies. *Mol Cancer* **5**: 24.

27. Yanaihara N, Caplen N, Bowman E *et al.* (2006) Unique microRNA molecular profiles in lung cancer diagnosis and prognosis. *Cancer Cell* **9**: 189–198.

28. Nagpal JK, Rani R, Trink B *et al.* (2010) Targeting miRNAs for drug discovery: A new paradigm. *Curr Mol Med* **10**: 503–510.

29. Nelson KM, Weiss GJ. (2008) MicroRNAs and cancer: Past, present, and potential future. *Mol Cancer Ther* **7**: 3655–3660.

30. Griffiths-Jones S, Saini HK, Van Dongen S *et al.* (2008) miRBase: Tools for microRNA genomics. *Nucleic Acids Res* **36**: D154–D158.

31. Hsu SD, Lin FM, Wu WY *et al.* (2011) miRTarBase: A database curates experimentally validated microRNA-target interactions. *Nucleic Acids Res* **39**: D163–D169.

32. Papadopoulos GL, Reczko M, Simossis VA *et al.* (2009) The database of experimentally supported targets: A functional update of TarBase. *Nucleic Acids Res* **37**: D155–D158.

33. Xiao F, Zuo Z, Cai G *et al.* (2009) miRecords: An integrated resource for microRNA-target interactions. *Nucleic Acids Res* **37**: D105–D110.

34. Jiang Q, Wang Y, Hao Y *et al.* (2009) miR2Disease: A manually curated database for microRNA deregulation in human disease. *Nucleic Acids Res* **37**: D98–D104.

35. Alexiou P, Vergoulis T, Gleditzsch M *et al.* (2010) miRGen 2.0: A database of microRNA genomic information and regulation. *Nucleic Acids Res* **38**: D137–D141.

36. Nam S, Kim B, Shin S *et al.* (2008) miRGator: An integrated system for functional annotation of microRNAs. *Nucleic Acids Res* **36**: D159–D164.

37. Wang X. (2008) miRDB: A microRNA target prediction and functional annotation database with a wiki interface. *RNA* **14**: 1012–1017.

38. Betel D, Wilson M, Gabow A *et al.* (2008) The microRNA.org resource: Targets and expression. *Nucleic Acids Res* **36**: D149–D153.

39. Hsu SD, Chu CH, Tsou AP *et al.* (2008) miRNAMap 2.0: Genomic maps of microRNAs in metazoan genomes. *Nucleic Acids Res* **36**: D165–D169.

40. Hsu PW, Huang HD, Hsu SD *et al.* (2006) miRNAMap: Genomic maps of microRNA genes and their target genes in mammalian genomes. *Nucleic Acids Res* **34**: D135–D139.

41. Weinstein JN, Pommier Y. (2006) Connecting genes, drugs and diseases. *Nat Biotech* **24**: 1365–1366.

42. Kitano H. (2003) Cancer robustness: Tumour tactics. *Nature* **426**: 125.
43. Kitano H. (2004) Cancer as a robust system: Implications for anticancer therapy. *Nat Rev* **4**: 227–235.
44. Brenner S, Johnson M, Bridgham J *et al.* (2000) Gene expression analysis by massively parallel signature sequencing (MPSS) on microbead arrays. *Nat Biotech* **18**: 630–634.
45. Shuen A, Foulkes WD. (2010) Clinical implications of next-generation sequencing for cancer medicine. *Curr Oncol* **17**: 39–42.
46. Sprenger RR, Jensen ON. (2010) Proteomics and the dynamic plasma membrane: Quo Vadis? *Proteomics* **10**: 3997–4011.
47. Wiśniewski JR. (2011) Tools for phospho- and glyco-proteomics of plasma membranes. *Amino Acids* **41**: 223–233.
48. Patrick A, B RR. (2006) Structural systems biology: Modelling protein interactions. *Nat Rev Mol Cell Biol* **7**: 188–197.
49. Huang T-C, Chang H-Y, Hsu C-H *et al.* (2008) Targeting Therapy for Breast Carcinoma by ATP Synthase Inhibitor Aurovertin B. *J Proteome Res* **7**: 1433–1444.
50. Chuang H-Y, Hofree M, Ideker T. (2010) A decade of systems biology. *Ann Rev Cell and Dev Biol* **26**: 721–744.

Chapter 2

Gene Network Construction
for Molecular Regulation

Hsueh-Fen Juan[*] and Hsuan-Cheng Huang[†]

1. Introduction

Networks are graphs of connected nodes and they are used in systems biology to represent different types of relationships between biological entities, including genes, proteins, chemical compounds, and transcription factors.[1] A meta-analysis by Chuang *et al.* of all systems biology publications recorded in PubMed over the past decade found that "network biology" showed a rise in publication rates,[2] suggesting that network biology is now a hot topic in systems biology studies.

Gene networks are graphical models that represent relationships among thousands of molecular and clinical traits (e.g. DNA variations, RNA levels, protein state, metabolite levels, and disease state) and provide a way to visualize complex relationships between genes.[3] Gene networks are composed of nodes that represent genes and edges between the nodes that represent relationships between genes (Fig. 1).[3] Gene expression data with time series combined with mathematical modeling are essential for constructing gene networks. To quantify the relative levels of gene expression in cells and test mathematical models of gene networks *in vivo*, researchers use modern assay techniques, such as microarray, flow cytometry, fluorescence microscopy, and microfluidic devices.

---

[*]Department of Life Science, Institute of Molecular and Cellular Biology, and Graduate Institute of Biomedical Electronics and Bioinformatics, National Taiwan University, Taipei 106, Taiwan. yukijuan@ntu.edu.tw

[†]Institute of Biomedical Informatics, Center for Systems and Synthetic Biology, National Yang-Ming University, Taipei 112, Taiwan. hsuancheng@ym.edu.tw

17

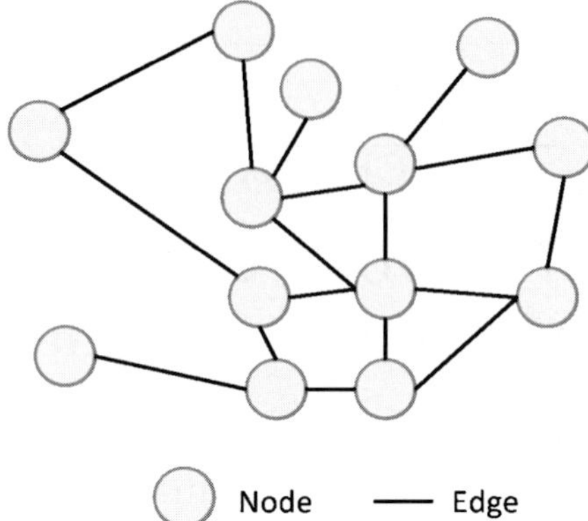

Fig. 1.   Gene networks. Gene networks comprise nodes (genes) and edges (relationships between genes).

Systems biology is a top-down approach, studying entire large-scale gene networks with the goal of obtaining an integrated understanding of genomic function.[4] In contrast, synthetic biology is a bottom-up approach, studying simplified gene networks consisting of just one or a few genes with the goal of uncovering the fundamental principles and mechanisms that govern gene regulation.[4]

In this chapter, we first introduce the assay techniques for quantifying gene activity, adopt a few mathematical methods to model gene networks, and then discuss the emerging field of synthetic biology as well as give examples of recent real-life applications of gene network construction.

2. Assay Techniques for Quantifying Gene Activity

Network analysis is broadly applicable through the drug discovery and development pipeline, from the biological mechanisms of disease to the pharmacological interactions of drugs.[5] Quantifying gene activity is the key to network analysis. Here, we will introduce four techniques: (1) DNA microarray, (2) flow cytometry, (3) fluorescence microscopy, and (4) microfluidic devices, to quantify gene expression/activity. A comparison of these techniques is summarized in Table 1.

2.1. *DNA microarray*

DNA microarray consists of an arrayed series of thousands of microscopic spots of DNA oligonucleotides. The microarray technique is a powerful

Table 1. Comparison of the four assay techniques.

	DNA Microarray	Flow cytometry	Fluorescence microscopy	Microfluidic devices
Easy-to-use		V		V
High-throughput	V	V		V
Individual cells can be imaged mutiple times			V	V
Quntifying the fluorescence of thousands of cells simultaneously		V		V
Detecting gene expression at the level of organisms and populations	V			V
Detecting gene expression at the level of a single cell		V	V	V

high-throughput method for accurately determining changes in global gene expressions.[6,7] It can measure the expression levels of tens of thousands of discrete gene sequences in a single array.[8] This technique has been used for novel gene discoveries,[9] gene function determinations, and pathway dissections.[10] Genes with a similar expression pattern often function in the same biological processes.[11] Likewise, genes on the same pathways or in the same functional complex often exhibit similar expression patterns under diverse temporal and physiological conditions.[12] Using microarray to qualify gene expression or activity is very useful in the study of both gene network construction and disease therapies.

Lin *et al.*, for example, developed a network-based comparative analysis approach to integrate gene expression profiles derived from microarray data with protein–protein interactions and biological function annotations to reveal dynamic functional modules under different biological states.[11] They successfully revealed network modules closely related to heart failure; more importantly, these network dynamics provided new insights into the cause of dilated cardiomyopathy. The molecular modules revealed by this work could be used as potential drug targets and provide new directions for heart failure therapy.[11] For a more detailed explanation of this work, see Chapter 1, Sec. 2.3 and Chapter 4.

In another example of DNA microarray data emerging from time-course gene expression measurements of an identified pathway, Cheng *et al.* reconstructed a plausible regulatory network of the genes involved in cell death using a reverse-engineering computational approach (for a more detailed

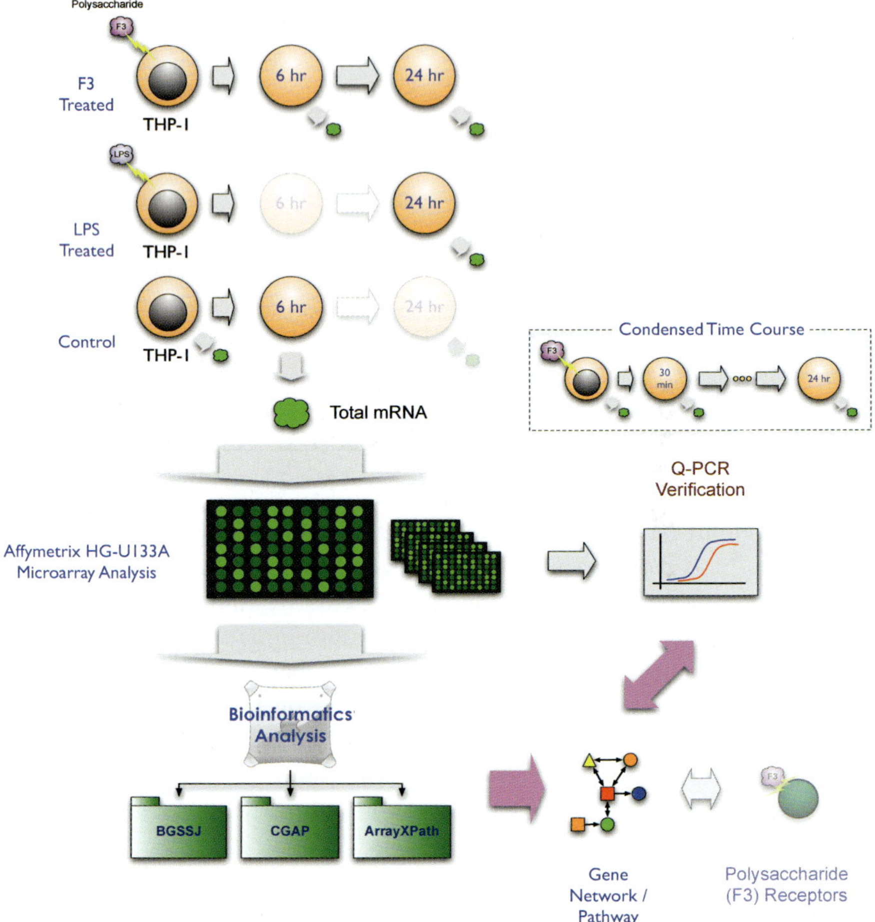

Fig. 2. A method of gene expression to network construction. This approach integrates time-course DNA microarray analysis, quantitative PCR assays, and bioinformatics techniques to study the F3-induced effects in human monocytic leukemia THP-1 cells. LPS, lipopolysaccharide; F3, a specific kind of polysaccharide originally isolated and purified from *G. lucidum.*

explanation, see Chapter 5), which revealed the efficacy of antitumor drugs, such as *Ganoderma lucidum* (*G. lucidum*) polysaccharides, F3, on tumor growth (Fig. 2).[13] This approach reveals a different way of elucidating the molecular mechanisms for antitumor compounds or drugs in cancer cells and provide a powerful tool for cancer diagnosis and therapy.

2.2. *Flow cytometry*

For biologists, the main tools for quantifying gene activity *in vivo* are fluorescent proteins. The most popular method for measuring the activities of fluorescent proteins in living cells is flow cytometry.[4] This is a powerful technique for counting and examining microscopic particles, such as cells and chromosomes, by suspending them in a stream of fluid and passing them through an electronic detection apparatus.[14] It can also be used to separate cells according to subtype or epitope expression for further biological studies. Researchers can easily and quickly analyze hundreds of thousands of cells as well as generate a snapshot of the distribution of the activity of a gene in a population of cells.[4]

2.3. *Fluorescence microscopy*

Another widely used technique for quantifying gene activity is fluorescence microscopy.[4] This is based on the phenomenon that certain materials emit energy detectable as visible light when irradiated with light of a specific wavelength. The sample may either fluoresce in its natural form, such as green fluorescent protein (GFP), chlorophyll, and some minerals, or may be treated with fluorescing chemicals. Fluorescence microscopy is particularly useful for obtaining subcellular structures and localization patterns of proteins and DNA.[15] When studying the dynamics of gene expression in single cells, fluorescence data can be invaluable.

2.4. *Microfluidic devices*

A flow cytometer is a kind of high-throughput microfluidic device. The first functional microfluidic devices were made from elastomers and used to perform DNA analysis[16,17] and cell sorting.[18] Microfluidic devices offer enhanced reaction times, sensitivity, and reliability, enabling researchers to address DNA sequencing and gene expression analysis from lower amounts of nucleic acid material at a reduced cost.[19] Maerkl and Quake developed a high-throughput microfluidic platform capable of detecting low-affinity transient-binding events based on the mechanically induced trapping of molecular interactions, which eliminates the off-rate (also called dissociation) problem facing the current array platforms and allows for absolute affinity measurements.[20] This platform was used to characterize DNA-binding energy landscapes for four eukaryotic transcription factors; these landscapes were used to test basic assumptions about transcription factor binding and predict their *in vivo* function.[20] Recently, Quake's group developed an *in vitro* microfluidic platform for high-throughput screening of protein interactions, called the protein interaction network generator (PING), which combines on-chip *in vitro* protein synthesis with an *in situ* microfluidic affinity assay.[21] Using microfluidic "lab-on-chip" devices, researchers can track the dynamics of gene networks in single cells under various environmental conditions.

3. Mathematical Methods

Various reverse engineering algorithms have been used to model gene regulatory networks.[22] A gene regulation matrix is a rectangular array of various gene expression levels and perturbations, such as time series or drug treatments. Here, we will discuss the five different inference models currently used to extract the gene regulation matrix from gene expression data and compare them: (1) a linear model is a continuous method that uses linear ordinary differential equations to describe the system[23]; (2) an ordinary differential equation (ODE) is a relation that contains functions of only one independent variable and one or more of their derivatives with respect to that variable; (3) a dynamic Bayesian network stochastically models causality between genes over time-series data[22]; (4) a Boolean network is a logical description in which variables and functions (the relationships between the components) are simply presented as ON or OFF; and (5) Petri nets are basic models of parallel and distributed systems.

Genetic algorithms are search heuristics routinely used to generate useful solutions to optimization and search problems by modeling aspects of natural evolution such as inheritance, mutation, selection, and crossover. Genetic algorithms belong to the larger class of evolutionary algorithms that are subsets of evolutionary computation and are generic, population-based, metaheuristic optimization algorithms. For the latter four models, genetic algorithms can be applied to effectively search for the optimal point in the large solution space and learn network structure.[24]

3.1. *Linear model*

The definition of a linear model is that each regulator (gene) contributes to the input of the regulation function independently of the other regulators (genes), in an additive manner (Fig. 3).[25] Time-series data usually contain many more genes than time points. This presents a difficulty in reverse

$$\frac{\Delta y_i(t)}{\Delta t} = \sum_j w_{ij}\, y_j(t) + b_i$$

(a)

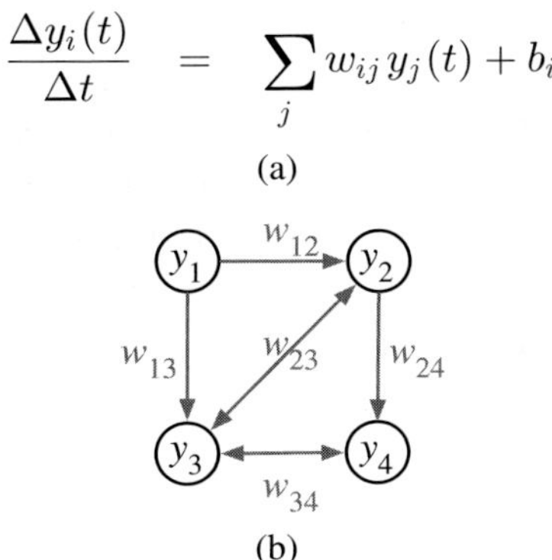

(b)

Fig. 3.   Linear model. The change in the level of each entity depends on a weight linear sum of the levels of its regulators. (a) $y_i(t)$ is the expression level of gene i at time t; w_{ij} indicates how much the expression level of gene i influences gene j; and b_i is a constant bias factor to model the activation level of a gene in the absence of any other regulator inputs. (b) An example of the linear model.

engineering a network's structure and regulation functions. The general linear model is probably unsuitable for capturing complex phenomena. Recently, Kabir *et al.* proposed a new approach for inferring gene regulatory networks from time-series gene expression data using linear time-variant model and self-adaptive differential evolution, a versatile and robust evolutionary algorithm, as the learning paradigm.[26] They applied the proposed approach to a simulated expression dataset of cAMP oscillations in *Dictyostelium discoideum*, a species of soil-living amoeba, which proved its strength in identifying methods of gene regulation. This approach has also been verified by analyzing the real expression dataset of the SOS DNA repair system in *Escherichia coli* (*E. coli*) and has succeeded in finding more correct and reasonable regulations as compared to various traditional methods.[26]

3.2. *Ordinary differential equation*

An equation that involves derivatives of the dependent variable is called a differential equation. Differential equations are widely used in biological modeling to describe dynamic processes in terms of rates of change.[27,28] If a differential equation involves only one independent variable and therefore contains only full derivatives, it is called an ordinary differential equation (ODE). If it involves more than one independent variable and, therefore, contains partial derivatives, it is called a partial differential equation (PDE). The order of a differential equation is the order of the highest derivative that appears in the equation. ODEs have been applied to many biological studies; for example, they are useful to describe the essential features of the dynamics of gene regulatory networks[29] and model the dynamics of biological processes.[30]

3.3. *Dynamic Bayesian network*

Dynamic Bayesian network is a general state-space model to describe a stochastic dynamic system and is a Bayesian network that represents sequences of variables. While the linear model cannot provide answers as to how the affinity of a transcription factor to its target promoter affects the network, time-dependent transcription factor activities can be inferred from microarray time-series data using dynamic Bayesian network (Fig. 4).[25] If we have n as the number of microarrays and p as the number of genes, then in a dynamic Bayesian network, we let $\mathbf{X}$ be $n \times p$, the microarray gene expression data matrix.

In the context of Bayesian networks, a gene is considered as a random variable. When we model a gene network by using statistical models described by the density or probability function, the statistical model should include p random variables. However, we have only n samples, and n is usually much smaller than p. In such cases, the inference of the model is quite difficult or sometimes impossible because the model has many

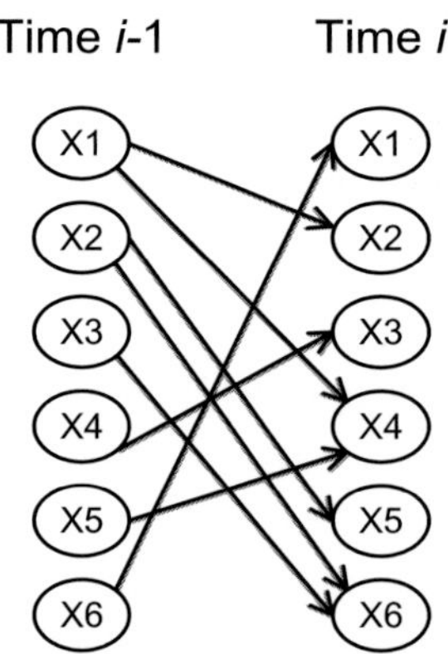

Fig. 4. Graphical view of a dynamic Bayesian network model. $X1$–$X6$ indicate gene expression levels at time $i - 1$ or i. The arrows indicate the regulatory directions.

parameters and the number of samples is not enough for estimating the parameters.[31] A Bayesian network model is able to overcome this problem because it is efficient at both inference and learning. It has been shown previously to perform well in integrating heterogeneous biological data. Jia and Huan investigated a novel nonstationary dynamic Bayesian network method with a potential regulator detection technique and a flexible lag choosing mechanism. They successfully applied this approach to the gene regulatory network inference on three nonstationary time series data: macrophages, *Arabidopsis*, and *Drosophila*.[32] Klinke also successfully used an empirical Bayesian approach to model an epidermal growth factor (EGF) signaling network.[33] Tonikian *et al.* chose the dynamic Bayesian network approach for modeling yeast SH3 domain interactome and predicting spatiotemporal dynamics of endocytosis proteins.[34]

3.4. *Boolean network*

Boolean networks were first proposed by Kauffman in 1969.[35] Two Boolean models, Boolean network and probabilistic Boolean network, are discussed here. In Boolean network, each node (gene) can attain two alternative levels: active (ON; 1) or inactive (OFF; 0). It is assumed to change synchronously, such that at every time step, the level entity is determined according to the levels of its regulators at the previous time step and the regulation function.[25] A detailed explanation is shown in Fig. 5(a). Despite the simplification of regulatory dynamics and interactions, this approach has the advantages of robustness and ease of interpretation.[36] In probabilistic Boolean network, the way genes interact with each other is formulated by standard logic functions.[37] The probabilistic Boolean network generates a sequence of global states that constitutes a Markov chain, a discrete set of times. Xiao provides an analysis of and simulation issues with Boolean networks and probabilistic Boolean networks for gene regulatory networks.[37] Teschendorff *et al.* used a modified Boolean network for pathway activity estimation in tumors and showed that pathway modules antagonize or synergize to delineate novel prognostic subtypes.[38]

3.5. *Petri nets*

Petri nets were designed by Carl Adam Petri in 1962.[39] A Petri net is a graphical and mathematical modeling tool and an alternative to

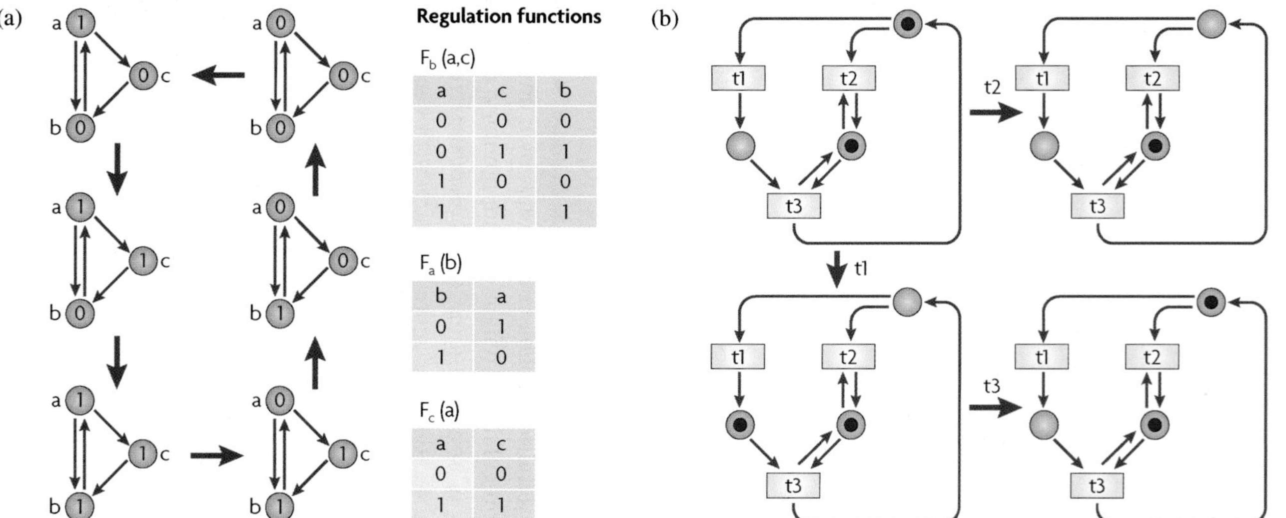

Fig. 5. (a) A Boolean network. Each of the entities a, b, and c in the network can be in state 0 or 1 (OFF or ON). State transitions obey the regulation functions shown on the right, which describe the rules of the model. For example, if a is in state 1 and c is in state 1, at the next time step, the state of b will be 1. Thin arrows indicate the regulators of each node. Time steps are represented by thick arrows. The global state of the model is the combination of the three entity states. The system cycles through the six global states. (b) A Petri net. The net contains *places* (light blue circles) that are the model's entities and *transitions* (rectangles) that constitute the regulation functions and define the model's dynamics. *Arcs* connect input places to transitions and transitions to their output places. Places that receive discrete values are called *tokens* (dark blue dots). A transition that is activated, or "fired," reduces the tokens in its input places and increases the number of tokens in each of its output places. At any time step, every transition that has enough tokens in its input places may be fired. In the example, every transition consumes one token from every input place and produces one token at every output place. Labels at thick arrows indicate which transition is fired. Transitions $t1$ and $t3$ can be fired in alternation indefinitely, whereas no other transition can be fired after $t2$ has been fired. The figures are adapted from Fig. 1 from Karlebach *et al.*[25]

ODEs for the simulation of time-dependent processes. It consists of *places* and *transitions*, represented graphically by circles and rectangles, respectively, and arcs that connect them. A detailed explanation is shown in Fig. 5(b).

Petri nets have proven useful for modeling, analysis, and simulation of a diversity of biological networks,[40–42] covering discrete,[43] stochastic,[44] continuous, and hybrid models.[45–47] Steggles *et al.* successfully used Petri nets to model the genetic regulatory network controlling sporulation in the bacterium *Bacillus subtilis*.[48] Using Petri nets, Simão *et al.* modeled the biosynthesis of tryptophan (Trp) in *E. coli*, taking into account two types of regulatory feedbacks: the direct inhibition of the first enzyme of the pathway by the final product of the pathway and the transcriptional inhibition of the Trp operon by the Trp-repressor complex.[49] The deployment of Petri nets to study biological applications has not only generated original models but also motivated fundamental research. Because of their complexity, size, and heterogeneity, biological networks raise distinctive challenges to the modelers.

4. Synthetic Biology

Systems biology is creating a new generation of biological research enabled by the genome revolution with a systems-level approach to understand organisms and the functional activities of their components by studying the underlying complex interactions. The payoff for systems biology research is not only abstract theoretical understanding but also empowerment to design new and improved biological function via "synthetic biology."[50] Combination of well-characterized biological parts to create synthetic wholes accelerates potential applications and finesses of natural systems.

4.1. *Engineering gene networks*

Synthetic biology is a new research area that combines the investigative nature of biology with the constructive nature of engineering.[51,52] The term "synthetic biology" was first introduced by the Polish geneticist Waclaw Szybalski in 1974.[53] In 1978, Szybalski and Skalka wrote an editorial comment for the discovery of restriction enzymes by Arber, Nathans, and Smith, who were awarded the Nobel Prize in Physiology or Medicine (1978). Szybalski and Skalka described that the work on restriction nucleases not only permits us easily to construct recombinant DNA molecules

and to analyze individual genes but also leads us into the new era of
"synthetic biology," in which existing genes are described and analyzed
and new gene arrangements can be constructed and evaluated.[51] Synthetic
biology aims to design modules and devices, based either on natural or
artificially modified macromolecular parts, enabling novel functionalities
in reconstructed biological systems, or in those built *de novo*[52,54] One
of the major objectives of bottom-up synthetic biology is to engineer
extant biological molecules to implement novel functionalities in living
systems.[52,54]

Engineering novel reusable gene networks to provide a greater control
over cellular processes is one of the goals of the emerging discipline of
synthetic biology.[55–57] Gene network engineers work with a framework in
which a network is designed based on a simplified mathematical model,
constructed using tools from molecular biology, deployed within a suitable
host, and experimentally validated against the mathematical abstraction.[56]
We will give some examples of how this has been implemented in the
next section. Engineered gene networks have numerous applications in
diverse biotechnological endeavors, such as metabolic engineering,[58,59] gene
oscillators,[55,60,61] as well as cell-based computers.[62,63] It certainly seems
that in the next few years engineering gene networks will have a variety
of solutions in agricultural, pharmaceutical, environmental, industrial, and
energy-related problems.

4.2. *The applications of synthetic biology*

The synthetic biology approach has been applied to cancer therapy.[52]
For example, Anderson *et al.* proposed an approach that could be used
to engineer bacteria to sense the microenvironment of a tumor and
respond by invading cancerous cells and releasing a cytotoxic agent to
kill tumorous cells.[64,65] The *inv*[+] *E. coli,* which was transferred by the
inv gene encoding invasin from *Yersinia pseudotuburculosis,* is therapeutic
bacteria for the treatment of cancer. By restricting the expression of *inv*
to tumor sites, invasion could be confined to malignant cells. The hypoxic
environment could provide a cue for detection of tumors and the induction
of cancer cell invasion.[64] The other application of synthetic biology is to
engineer parallel metabolic systems that interface with cellular metabolic
machinery to provide cost-effective chemical and drug synthesis.[66] For
example, Martin *et al.* successfully engineered a synthetic metabolic
pathway based on the isoprenoid pathway of *S. cerevisiae* into *E. coli.*

Isoprenoid is a viable terpenoid precursor that is used for the synthesis of various drugs.[67] By modifying their isoprenoid systems, researchers constructed an artemisinin (its derivatives are a group of drugs) biosynthetic pathway in yeast, potentially providing an affordable and reliable source of highly potent antimalarial drugs, since artemisinin and its derivatives are used to produce antimalarial drugs.[68,69] Ro *et al.* showed that yeast can be engineered with a plant metabolic pathway to efficiently produce a drug precursor that is difficult and expensive to obtain using traditional methods.[69]

5. Summary and Future Perspectives

Network analysis is broadly applicable through the drug discovery and development pipeline.[5] As large molecular datasets are processed with the help of network analysis, a growing set of reconstructed pathways and networks will emerge. These will be very useful for discovery of drug targets,[70] elucidating molecular regulation of drugs in diseases such as cancer[13,71] and cardiovascular diseases,[11] understanding toxin action, and modeling biological processes among other applications.

Acknowledgments

This work was supported by National Research Program for Genomic Medicine, Department of Health (DOH), National Science Council of Taiwan, and the National Taiwan University Frontier and Innovative Research Program.

References

1. Pujol A, Mosca R, Farres J *et al.* (2010) Unveiling the role of network and systems biology in drug discovery. *Trends Pharmacol Sci* **31**: 115–123.
2. Chuang HY, Hofree M, Ideker T. (2010) A decade of systems biology. *Annu Rev Cell Dev Biol* **26**: 721–744.
3. Schadt EE, Friend SH, Shaywitz DA. (2009) A network view of disease and compound screening. *Nat Rev Drug Discov* **8**: 286–295.
4. Bennett MR, Hasty J. (2009) Microfluidic devices for measuring gene network dynamics in single cells. *Nat Rev Genet* **10**: 628–638.
5. Nikolsky Y, Nikolskaya T, Bugrim A. (2005) Biological networks and analysis of experimental data in drug discovery. *Drug Discov Today* **10**: 653–662.

6. Yue H, Eastman PS, Wang BB *et al.* (2001) An evaluation of the performance of cDNA microarrays for detecting changes in global mRNA expression. *Nucleic Acids Res* **29**: E41-1.

7. Ideker T, Thorsson V, Ranish JA *et al.* (2001) Integrated genomic and proteomic analyses of a systematically perturbed metabolic network. *Science* **292**: 929–934.

8. Hughes TR, Marton MJ, Jones AR *et al.* (2000) Functional discovery via a compendium of expression profiles. *Cell* **102**: 109–126.

9. Seo J, Kim M, Kim J. (2000) Identification of novel genes differentially expressed in PMA-induced HL-60 cells using cDNA microarrays. *Mol Cells* **10**: 733–739.

10. Le Naour F, Hohenkirk L, Grolleau A *et al.* (2001) Profiling changes in gene expression during differentiation and maturation of monocyte-derived dendritic cells using both oligonucleotide microarrays and proteomics. *J Biol Chem* **276**: 17920–17931.

11. Lin CC, Hsiang JT, Wu CY *et al.* (2010) Dynamic functional modules in co-expressed protein interaction networks of dilated cardiomyopathy. *BMC Syst Biol* **4**: 138.

12. Ruan J, Dean AK, Zhang W. (2010) A general co-expression network-based approach to gene expression analysis: Comparison and applications. *BMC Syst Biol* **4**: 8.

13. Cheng KC, Huang HC, Chen JH *et al.* (2007) *Ganoderma lucidum* polysaccharides in human monocytic leukemia cells: From gene expression to network construction. *BMC Genomics* **8**: 411.

14. Bergquist PL, Hardiman EM, Ferrari BC *et al.* (2009) Applications of flow cytometry in environmental microbiology and biotechnology. *Extremophiles* **13**: 389–401.

15. Boland MV, Markey MK, Murphy RF. (1998) Automated recognition of patterns characteristic of subcellular structures in fluorescence microscopy images. *Cytometry* **33**: 366–375.

16. Effenhauser CS, Bruin GJM, Paulus A *et al.* (1997) Integrated capillary electrophoresis on flexible silicone microdevices: Analysis of DNA restriction fragments and detection of single DNA molecules on microchips. *Analytical chem* **69**: 3451–3457.

17. Chou HP, Spence C, Scherer A *et al.* (1999) A microfabricated device for sizing and sorting DNA molecules. *Proc Natl Acad Sci USA* **96**: 11–13.

18. Fu AY, Spence C, Scherer A *et al.* (1999) A microfabricated fluorescence-activated cell sorter. *Nat Biotechnol* **17**: 1109–1111.

19. Szita N, Polizzi K, Jaccard N *et al.* (2010) Microfluidic approaches for systems and synthetic biology. *Curr Opin Biotechnol* **21**: 517–523.

20. Maerkl SJ, Quake SR. (2007) A systems approach to measuring the binding energy landscapes of transcription factors. *Science* **315**: 233–237.

21. Gerber D, Maerkl SJ, Quake SR. (2009) An *in vitro* microfluidic approach to generating protein-interaction networks. *Nat Methods* **6**: 71–74.

22. De Jong H. (2002) Modeling and simulation of genetic regulatory systems: A literature review. *J Comput Biol* **9**: 67–103.

23. D'haeseleer P, Wen X, Fuhrman S *et al.* (1999) Linear modeling of mRNA expression levels during CNS development and injury. *Pac Symp Biocomput*: 41–52.
24. Repsilber D, Liljenstrom H, Andersson SG. (2002) Reverse engineering of regulatory networks: Simulation studies on a genetic algorithm approach for ranking hypotheses. *Biosystems* **66**: 31–41.
25. Karlebach G, Shamir R. (2008) Modelling and analysis of gene regulatory networks. *Nat Rev Mol Cell Biol* **9**: 770–780.
26. Kabir M, Noman N, Iba H. (2010) Reverse engineering gene regulatory network from microarray data using linear time-variant model. *BMC Bioinformatics* **11**(Suppl 1): S56.
27. Aldridge BB, Burke JM, Lauffenburger DA *et al.* (2006) Physicochemical modelling of cell signalling pathways. *Nat Cell Biol* **8**: 1195–1203.
28. Kim KA, Spencer SL, Albeck JG *et al.* (2010) Systematic calibration of a cell signaling network model. *BMC Bioinformatics* **11**: 202.
29. Ironi L, Panzeri L. (2009) A computational framework for qualitative simulation of nonlinear dynamical models of gene-regulatory networks. *BMC Bioinformatics* **10**(Suppl 12): S14.
30. Donze A, Clermont G, Langmead CJ. (2010) Parameter synthesis in nonlinear dynamical systems: Application to systems biology. *J Comput Biol* **17**: 325–336.
31. Kim S, Imoto S, Miyano S. (2004) Dynamic Bayesian network and nonparametric regression for nonlinear modeling of gene networks from time series gene expression data. *Biosystems* **75**: 57–65.
32. Jia Y, Huan J. (2010) Constructing non-stationary Dynamic Bayesian Networks with a flexible lag choosing mechanism. *BMC Bioinformatics* **11**(Suppl 6): S27.
33. Klinke DJ, 2nd. (2009) An empirical Bayesian approach for model-based inference of cellular signaling networks. *BMC Bioinformatics* **10**: 371.
34. Tonikian R, Xin X, Toret CP *et al.* (2009) Bayesian modeling of the yeast SH3 domain interactome predicts spatiotemporal dynamics of endocytosis proteins. *PLoS Biol* **7**: e1000218.
35. Kauffman SA. (1969) Metabolic stability and epigenesis in randomly constructed genetic nets. *J Theor Biol* **22**: 437–467.
36. Shmulevich I, Kauffman SA. (2004) Activities and sensitivities in Boolean network models. *Phys Rev Lett* **93**: 048701.
37. Xiao Y. (2009) A tutorial on analysis and simulation of Boolean gene regulatory network models. *Curr Genomics* **10**: 511–525.
38. Teschendorff AE, Gomez S, Arenas A *et al.* (2010) Improved prognostic classification of breast cancer defined by antagonistic activation patterns of immune response pathway modules. *BMC Cancer* **10**: 604.
39. Petri CA. (1962) Kommunikation mit Automaten. *Schriften des Institutes Institutes für Instrumentelle Mathematik.*
40. Peleg M, Yeh I, Altman RB. (2002) Modelling biological processes using workflow and Petri net models. *Bioinformatics* **18**: 825–837.

41. Chaouiya C. (2007) Petri net modelling of biological networks. *Brief Bioinform* **8**: 210–219.
42. Breitling R, Gilbert D, Heiner M *et al.* (2008) A structured approach for the engineering of biochemical network models, illustrated for signalling pathways. *Brief Bioinform* **9**: 404–421.
43. Materi W, Wishart DS. (2007) Computational systems biology in drug discovery and development: Methods and applications. *Drug Discov Today* **12**: 295–303.
44. Mura I, Csikasz-Nagy A. (2008) Stochastic Petri net extension of a yeast cell cycle model. *J Theor Biol* **254**: 850–860.
45. Goss PJ, Peccoud J. (1998) Quantitative modeling of stochastic systems in molecular biology by using stochastic Petri nets. *Proc Natl Acad Sci USA* **95**: 6750–6755.
46. Matsuno H, Tanaka Y, Aoshima H *et al.* (2003) Biopathways representation and simulation on hybrid functional Petri net. *In Silico Biol* **3**: 389–404.
47. Wu J, Voit E. (2009) Hybrid modeling in biochemical systems theory by means of functional Petri nets. *J Bioinform Comput Biol* **7**: 107–134.
48. Steggles LJ, Banks R, Shaw O *et al.* (2007) Qualitatively modelling and analysing genetic regulatory networks: A Petri net approach. *Bioinformatics* **23**: 336–343.
49. Simao E, Remy E, Thieffry D *et al.* (2005) Qualitative modelling of regulated metabolic pathways: Application to the tryptophan biosynthesis in *E.coli*. *Bioinformatics* **21**(Suppl 2): ii190–ii196.
50. Silver P, Way J. (2004) Cells by design. *The Scientist* **18**: 30–31.
51. Szybalski W, Skalka A. (1978) Nobel prizes and restriction enzymes. *Gene* **4**: 181–182.
52. Purnick PE, Weiss R. (2009) The second wave of synthetic biology: From modules to systems. *Nat Rev Mol Cell Biol* **10**: 410–422.
53. Szybalski W. (1974) *In Vivo* and *in vitro* Initiation of Transcription. *Control of Gene Expression*, Plenum Press, New York, pp. 23–24.
54. Giraldo R. (2010) Amyloid assemblies: Protein legos at a crossroads in bottom-up synthetic biology. *Chembiochem* **11**: 2347–2357.
55. Stricker J, Cookson S, Bennett MR *et al.* (2008) A fast, robust and tunable synthetic gene oscillator. *Nature* **456**: 516–519.
56. Bhalerao KD. (2009) Synthetic gene networks: The next wave in biotechnology? *Trends Biotechnol* **27**: 368–374.
57. Bansal K, Yang K, Nistala GJ *et al.* (2010) A positive feedback-based gene circuit to increase the production of a membrane protein. *J Biol Eng* **4**: 6.
58. Atsumi S, Liao JC. (2008) Metabolic engineering for advanced biofuels production from *Escherichia coli*. *Curr Opin Biotechnol* **19**: 414–419.
59. Lee SK, Chou H, Ham TS *et al.* (2008) Metabolic engineering of microorganisms for biofuels production: From bugs to synthetic biology to fuels. *Curr Opin Biotechnol* **19**: 556–563.
60. Elowitz MB, Leibler S. (2000) A synthetic oscillatory network of transcriptional regulators. *Nature* **403**: 335–338.

61. Fung E, Wong WW, Suen JK *et al.* (2005) A synthetic gene-metabolic oscillator. *Nature* **435**: 118–122.
62. Dueber JE, Yeh BJ, Bhattacharyya RP *et al.* (2004) Rewiring cell signaling: The logic and plasticity of eukaryotic protein circuitry. *Curr Opin Struct Biol* **14**: 690–699.
63. Balagadde FK, Song H, Ozaki J *et al.* (2008) A synthetic *Escherichia coli* predator–prey ecosystem. *Mol Syst Biol* **4**: 187.
64. Anderson JC, Clarke EJ, Arkin AP *et al.* (2006) Environmentally controlled invasion of cancer cells by engineered bacteria. *J Mol Biol* **355**: 619–627.
65. Anderson JC, Voigt CA, Arkin AP. (2007) Environmental signal integration by a modular AND gate. *Mol Syst Biol* **3**: 133.
66. Ajikumar PK, Tyo K, Carlsen S *et al.* (2008) Terpenoids: Opportunities for biosynthesis of natural product drugs using engineered microorganisms. *Mol Pharm* **5**: 167–190.
67. Martin VJ, Pitera DJ, Withers ST *et al.* (2003) Engineering a mevalonate pathway in *Escherichia coli* for production of terpenoids. *Nat Biotechnol* **21**: 796–802.
68. Hale V, Keasling JD, Renninger N *et al.* (2007) Microbially derived artemisinin: A biotechnology solution to the global problem of access to affordable antimalarial drugs. *Am J Trop Med Hyg* **77**: 198–202.
69. Ro DK, Paradise EM, Ouellet M *et al.* (2006) Production of the antimalarial drug precursor artemisinic acid in engineered yeast. *Nature* **440**: 940–943.
70. Chu LH, Chen BS. (2008) Construction of a cancer-perturbed protein–protein interaction network for discovery of apoptosis drug targets. *BMC Syst Biol* **2**: 56.
71. Huang TC, Huang HC, Chang CC *et al.* (2007) An apoptosis-related gene network induced by novel compound-cRGD in human breast cancer cells. *FEBS Lett* **581**: 3517–3522.

Chapter 3

MicroRNA Regulation in Cellular Networks

Hsueh-Fen Juan[*] and Hsuan-Cheng Huang[†]

1. Introduction

MicroRNAs (miRNAs) are a class of endogenous, small and highly conserved noncoding RNAs that control gene expression post-transcriptionally, either by degradation of target mRNAs or by inhibition of protein translation (Fig. 3 in Chapter 1). A single miRNA can provoke a chain reaction and feedback pathways involving multiple miRNAs and affect multiple target genes of the same or different pathways. The dysregulation of a miRNA is sufficient to trigger global alterations of genetic programs implicated in cell proliferation, differentiation, survival, or invasiveness.[1]

A crucial task in molecular biology is to understand gene regulation in the context of biological networks.[2] Many reports have identified the role of miRNAs in the regulation of cellular networks.[2-4] Cellular networks represent different types of relationships including protein–protein interactions (PPIs) (see also Chapter 4), metabolites in the same biochemical reactions (see also Chapter 5), regulatory relationships between a transcription factor and a gene (called transcriptional regulatory networks, see Sec. 3 and also Chapter 7), signal transduction between a kinase and its substrates, and interactions between regulatory RNAs, such as miRNAs and small

[*]Department of Life Science, Institute of Molecular and Cellular Biology, and Graduate Institute of Biomedical Electronics and Bioinformatics, National Taiwan University, Taipei 106, Taiwan. yukijuan@ntu.edu.tw
[†]Institute of Biomedical Informatics, Center for Systems and Synthetic Biology, National Yang-Ming University, Taipei 112, Taiwan. hsuancheng@ym.edu.tw

interfering RNAs (siRNAs). In this chapter, we will describe in detail miRNA regulation in two kinds of cellular networks: PPI networks (PPINs) and transcriptional regulatory networks.

2. MicroRNA Regulation in PPI Networks

Protein–protein interactions (PPIs) are critical to almost every biological process; thus, a better understanding of PPI and the networks of interacting proteins they form is important for investigating cellular mechanisms.[5] PPINs can be characterized by topological properties such as degree, clustering coefficient, betweenness centrality, and closeness centrality.[6] *Degree* is defined as the number of edges per node; that is, the number of interacting partners. A node with high degree is called a *network hub*, which represents a protein with many interacting partner proteins. *Clustering coefficient* is defined as the ratio between the number of edges linking adjacent nodes and the total number of possible edges among them. *Betweenness centrality* measures the number of nonredundant shortest paths going through a node. A node with high betweenness centrality is called a *network bottleneck*, as they have many shortest paths going through them, analogous to major bridges and tunnels on a highway map.[7] *Closeness centrality* of a node is defined as the reciprocal of the sum of shortest paths to all other nodes. Intuitively, we say two sets are close if they are arbitrarily near to each other. Since proteins fulfill their functions largely through PPIN, it is important to study these relationships and understand how the dynamics of PPIN is influenced by miRNAs.[2]

The global correlation between miRNA regulation and PPIN has been well documented.[2,5] Liang and Li suggested that intermodular hub proteins are more likely to be under miRNA regulation than intramodular hubs (Fig. 1). This can be attributed to the functional difference between these two types of hub proteins: intramodular hub proteins interact with most of their partners simultaneously, whereas intermodular hubs tend to bind different partners at separate times and places.

Hsu *et al.* showed that target proteins of individual miRNAs tend to interact with more proteins than other non-miRNA targets and have a higher betweenness centrality, indicating that many miRNA targets are network hubs and bottlenecks[5] (Fig. 2). Both hubs and bottlenecks are known to represent important nodes in biological networks.[7,8] When these genes are inhibited by an individual miRNA, they may consequently affect a large number of interacting proteins, indicating that miRNAs

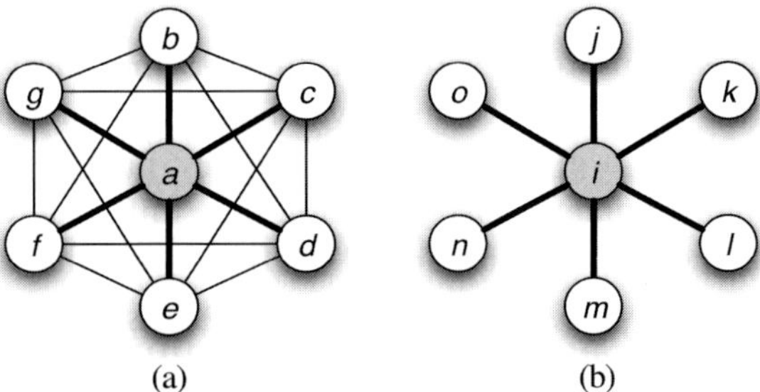

(a)   (b)

Fig. 1. Different types of hub proteins. (a) An example of an intramodular hub. (b) An example of an intermodular hub. The interactions among proteins (nodes, shown as small circles) are shown as lines (edges). The hub proteins are shown as gray nodes. The different letters represent the different proteins.

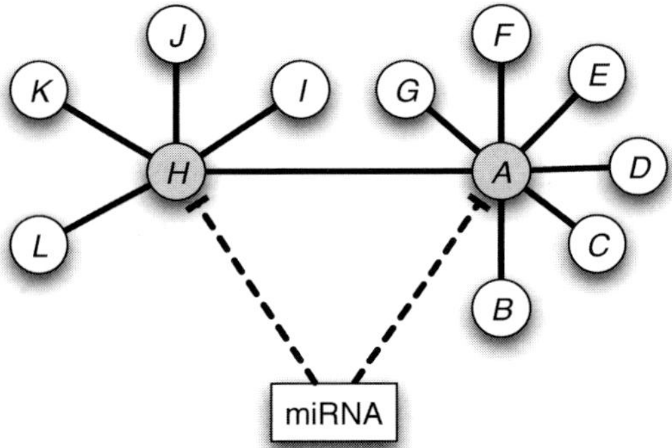

Fig. 2. An example subnetwork of miRNA-regulation in a protein–protein interaction network. The subnetwork consists of miRNA targets (gray nodes) and their interacting partner proteins (open nodes). The interactions among proteins are shown as solid lines. The regulatory interactions between the miRNA and its targets are shown as dashed lines. The different letters represent the different proteins.

may selectively target key components of the PPI network. On the other hand, the characteristic path lengths between the target genes of the same miRNA are significantly shorter, implying that the target genes of individual miRNAs are closer to each other in the network. By selectively targeting the hub proteins and bottleneck proteins, miRNA may regulate the protein interaction network on a larger scale. Since the target genes and their interacting partners might form functional modules, miRNAs could influence specific biological functions through regulating a smaller number of selected genes.[5,9]

key players in tumor development by placing miRNAs in pivotal roles in a well-known tumor-suppressor network.

5. Summary and Future Perspectives

A miRNA can affect many downstream targets that in turn form a complicated network. Therefore, there is considerable potential for the miRNA-regulated network as a novel therapeutic target. The therapeutic strategy could be used for cancer and the other diseases such as neurological and cardiovascular disorders. However, the use of RNA interference and miRNA oligonucleotides into clinic is needed especially with regard to unwanted off-target effects and side effects. Using miRNA-regulated network as a novel therapeutic target could help reduce off-target effects and alleviate the risk of side effects.

Acknowledgments

This work was supported by the National Research Program for Genomic Medicine, Department of Health (DOH), National Science Council of Taiwan, and the National Taiwan University Frontier and Innovative Research Program.

References

1. Zhang H, Li Y, Lai M. (2010) The microRNA network and tumor metastasis. *Oncogene* **29**: 937–948.
2. Liang H, Li WH. (2007) MicroRNA regulation of human protein–protein interaction network. *RNA* **13**: 1402–1408.
3. Cui Q, Yu Z, Pan Y *et al.* (2007) MicroRNAs preferentially target the genes with high transcriptional regulation complexity. *Biochem Biophys Res Commun* **352**: 733–738.
4. Cui Q, Yu Z, Purisima EO *et al.* (2006) Principles of microRNA regulation of a human cellular signaling network. *Mol Syst Biol* **2**: 46.
5. Hsu CW, Juan HF, Huang HC. (2008) Characterization of microRNA-regulated protein–protein interaction network. *Proteomics* **8**: 1975–1979.
6. Zhang S, Jin G, Zhang XS *et al.* (2007) Discovering functions and revealing mechanisms at molecular level from biological networks. *Proteomics* **7**: 2856–2869.
7. Yu H, Kim PM, Sprecher E *et al.* (2007) The importance of bottlenecks in protein networks: Correlation with gene essentiality and expression dynamics. *PLoS Comput Biol* **3**: e59.

8. Lee Y, Yang X, Huang Y *et al.* (2010) Network modeling identifies molecular functions targeted by miR-204 to suppress head and neck tumor metastasis. *PLoS Comput Biol* **6**: e1000730.

9. Cheng J, Zhou L, Xie QF *et al.* (2010) The impact of miR-34a on protein output in hepatocellular carcinoma HepG2 cells. *Proteomics* **10**: 1557–1572.

10. Baskerville S, Bartel DP. (2005) Microarray profiling of microRNAs reveals frequent coexpression with neighboring miRNAs and host genes. *RNA* **11**: 241–247.

11. Yu J, Wang F, Yang GH *et al.* (2006) Human microRNA clusters: Genomic organization and expression profile in leukemia cell lines. *Biochem Biophys Res Commun* **349**: 59–68.

12. Yuan X, Liu C, Yang P *et al.* (2009) Clustered microRNAs' coordination in regulating protein–protein interaction network. *BMC Syst Biol* **3**: 65.

13. Shen-Orr SS, Milo R, Mangan S *et al.* (2002) Network motifs in the transcriptional regulation network of *Escherichia coli. Nat Genet* **31**: 64–68.

14. Lee TI, Rinaldi NJ, Robert F *et al.* (2002) Transcriptional regulatory networks in *Saccharomyces cerevisiae. Science* **298**: 799–804.

15. Yu H, Gerstein M. (2006) Genomic analysis of the hierarchical structure of regulatory networks. *Proc Natl Acad Sci USA* **103**: 14724–14731.

16. Hobert O. (2008) Gene regulation by transcription factors and microRNAs. *Science* **319**: 1785–1786.

17. Shalgi R, Lieber D, Oren M *et al.* (2007) Global and local architecture of the mammalian microRNA-transcription factor regulatory network. *PLoS Comput Biol* **3**: e131.

18. Zhou Y, Ferguson J, Chang JT *et al.* (2007) Inter- and intra-combinatorial regulation by transcription factors and microRNAs. *BMC Genomics* **8**: 396.

19. Milo R, Shen-Orr S, Itzkovitz S *et al.* (2002) Network motifs: Simple building blocks of complex networks. *Science* **298**: 824–827.

20. Mangan S, Alon U. (2003) Structure and function of the feed-forward loop network motif. *Proc Natl Acad Sci USA* **100**: 11980–11985.

21. Tsang J, Zhu J, Van Oudenaarden A. (2007) MicroRNA-mediated feedback and feedforward loops are recurrent network motifs in mammals. *Mol Cell* **26**: 753–767.

22. Brosh R, Shalgi R, Liran A *et al.* (2008) p53-Repressed miRNAs are involved with E2F in a feed-forward loop promoting proliferation. *Mol Syst Biol* **4**: 229.

23. Chen C-Y, Chen S-T, Fuh C-S *et al.* (2011) (in press) Coregulation of transcription factors and microRNAs in human transcriptional regulatory network. *BMC Bioinformatics* **12**(Suppl 1): S41.

24. Cui Q, Purisima EO, Wang E. (2009) Protein evolution on a human signaling network. *BMC Syst Biol* **3**: 21.

25. Qiu C, Wang J, Yao P *et al.* (2010) microRNA evolution in a human transcription factor and microRNA regulatory network. *BMC Syst Biol* **4**: 90.

26. Hermeking H. (2010) The miR-34 family in cancer and apoptosis. *Cell Death Differ* **17**: 193–199.

27. He L, He X, Lowe SW *et al.* (2007) MicroRNAs join the p53 network — another piece in the tumour-suppression puzzle. *Nat Rev Cancer* **7**: 819–822.

28. Emmrich S, Putzer BM. (2010) Checks and balances: E2F-microRNA crosstalk in cancer control. *Cell Cycle* **9**(13): 2555–2567.

29. Volinia S, Galasso M, Costinean S *et al.* (2010) Reprogramming of miRNA networks in cancer and leukemia. *Genome Res* **20**: 589–599.

30. Bommer GT, Gerin I, Feng Y *et al.* (2007) p53-mediated activation of miRNA34 candidate tumor-suppressor genes. *Curr Biol* **17**: 1298–1307.

31. Chang TC, Wentzel EA, Kent OA *et al.* (2007) Transactivation of miR-34a by p53 broadly influences gene expression and promotes apoptosis. *Mol Cell* **26**: 745–752.

32. He L, He X, Lim LP *et al.* (2007) A microRNA component of the p53 tumour suppressor network. *Nature* **447**: 1130–1134.

33. Raver-Shapira N, Marciano E, Meiri E *et al.* (2007) Transcriptional activation of miR-34a contributes to p53-mediated apoptosis. *Mol Cell* **26**: 731–743.

34. Tarasov V, Jung P, Verdoodt B *et al.* (2007) Differential regulation of microRNAs by p53 revealed by massively parallel sequencing: miR-34a is a p53 target that induces apoptosis and G1-arrest. *Cell Cycle* **6**: 1586–1593.

Chapter 4

Disease Modules in Protein–Protein Interaction Networks

Hsueh-Fen Juan[*] and Hsuan-Cheng Huang[†]

1. Introduction

Protein–protein interactions (PPIs) are critical to almost any biological process, and thus a better understanding of PPI and networks of interacting proteins is important in investigating cellular mechanisms. High-throughput experiments on PPIs allow us to build an interaction network and gain insight into the underlying cellular mechanisms.[1] PPI networks are conveniently modeled by graphs, where the nodes are proteins and two nodes are connected by a link if the corresponding proteins physically bind to each other (Fig. 1). This representation of complex cellular networks as graphs has made it possible to systematically investigate the topology and function of these networks using well-established mathematical and computational methods to predict the structural and dynamical properties of the underlying network.[2]

Modularity is recognized as a design principle in biological systems or as an important structural feature of biological networks.[3,4] Most biological functions are carried out by specific genes or proteins, and in this way one can separate the network structure into functional modules. If a gene or protein is involved in a specific biochemical process or disease, its direct interactors in PPI networks might also have some role in the same process.

[*]Department of Life Science, Institute of Molecular and Cellular Biology, and Graduate Institute of Biomedical Electronics and Bioinformatics, National Taiwan University, Taipei 106, Taiwan. yukijuan@ntu.edu.tw

[†]Institute of Biomedical Informatics, Center for Systems and Synthetic Biology, National Yang-Ming University, Taipei 112, Taiwan. hsuancheng@ym.edu.tw

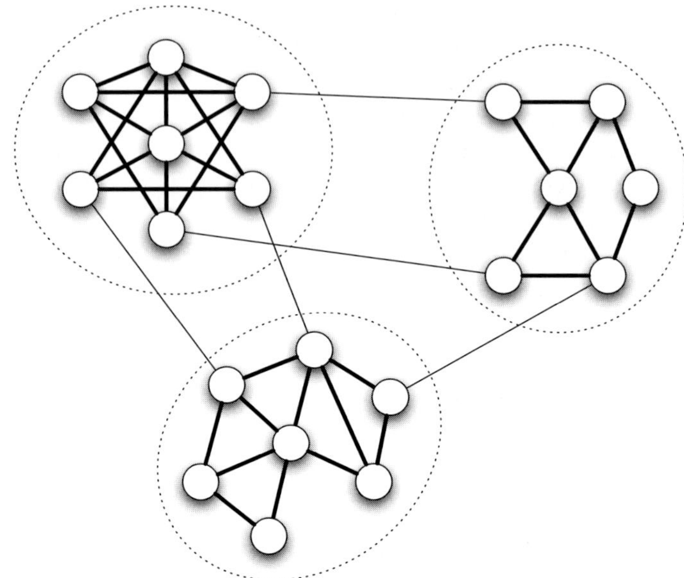

Fig. 1. A simple network with three modules, enclosed by the dashed circles. A module consists of densely connected nodes with links shown as thick lines. Thin lines represent the loose connections between modules. In a PPI network, a node represents a single protein and links represent physical interactions between proteins. A PPI network module might indicate a group of proteins carrying out some specific biological function.

We expect that each disease can be linked to some neighborhood in the network as a disease module.

2. Network Modules

Network modules are not rigorously defined. A module is generally considered to be a group of nodes that are densely connected amongst each other but loosely connected to other nodes outside the module (Fig. 1). Identification of modules within a complex network is computationally challenging. This computational problem is closely related to the clustering of elements, that is, the assignment of "entities" into distinct categories based on a suitably defined notion of similarity. In the context of complex network analysis, the property common to all nodes, which identifies them as being part of a specific module, could be defined in different ways. For example, the shortest path between two nodes or the total number of paths between nodes (weighted according to path length) are among several other possibilities. Likewise, and again resembling the situation in clustering algorithms, modules can be detected using agglomerative or divisive methods, that is, by grouping similar nodes into larger units (agglomerative methods) or by iteratively breaking down a complex network into smaller and smaller subsets (divisive methods).[5] In a seminal paper published in 2002, Girvan and Newman proposed a new algorithm, aiming at the identification of links lying between modules and their successive removal, a procedure that, after some iterations, leads to the isolation of the modules.[6] The intermodule links are detected according to the values of a centrality measure that expresses the importance of the role of the links in processes whereby signals are transmitted across the network following paths of minimal length. This paper triggered a surge in activity in the field, and a wide array of new methods and tools have emerged over the past few years.[7] Some examples are described in the following.

The Markov clustering algorithm (MCL)[8] simulates random walks in PPI networks. The MCL manipulates the weighted or unweighted adjacency matrix with two operators called expansion and inflation. The expansion operator assigns new probabilities for all pairs of nodes, while the inflation operator changes the probabilities for all these walks in the graph, boosting the probabilities of intracluster walks and demoting intercluster walks. Iterative expansion and inflation will separate the PPI network into many clusters (modules).

CFinder[9,10] uses the clique percolation method to locate the k-clique percolation clusters of a network interpreted as modules. A k-clique is a complete subgraph on k nodes ($k = 3, 4, \ldots$), and two k-cliques are said to be adjacent if they share exactly $k - 1$ nodes. A k-clique percolation cluster consists of a union of all k-cliques that can be reached from each other through a series of adjacent k-cliques (Fig. 2). Note that larger values of k correspond to a higher stringency during the identification of dense groups and provide smaller groups with a higher density of links inside them. The advantage of this method is that it can identify overlapping and nested modules.

Recently, Ahn *et al.* proposed to redefine modules as groups of links rather than nodes.[11] They used hierarchical clustering with a similarity between links to build a dendrogram where each leaf is a link from the original network and branches represent modules of links (Fig. 3). In this dendrogram, links occupy unique positions whereas nodes naturally occupy multiple positions, owing to their links. They extracted link communities

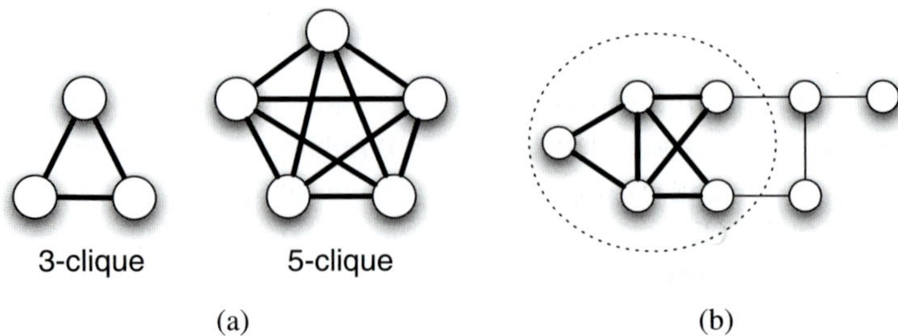

Fig. 2. (a) Examples of k-clique ($k = 3$ and 5). (b) A 3-clique percolation cluster, enclosed by the dashed circle. Thick lines represent the links within the module, and thin lines represent the links outside the module.

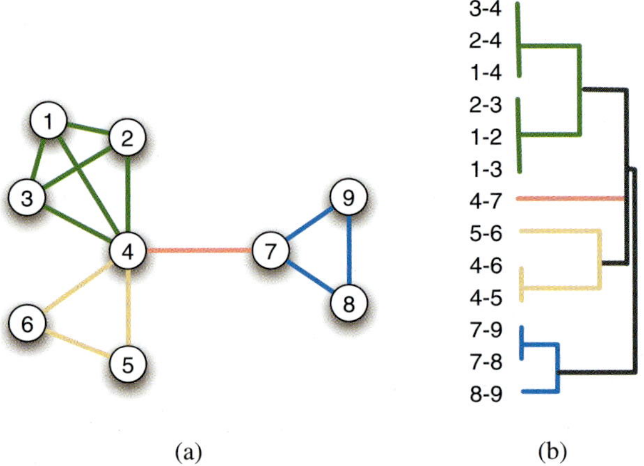

Fig. 3. (a) An example network showing link communities as represented by different colors. (b) A link dendrogram constructed by link similarities. (Modified from Ahn *et al.*[11])

at multiple levels by cutting this dendrogram at various thresholds. Each node inherits all memberships of its links and can thus belong to multiple overlapping modules. They showed that this unorthodox approach successfully reconciles the antagonistic organizing principles of overlapping modules and hierarchy.

3. Disease Subnetworks

To understand the molecular basis of diseases, it is important to discover their causal genes. Typically, a disease is associated with a linkage interval on the chromosome. These linkage intervals define a set of candidate disease-causing genes. Genes related to the same disease are also known to have protein products that physically interact, which form subnetworks in a PPI network. Subnetworks are significant because, in contrast to individual proteins, they can provide hypotheses as to the molecular complexes, signaling pathways, and other mechanisms that impact the disease outcome.[12–15]

Huntington's disease is an autosomal dominant neurodegenerative disease, which is caused by the repeat expansion of the trinucleotide CAG in the Huntingtin (Htt) gene. This expansion causes aggregation of the mutant Htt protein into insoluble neuronal inclusion bodies, which consequently leads to neuronal degeneration. Goehler *et al.* reported an experimental strategy to construct a PPI subnetwork around Htt, revealing many new interactions and permitting the functional annotation of several uncharacterised proteins.[16] Each direct interactor of HTT was tested for its ability to enhance Htt aggregation. This screen identified a new enhancer of Htt aggregation, GIT1 (a GTPase-activating protein), and additional tests verified its role in disease pathogenesis.

Lim *et al.* used yeast two-hybrid screens to construct a PPI subnetwork around 23 proteins involved in inherited ataxias.[17] The resulting network was expanded by previously known interactions and contained a total of 7000 interactions among 3500 proteins. They found that most of the proteins related to ataxias interact with each other. The network was highly connected, and the mean distance between the ataxia-causing proteins in the network was much lower than in a network nucleated around 30 disease proteins sampled independent of phenotypes. They demonstrated that networks can be utilized to uncover novel ataxia-causing genes and genetic modifiers for ataxia.

4. Network-Based Disease Classification

An emerging application of molecular network analysis is the use of networks to improve the task of disease classification. Ma *et al.* used the concept that candidate genes associated with specific disease phenotypes fall into the same network neighborhoods, to identify genes predictive of Alzheimer's disease. Their method uses the PPI network to define a Markov random field, in which the score of association between a gene's expression level and the disease class depends on the scores of its network neighbors. It was reported to perform better than when gene-expression data were used as the sole source of information.[18]

Chuang *et al.* applied a network-based classifier to the prognosis of breast cancer metastasis.[19] According to their method, the gene-expression profiles of metastatic and nonmetastatic patients are superimposed on a human PPI network. Subnetworks are identified whose expression levels correlate with metastasis, where the expression level of a subnetwork is defined as a function of the expression levels of its member genes. They found that subnetwork markers were more reproducible than individual marker genes selected without network information and that they achieved higher accuracy in the classification of metastatic versus nonmetastatic tumors.

Taylor *et al.* showed that the dynamic structure of PPI networks can be used to predict breast cancer outcome.[20] Changes in the biochemical wiring of oncogenic cells drives phenotypic transformations that directly affect disease outcome. They searched for changes in the global modularity of the human interactome that indicated altered organization and information flow. Other mechanisms that affect network connectivity, such as alterations in protein stability and post-translational modification, may also influence network modularity on a global scale during cancer progression. Their studies suggest the need to search for multimodal therapies that target hubs in networks with altered modularity in disease. Furthermore, the favorable performance of their classification algorithms suggests that changes in network modularity may be a defining feature of tumor phenotype that in turn determines patient prognosis.

Lin *et al.* developed a network-based comparative analysis approach that integrates PPIs with gene expression profiles and biological function annotations to reveal dynamic functional modules under different biological states.[21] They found that hub proteins in condition-specific co-expressed PPI networks tended to be differentially expressed between biological states. Applying this method to a cohort of heart failure patients, they identified two functional modules that significantly emerged from the interaction networks. The dynamics of these modules between normal and disease states further suggests a potential molecular model of dilated cardiomyopathy and provides new insights into its cause. The revealed molecular modules might be used as potential drug targets and provide new directions for heart failure therapy.

5. Summary and Perspectives

Identification and analysis of disease modules in PPI networks offer a powerful means to elucidate the molecular mechanisms of human disease.

However, human PPI network data remain sparse, and many important types of networks, such as networks of regulatory and synthetic–lethal or chemical–genetic interactions, are still forthcoming. Collecting and interrogating these data to perform a comprehensive network-level investigation are pressing issues for systems biology research. On the other hand, with more and more high-throughput experimental data sets that provide snapshots of biological systems at different times and conditions, it is increasingly permissible to infer and elucidate the interaction network dynamics with advances in computational methods. Although this remains a challenging research area, we can foresee that integrative dynamic network analysis will further our understanding of cellular processes and disease behavior. It will eventually revolutionize our view of disease toward network and systems medicine.

Acknowledgments

This work was supported by the National Research Program for Genomic Medicine, Department of Health (DOH), National Science Council of Taiwan, and the National Taiwan University Frontier and Innovative Research Program.

References

1. Rual JF, Venkatesan K, Hao T *et al.* (2005) Towards a proteome-scale map of the human protein–protein interaction network. *Nature* **437**: 1173–1178.
2. Aittokallio T, Schwikowski B. (2006) Graph-based methods for analysing networks in cell biology. *Brief Bioinform* **7**: 243-255.
3. Hartwell LH, Hopfield JJ, Leibler S *et al.* (1999) From molecular to modular cell biology. *Nature* **402**: C47–C52.
4. Wang X, Dalkic E, Wu M *et al.* (2008) Gene module level analysis: Identification to networks and dynamics. *Curr Opin Biotechnol* **19**: 482–491.
5. Junker BH, Schreiber F. (2008) *Analysis of Biological Networks*. John Wiley & Sons, Inc.
6. Girvan M, Newman ME. (2002) Community structure in social and biological networks. *Proc Natl Acad Sci USA* **99**: 7821–7826.
7. Fortunato (2010) Community detection in graphs. *Phys Rep* **486**: 75-174.
8. Enright AJ, Van Dongen S, Ouzounis CA. (2002) An efficient algorithm for large-scale detection of protein families. *Nucleic Acids Res* **30**: 1575–1584.
9. Palla G, Derenyi I, Farkas I *et al.* (2005) Uncovering the overlapping community structure of complex networks in nature and society. *Nature* **435**: 814–818.

10. Adamcsek B, Palla G, Farkas IJ *et al.* (2006) CFinder: Locating cliques and overlapping modules in biological networks. *Bioinformatics* **22**: 1021–1023.

11. Ahn YY, Bagrow JP, Lehmann S. (2010) Link communities reveal multiscale complexity in networks. *Nature* **466**: 761–764.

12. Goh KI, Cusick ME, Valle D *et al.* (2007) The human disease network. *Proc Natl Acad Sci USA* **104**: 8685-8690.

13. Kann MG. (2007) Protein interactions and disease: Computational approaches to uncover the etiology of diseases. *Brief Bioinform* **8**: 333–346.

14. Ideker T, Sharan R. (2008) Protein networks in disease. *Genome Res* **18**: 644–652.

15. Navlakha S, Kingsford C. (2010) The power of protein interaction networks for associating genes with diseases. *Bioinformatics* **26**: 1057–1063.

16. Goehler H, Lalowski M, Stelzl U *et al.* (2004) A protein interaction network links GIT1, an enhancer of huntingtin aggregation, to Huntington's disease. *Mol Cell* **15**: 853–865.

17. Lim J, Hao T, Shaw C *et al.* (2006) A protein–protein interaction network for human inherited ataxias and disorders of Purkinje cell degeneration. *Cell* **125**: 801–814.

18. Ma X, Lee H, Wang L *et al.* (2007) CGI: A new approach for prioritizing genes by combining gene expression and protein–protein interaction data. *Bioinformatics* **23**: 215–221.

19. Chuang HY, Lee E, Liu YT *et al.* (2007) Network-based classification of breast cancer metastasis. *Mol Syst Biol* **3**: 140.

20. Taylor IW, Linding R, Warde-Farley D *et al.* (2009) Dynamic modularity in protein interaction networks predicts breast cancer outcome. *Nat Biotechnol* **27**: 199–204.

21. Lin CC, Hsiang JT, Wu CY *et al.* (2010) Dynamic functional modules in co-expressed protein interaction networks of dilated cardiomyopathy. *BMC Syst Biol* **4**: 138.

Chapter 5

Biomolecular Pathway Modeling

Feng-Sheng Wang[*] and Wu-Hsiung Wu[†]

1. Introduction

Systems biology aims at a holistic analysis of biological networks. Predicting the behavior of large-scale biomolecular networks, such as gene networks, protein interaction networks, metabolic networks and signaling networks, represents one of the greatest challenges of bioinformatics and computational systems biology.[1–4] To understand biological networks as systems, we must investigate four areas: system structure identification, system behavior analysis, system control and system design.[1] Mathematical modeling and dynamic simulation of genetic networks, metabolic networks and signal transduction cascades are central tools in systems biology to understand these areas and are increasingly attracting attention in the post-genomic era.[1,5–9] Used appropriately, mathematical models can represent pathways in a physically and biologically realistic manner, incorporate a wide variety of empirical observations, and generate novel and useful hypotheses. The ultimate goal of mathematical modeling is to obtain expressions that quantitatively describe every detail and principle of biological systems. A major current challenge in biological modeling is to clarify the relationship between structure, function and regulation in complex networks that can be reconstructed from genomic or biochemical data. Biological models can be classified into two groups: *first-principle* (fundamental) models and *data-based* (empirical, black box) models. First-principle models are based

[*]Department of Chemical Engineering, National Chung Cheng University, 168 University Road, Minhsiung Township, Chiayi County 62102, Taiwan. chmfsw@ccu.edu.tw

[†]Department of Computer Science and Information Engineering, National Chung Cheng University, 168 University Road, Minhsiung Township, Chiayi County 62102, Taiwan. wwh@cs.ccu.edu.tw

on fundamental physical/chemical principles, such as the conservation of mass and energy. One of the most important reasons for using fundamental models is the analytical expressions they provide, relating key features of the physical system to its dynamic behavior. Data-based models provide relations between measured inputs and outputs that describe how the biological system responds to changes in various inputs. They can be developed much faster than first-principle models, but their accuracy, robustness and utility are limited. They provide an inexpensive alternative to fundamental models in most monitoring, diagnosis and control tasks.

Building a mathematical model primarily involves two steps: (1) deciding on the model structure and (2) estimating the involved parameter values. Biological processes and interactions are highly nonlinear and complex. A linear model is usually not very well-suited for describing such dynamic behavior. The dynamic responses of biological systems are of particular interest. Therefore, a suitable mathematical model will have to be time-dependent, which almost always requires formulation as a set of differential equations. Given a model structure, parameter estimation remains the limiting step in the modeling and simulation of biological systems. There have been in-depth studies regarding parameter estimation for nonlinear dynamic systems[10–16] and for steady-state systems[17,18]; however, difficulties still remain because of measurement noises and the escalating complexity resulting from the increasing number of parameters to be estimated.[10,19]

Determining the parameters for a mathematical model from quantitative measurements is the main bottleneck in modeling biological systems. Two major challenges for parameter estimation using dynamic observations have to be surmounted to be able to efficiently estimate parameter values of nonlinear dynamic biological systems. The first challenge is to develop a powerful numerical integration method for solving differential equations. In conventional parameter estimation, numerical integration is required to yield dynamic profiles for the model. Numerical integration failure is the major problem in the optimization search. In addition, numerical integration is time-consuming. Time-course slope information for a dynamic system has been applied to avoid numerical integration; using the dynamic model to compute the slope error information can alleviate the computational burden.[9,10,20,21] The decomposition strategy has previously been applied to solve differential equations in parallel to yield dynamic profiles.[19,22] Hybrid differential evolution with the modified collocation method has been applied to determine model parameters for biological systems.[13,15,23] The second challenge is to develop an efficient optimization method to find the global estimates. Several conventional methods, such as the graphical method and gradient-based nonlinear optimization methods,[14] have been employed as estimation techniques to obtain parameter values. The graphical method can be applied only to those problems that can be converted to linear regression problems. Basically, a gradient-based nonlinear optimization method does not have such a restriction, but it requires gradient information about the error function with respect to the parameter estimates and often converges to local minima. Evolutionary algorithms can be applied to overcome such drawbacks.

Parameter sensitivity analysis is sometimes used as a tool for assessing model accuracy. This type of analysis can be described as the study of the behavior of dynamic systems with small perturbations in system parameters. Specific experiments have been proposed to obtain sensitivity measures of the system to validate a model. Fell[24] reviewed experimental techniques for estimating elasticities and control coefficients that are based on changing enzyme activities while only minimally affecting other system properties.[25,26] The two prominent frameworks for metabolic analysis, biochemical systems theory (BST)[9,27–30] and metabolic control analysis (MCA),[24,31,32] use sensitivity coefficients in the form of logarithmic derivatives to characterize systemic and local properties.[17,32–34] The systemic sensitivity coefficients are known as control and response coefficients in MCA and as logarithmic gains in BST, while the local sensitivity coefficients are called elasticity coefficients in MCA and kinetic orders in BST. These coefficients are the basis of two important properties of steady-state metabolic systems, namely the summation and the connectivity relationships, which were discovered in MCA and similarly hold in BST for the majority of practical examples (see Kimura *et al.*[19] for exceptions). They are intrinsic features of metabolic systems in which the enzymes affect reactions in a linear fashion. The summation relationship is a local property that states that the sum of all sensitivities of a particular flux with respect to all rate constants is always equal to one. A particularly useful and important feature of the connectivity relationship is that it relates the kinetic properties of the individual reactions (local properties) to (global) properties of the intact pathway. BST and MCA have achieved great success in addressing steady-state sensitivities. However, some biological systems, such as signal transduction or cell-cycle regulation, do not have steady-state responses and there is a need to perform sensitivity analysis on their dynamic oscillation behaviors. Understanding transient dynamics is becoming possible because advanced experimental techniques and mathematical tools have been developed. Dynamic sensitivity analysis can be used to analyze interaction effects of cellular mechanisms to understand the dynamic behavior of biological systems, such as oscillations of cell-cycle control.

2. Modeling of Biochemical Networks

The correct mathematical form for a physicochemical model depends on the properties of the system being studied and the goals of the modeling effort. Physicochemical modeling seeks to describe biomolecular transformations (such as covalent modification, intermolecular association, and intracellular localization) in terms of equations derived from established physical and chemical theory. These "kinetic" or "reaction" models use prior knowledge to make specific molecular predictions and work best with pathways in which components and connectivity are relatively well established. When prior knowledge is sparse, data-driven statistical models are more appropriate. A biochemical network, such as metabolic pathways, gene regulation networks and protein interaction pathways, can be formulated through a set of coupled ordinary differential equations (ODEs). ODEs are most commonly cast in either deterministic or stochastic form. Stochastic equations include effects arising from random fluctuations around the average behavior. ODE networks represent the rates of production and consumption of individual biomolecular species, dx_i/dt, in terms of mass action kinetics — an empirical law stating that rates of a reaction are proportional to the concentrations of the reacting species. Each biochemical transformation is therefore represented by an elementary reaction with forward and reverse rate constants. Some models utilize partial differential equations (PDEs) to represent biological systems if space or mechanics are involved. Changes in localization of PDEs, a central feature of biological pathways, can be represented by compartmentalization. As a result, PDEs can be reformulated as a set of compartmentalized ODEs. Two fundamental assumptions of the compartmentalized ODE formalism are that: first, within a compartment, the concentration of each species is high and transport essentially instantaneous, that is, the compartment is well mixed; and second, between compartments, transport is slower and associated with an observable rate. If these assumptions are not satisfied, then it is necessary to model changes in species concentrations explicitly with respect to space.

The dynamics of a biomolecular reaction network can be represented generically using a set of nonlinear ODEs with the following structure:

$$\frac{\mathrm{d}x_i}{\mathrm{d}t} = f_i(\mathbf{x}, \theta) = \sum_{j=1}^{r} N_{ij} v_j, \quad i = 1, \ldots, n, \tag{1}$$

where n is the number of the biomolecular species or component, x_i is the concentration for the ith biomolecular species or pool, v_j is the reaction rate for the jth pathway, N_{ij} are the coefficients of the stoichiometric matrix $\mathbf{N}$, $\mathbf{x}$ is the $n \times 1$ state vector consisted of $x_i, i = 1, \ldots, n$, and θ is the $n_p \times 1$ parameter vector included in the reaction rates. The stoichiometric matrix serves as a connectivity matrix and represents a network. This network is represented by a map. Each species (or node) in the map corresponds to a row in the matrix, and each column corresponds to a link in the map. The stoichiometric property of a network is typically considered time-invariant, while kinetic aspects are used to capture the dynamics of a system and are driven by the state of the system and may change rather quickly. The stoichiometry of a biochemical pathway determines the wiring diagram of the network and describes which fluxes enter or leave which pool and ensures that mass is conserved in the process. The translation of a topological wiring diagram into a matrix equation is straightforward.

Many mathematical formalisms are available to express reaction rates.[9,35–37] Michaelis–Menten-based rate law is well known in the field of biological systems and is expressed as

$$v_j = \frac{V_{j,max} x_j}{K_{j,M} + x_j}. \tag{2}$$

This model is derived by making quasi steady-state assumptions for some components. More complex rate law expressions can be derived using different versions of the quasi steady-state approximation to describe more complex protein-catalyzed dynamics, such as allosteric processes, competitive inhibition and cooperativity, and can be formulated as

$$v_j = \frac{V_{j,max} \prod_k x_k^{m_{jk}}}{K_{j,M} + \prod_l x_l^{m_{jl}}}, \tag{3}$$

where m_{jk} are the kinetic orders. While the quasi-stationary approximation is generally accepted for systems operating in a preferential steady state, e.g. metabolic pathways, these assumptions are questionable in the context of signal transduction pathways where the focus is on transient behavior.[38]

Savageau[27–30] proposed approximating kinetic rate laws with multivariate power-law functions. This representation results from Taylor's theorem in logarithmic coordinates. It offers striking procedural, mathematical and computational benefits and is the basis of a conceptual framework known as BST. The two most important variants within BST are generalized mass action (GMA) and S-system. GMA focuses on processes such that each enzyme reaction or transport step is modeled individually as a product of power-law functions. In the GMA representation, each flux entering or leaving the variable x_i is expressed as a product of power-law functions. Each rate law in Eq. (1) is therefore expressed as

$$v_j = \alpha_j \prod_{k=1}^{n} x_k^{g_{jk}}, \tag{4}$$

where α_j is the rate constant for the jth rate law and g_{jk} are kinetic orders. The S-system focuses on component pools. It collects all influxes into one term V_i^+, which is subsequently represented collectively with one product of power-law functions, and it does the same for all effluxes, resulting in V_i^-. Each S-system equation thus always consists of only one difference between two power-law terms and the generic S-system is given by

$$\frac{\mathrm{d}x_i}{\mathrm{d}t} = V_i^+ - V_i^- = \alpha_i \prod_{j=1}^{n} x_j^{g_{ij}} - \beta_i \prod_{j=1}^{n} x_j^{h_{ij}}, \quad i = 1, \ldots, n. \tag{5}$$

The multipliers α_i and β_i are rate constants that quantify the turnover rate of production or degradation, respectively. The exponents g_{ij} and h_{ij} are real-valued kinetic orders and quantify the effect of x_j on the production or degradation of x_i, respectively. The modeled effect is activating or augmenting if the kinetic order is positive; if it is negative the effect is inhibiting. If a variable x_j does not have an effect on x_i, the kinetic order is zero.

The log-lin and lin-log models[39–41] were introduced as further model representations. These representations are also based on ordinary differential equations that consist of a combination of logarithmic and additive components and share much of the modeling philosophy of BST. Both log-lin and lin-log models may be derived as Taylor approximations and have structural features in common with BST. However, it is more typical to formulate them directly based on the topology of the biochemical reaction networks by including in each process description each enzyme and each contributing variable in relation to a chosen reference steady state.

Specifically, the rates v_i in a log-lin model are represented as

$$\frac{v_i}{v_i^0} = 1 + \ln\left(\frac{e_i}{e_i^0}\right) + \sum_{j=1}^{n} \varepsilon_{ij}^0 \ln\left(\frac{x_j}{x_j^0}\right),\tag{6}$$

while the rates v_i in a lin-log model are given as

$$\frac{v_i}{v_i^0} = \frac{e_i}{e_i^0}\left\{1 + \sum_{j=1}^{n} \varepsilon_{ij}^0 \ln\left(\frac{x_j}{x_j^0}\right)\right\}.\tag{7}$$

In both cases, v_i^0 is the ith reference steady-state flux rate, x_j^0 is the reference value of a dependent or independent variable, e_i^0 is the reference level of the ith enzyme activity, and ε_{ij}^0 is the reference elasticity, which is analogous to kinetic orders in BST. If the enzyme levels do not change during an experiment, the two models (6) and (7) are equivalent. No matter which estimation method is used, it is preferred to avoid redundancies that occur if parameters always appear in the same combination, such as a sum or a product. This issue is pertinent for S-systems and log-lin/lin-log models. Wang *et al.*[16] have shown that S-system and log-lin/lin-log models have identical characteristic behaviors in a steady state.

3. Parameter Estimation

Parameter estimation for biological models is the bridge connecting the experimental results with the computational models. Determining the parameters of a mathematical model from quantitative measurements is the main impediment in modeling biological systems.[10] Parameter values can be estimated from steady-state or dynamic data. The nature of suitable data for these two types of estimation is rather different. For instance, estimations of parameter values from steady-state data in pathway models, such as rate constants and kinetic orders, are generally based on experiments that measure how a biochemical system responds to small perturbations around the steady state. In contrast, parameter estimation from dynamic data requires time-series measurements for all dependent variables. Some new experiments should then be carried out to validate the estimated model. As mentioned in the introduction, there are still many challenges in using time-series data to solve parameter estimation problems for nonlinear biological systems. Inconsistencies may lead to improved second-round estimates that again must be validated.

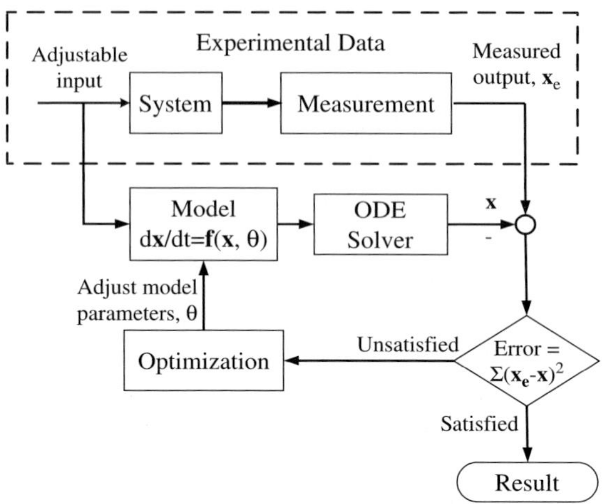

Fig. 1. A conventional scheme of parameter estimation.

Parameter estimation is generally formulated as an optimization problem, which is referred to as an inverse problem. Several conventional methods, such as graphical methods and gradient-based optimization methods, have been applied to parameter estimation problems. The graphical method is only suited for special cases such as the Michaelis–Menten model. Figure 1 shows the concepts of the computational technique for parameter estimation. As shown in the dash box of Fig. 1, independent inputs are provided to perform experiments to obtain measurable output data. Such inputs are also provided for the proposed model to compute the system output data. Since there is no information on most of the model parameters, they must be inferred from the data. Therefore, we optimize an error criterion that measures the discrepancy of simulated data from the real data. The error criterion in conventional parameter estimation techniques is evaluated from the measured concentrations and dynamic profiles computed using ordinary differential equations. Such a criterion is expressed as

$$J_{1l} = \frac{1}{nN_s} \sum_{i=1}^{n} \sum_{s=1}^{N_s} \frac{(x_{e_i}(t_s) - x_i(t_s))^2}{x_{e_i,max}^2}, \quad l = 1, \ldots, N_{exp}, \qquad (8)$$

where $x_{e_i}(t_s)$ is the measured data for the ith component at $t = t_s$, $x_i(t_s)$ is the computed concentration for the ith component at $t = t_s$ and $x_{e_i,max}$ is the maximum measured concentration of the ith component. Here, N_s is the number of sampled data points and N_{exp} is the number of experiments. Meanwhile, an optimization method numerically generates model parameters for the model. Differential equations including these model parameters are numerically solved to yield dynamic profiles. A least squared regression sums up every observed error in (8) to yield a criterion to inspect how close the experimental data are to the computed profiles. If the error criterion is not satisfied, then new model parameters are generated using the optimization algorithm and this is repeated until a satisfied result is obtained.[10,42]

The quality of a solution to an inverse problem depends on the efficiency of the optimization method and differential equation solver, as mentioned above. Most nonlinear regressions are gradient-based optimization methods so that the solution quality strongly depends on the initial starting point. Moreover, gradient-based methods may yield a local minimum, not a global solution. Evolutionary algorithms can be applied to overcome such drawbacks. Numerical integration failure is a major problem during the evolutionary search process. In addition, numerical integration is time consuming. Many techniques have been employed to alleviate the numerical integration burden. Voit and Almeida[21] utilized a decoupling scheme to estimate the slopes of the dynamic processes. The decomposing method[22,43] is employed to convert ordinary differential equations into decoupled algebraic equations. Tsai and Wang[15] have introduced the modified collocation method[44] to approximate dynamic profiles so that experimental data could be directly applied to the parameter estimation problem. Figure 2 shows the computational concepts of parameter estimation with the modified collocation method. The approach is very similar to Fig. 1 except for the approximation procedure. The modified collocation method is applied to convert differential equations into algebraic equations. Such algebraic equations directly adopt the measured data to approximately yield dynamic profiles to evaluate the error criterion.

Time-course slopes can also be applied to avoid numerical integration. An artificial neural network is first applied to smooth the measured data to evaluate the time-course slopes for each component.[21,45] The evaluated slopes are compared with the slopes of the model to yield the error criterion in the form:

$$J_{2l} = \frac{1}{nN_s} \sum_{i=1}^{n} \sum_{s=1}^{N_s} \frac{(\dot{x}_{e_i}(t_s) - \dot{x}_i(t_s))^2}{\dot{x}_{e_i,max}^2}, \quad l = 1, \ldots, N_{exp}, \tag{9}$$

where $\dot{x}_{e_i}(t_s)$ is the measured slope for the ith component at $t = t_s$, $\dot{x}_i(t_s)$ is the computed slope for the ith component at $t = t_s$ and $\dot{x}_{e_i,max}$ is the

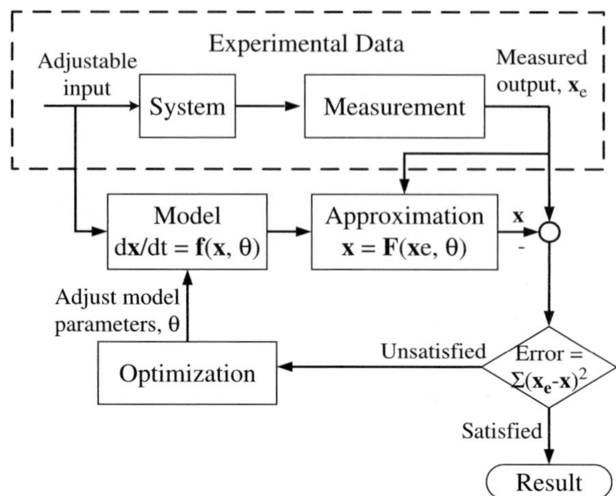

Fig. 2. Parameter estimation using the modified collocation approximation technique.

maximum slope of the ith component. Using the model slope to compute the error criterion can alleviate the computational burden. As discussed in the textbook by Voit,[9] a very small error value may be obtained by this slope approach. This implies that the slopes of the model are nearly identical to those of experiments. However, dynamic profiles of the model may not accurately relate to the experimental concentrations. Acceptable estimates can be obtained if the estimation problem in Eq. (9) is resolved with another initial guess.

A multi-objective optimization method can consider many criteria at the same time. Liu and Wang[13,46] have introduced an interactive inference algorithm to infer a realizable S-system structure for biochemical networks. The inference algorithm investigates the concentration error criterion, the slope error criterion and interaction measure, simultaneously. The concentration error criterion is employed to measure the goodness-of-fit of the model with respect to a given experimental time-course data set. The slope error criterion is used to judge the accuracy of the dynamic function, i.e. the net rate equation in (5). Each kinetic order, g_{ij} or h_{ij}, is applied to quantify the effect of x_j variable on the production or degradation of x_i. A smaller parameter value means less interaction between state variables, x_j and x_i. The interaction measure sums up the magnitude of kinetic orders that serves as an index to prune a skeletal structure of an S-system.

Various optimization algorithms, such as gradient-based methods,[14] genetic algorithms,[34,47–49] branch-and-bound methods,[50] Newton-flow analysis,[51] decomposition approaches,[52] multiple shooting methods,[53] alternating regression techniques[54] and stochastic optimization,[15,33,55,56] have been applied to determine parameters in biochemical network models. Most commercial nonlinear regression programs are gradient-based optimization methods where the quality of the solution strongly depends on the initial starting point. Evolutionary algorithms, which include genetic algorithms, evolutionary programming, evolution strategies, genetic programming and their variants[34,47–49] can be applied to overcome such drawbacks. Hybrid methods[15,57,58] can be applied using an evolutionary algorithm that acts as a global search engine to determine a satisfied solution and then uses this solution as the initial starting point for a gradient-based optimization method to obtain the refined solution.

Moles *et al.*[55] have reviewed and compared several evolutionary algorithms for parameter estimation of a three-step pathway. The three-enzyme pathway system with gene expression data shown in Fig. 3 is a challenging benchmark problem recently presented by Mendes,[59] Moles *et al.*[55] and Rodriguez-Fernandez *et al.*[58] The mathematical formulation of the "true" system is listed as follows:

$$\frac{dG_1}{dt} = \frac{V_{1,max}}{(1 + (P/K_{i_1})^{n_{i_1}} + (K_{a_1}/S)^{n_{a_1}})} - k_1 G_1,$$

$$\frac{dG_2}{dt} = \frac{V_{2,max}}{(1 + (P/K_{i_2})^{n_{i_2}} + (K_{a_2}/S)^{n_{a_2}})} - k_2 G_2,$$

$$\frac{dG_3}{dt} = \frac{V_{3,max}}{(1 + (P/K_{i_3})^{n_{i_3}} + (K_{a_3}/S)^{n_{a_3}})} - k_3 G_3,$$

$$\frac{dE_1}{dt} = \frac{V_{4,max} G_1}{K_4 + G_1} - k_4 E_1,$$

$$\frac{dE_2}{dt} = \frac{V_{5,max} G_2}{K_5 + G_2} - k_5 E_2,$$

$$\frac{dE_3}{dt} = \frac{V_{6,max} G_3}{K_6 + G_3} - k_6 E_3,$$

$$\frac{dM_1}{dt} = \frac{k_{cat_1} E_1 (S - M_1)/K_{m_1}}{1 + S/K_{m_1} + M_1/K_{m_2}} - \frac{k_{cat_2} E_2 (M_1 - M_2)/K_{m_3}}{1 + M_1/K_{m_3} + M_2/K_{m_4}},$$

$$\frac{dM_2}{dt} = \frac{k_{cat_2} E_2 (M_1 - M_2)/K_{m_3}}{1 + M_1/K_{m_3} + M_2/K_{m_4}} - \frac{k_{cat_3} E_3 (M_2 - P)/K_{m_5}}{1 + M_2/K_{m_5} + P/K_{m_6}}.$$

$$(10)$$

The parameter estimation problem consisted of the 36 parameters, which were divided into two different classes: Hill coefficients, allowed to vary

The local parameter sensitivity for the ith state variable x_i with respect to a change in the jth parameter θ_j is defined as

$$s_a(x_i, \theta_j) = \frac{\partial x_i}{\partial \theta_j} = \lim_{\Delta \theta_j \to 0} \frac{x_i(t; \theta_j + \Delta \theta_j) - x_i(t; \theta_j)}{\Delta \theta_j}, \qquad (12)$$

where $x_i(t; \theta_j + \Delta \theta_j)$ is the value of x_i with an increment $\Delta \theta_j$ on the jth parameter. In addition, the relative parameter sensitivity of x_i with respect to θ_j is defined as

$$s_r(x_i, \theta_j) = \frac{\partial \ln x_i(t)}{\partial \ln \theta_j} = \frac{\partial x_i}{\partial \theta_j} \frac{\theta_j}{x_i} = \frac{\theta_j}{x_i} s_a(x_i, \theta_j). \qquad (13)$$

The simplest way of calculating local sensitivities is to use the finite difference approximation, which is referred to as an indirect method. It is very easy to implement because it requires no extra code beyond an ODE solver. This numerical computation is only valid if we consider an infinitesimal variation (perturbation) for each parameter θ_j. Equation (12) shows that the application of the finite difference approximation method requires the solution $x_i(t; \theta_j)$ of the system in Eq. (1) using the nominal value of the parameters θ_j and n_p solutions $x_i(t; \theta_j + \Delta \theta_j)$ of the systems using the perturbed parameters $\theta_j + \Delta \theta_j$, where n_p is the number of parameters. The finite difference technique is too computationally expensive, especially in cases where the problem consists of many parameters. Therefore, Atherton *et al.*[62] developed the direct method for sensitivity analysis that is applicable to systems of ODEs. Taking a small perturbation for each parameter in Eq. (1), we yield the absolute sensitivity matrix ODEs as

$$\frac{d}{dt} \frac{\partial \mathbf{x}}{\partial \theta} = \frac{\partial \mathbf{f}}{\partial \mathbf{x}} \frac{\partial \mathbf{x}}{\partial \theta} + \frac{\partial \mathbf{f}}{\partial \theta}, \qquad (14)$$

where $\partial \mathbf{x} / \partial \theta$ is the parameter sensitivity matrix of dimension n by p and $\partial \mathbf{f} / \partial \mathbf{x}$ is the Jacobian matrix of dimension n by n. To evaluate the values of partial derivatives of function $\mathbf{f}$ ($\partial \mathbf{f} / \partial \mathbf{x}$ and $\partial \mathbf{f} / \partial \theta$), knowledge of the solution of Eq. (1) is required. A coupled direct method solves Eqs. (1) and (14), simultaneously. Such a method is the simplest one to code, but is the least economical and might cause numerical problems due to ill-conditioned Jacobian matrices.[63–65] Due to the efficiency, Dunker[66] proposed the decoupled direct method, in which the sensitivity equations are solved separately from the model equations. First, the model equations are solved to yield the dynamic profiles that are stored in a table. The sensitivity matrix ODEs are then solved using the dynamic profiles stored in the table to evaluate the Jacobian matrix $\partial \mathbf{f}/\partial \mathbf{x}$ and the partial derivative $\partial \mathbf{f}/\partial \theta$.

Some biological systems, such as glycolytic oscillations,[67–69] the cell cycle,[70] circadian rhythms[71] and periodic neuronal signals,[70] do not have steady-state responses and need to perform sensitivity analysis on their dynamic oscillation behaviors. In many practical applications, when performing the dynamic sensitivity analysis, we are interested in a specific characteristic of the system that is not defined in the system domain, e.g. the period feature in oscillating systems. As pointed out in previous papers,[72,73] dynamic sensitivities derived from oscillating systems must be interpreted with care. In biology, limit cycle systems are of particular interest, since some of them are the basis for various biological clocks, including the one governing the cell cycle. It has been recognized that the sensitivity of various features of oscillating systems (e.g. period and amplitude) could be estimated through simulation,[74] even if there was no general theory to provide a means of analytic computation. Buré and Rozenvasser[75] present a method for deriving the sensitivity of features in limit cycle trajectories to changes in parameters.

Wu *et al.*[76] have proposed an efficient algorithm, called adaptive modified collocation method (AMCM), with an adaptive step-size control to compute the solution and dynamic sensitivities of biological systems described by ODEs. The AMCM algorithm, similar to decoupled direct methods, is applied to solve dynamic equations and matrix sensitivity equations separately by using an adaptive step size to automatically ensure obtaining convergent solutions. The basic idea behind this algorithm is that the solution of Eq. (1) is approximated by the modified collocation method[44] with Lagrange polynomials as shape functions. Instead of computing by trial-and-error using error control, the adaptive step-size approach for the algorithm is obtained based on the Banach fixed-point theorem[77] and can be applied to solve both the dynamic model equations and dynamic matrix sensitivity equations with moderate accuracy even though sensitivity equations are more stiff than model equations. An automatic differentiation technique is embedded in the AMCM to compute the Jacobian matrix for solving the dynamic sensitivities of biological systems so that the user can provide the model equations only for computing dynamic sensitivities. In the following section, we will apply the AMCM algorithm to perform dynamic sensitivity analysis on the mitogen-activated protein kinase (MAPK) pathway.

4.1. *Brief description of the MAPK pathway*

In many human tumors, the MAPK pathway is frequently deregulated. The MAPK pathway transfers extracellular signals from ligand-bound cell-surface tyrosine kinase receptors, such as epidermal growth factor receptor (EGFR) and platelet-derived growth factor receptor (PDGFR), to the nucleus and is an important signal transduction pathway for the control of cell proliferation, differentiation, survival and apoptosis. Signal transduction of the MAPK pathway occurs through cascades of activation and inactivation of signaling proteins, including Ras, Raf, MEK, and extracellular signal-regulated kinase (ERK). It plays an essential role in tumor maintenance and is therefore an appropriate target for anticancer therapy.[78–80] Aberrant activation of individual components in the MAPK pathway is commonly observed in many types of cancer. Finding inhibitors of this pathway has been a new strategy for anticancer drug discovery[81] in recent years. Inhibitors of Raf and MEK kinases have been developed and many are currently in clinical trials.[82–87]

The topology of a signaling network is important for identification of oncogenes and tumor repressor genes. Feedback loops, in which a component of a pathway positively or negatively regulates the activity of upstream components of the pathway, can dramatically influence the output of a network. We will analyze the dynamics of the MAPK pathway with two feedback loops from the viewpoint of dynamic sensitivity in the following example.

4.2. *Kinetic model*

Shin *et al.*[88] constructed a mathematical model of the MAPK pathway based on experimental data obtained from mammalian COS-1 cells. It elucidates the dynamics of the Ras-Raf-MEK-ERK signaling pathway with two feedback loops, including the regulation of Raf kinase inhibitor protein (RKIP). The schematic diagram of the signaling pathway and its feedback regulation are shown in Fig. 4. Each protein in this model has only two possible states, active and inactive (e.g. Ras* and Ras), and the total amount is conserved. In the model, $u(t)$ denotes the activating stimulation and it activates the GDP-bound Ras protein by exchanging its own GDP with a GTP molecule. The GTP-bound Ras protein is deactivated by the hydrolysis of a bound GTP molecule into GDP. The process of exchanging the bound nucleotide is facilitated by guanine nucleotide

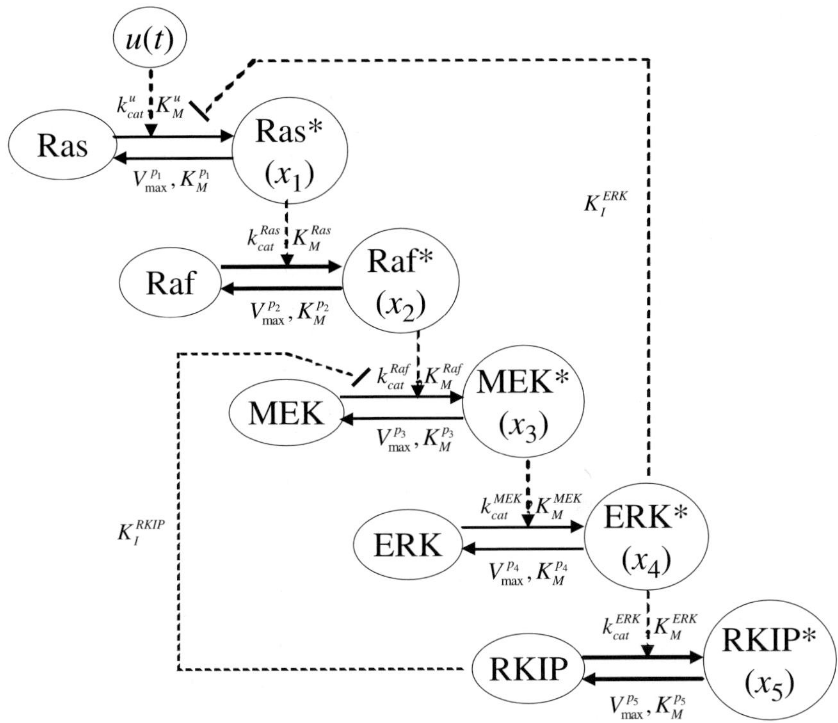

Fig. 4. Schematic diagram of the MAPK pathway. The solid lines indicate irreversible reactions and the dash lines denote regulation processes.

exchange factors (GEFs) and GTPase activating proteins (GAPs). The activated Ras protein activates a cascade of three kinases (Raf, MEK, ERK) where each kinase phosphorylates the next one in turn. The phosphorylation through each kinase is counteracted by dephosphorylation through its corresponding phosphatase, e.g. phosphorylation of Ras occurs via the activating stimulation and dephosphorylation is mediated by phosphatase 1. The GTP-bound Ras activates Raf, the first protein kinase of the MAPK cascade, by recruiting Raf to the plasma membrane, where Raf is phosphorylated at a tyrosine residue. When Raf is activated, it can activate the second kinase MEK by which a third kinase ERK is activated in turn through phosphorylation. Activated ERK dissociates from the Ras-Raf-MEK-ERK complex and phosphorylates more than 70 cytoskeletal proteins and nuclear transcription factors. The well-known function of the active ERK is to suppress the activation of Ras by the inactivation of guanosine nucleotide exchange factors and forms a negative feedback loop in the MAPK pathway. The other function of ERK is to inhibit the RKIP inhibitor by phosphorylation. As its name implies, the RKIP is an inhibitor that suppresses the function of Raf kinase and thereby suppresses the activation of MEK. So the phosphorylation of RKIP by active ERK inactivates the inhibitory function of RKIP and influences positively the activation of MEK. The downregulation of RKIP by active ERK induces the activation of MEK and forms a positive feedback from ERK to MEK in the MAPK pathway. In this computational model, the negative feedback is assumed to be allosteric inhibition and the RKIP in the positive feedback loop is modeled as a competitive inhibitor. All reactions in the pathway follow the Michaelis–Menten rate law and enzyme kinetics, e.g. the velocity for the reaction activating the formation of inactive Ras is given by $v = V_{max}x_1/(K_M + x_1)$, where x_1 is the concentration of active Ras, V_{max} is the maximum initial velocity of that reaction and K_M is the Michaelis constant. The ordinal differential equations describing the dynamic behavior of the MAPK pathway are expressed as follows:

$$\frac{\mathrm{d}x_1}{\mathrm{d}t} = \frac{k_{cat}^u u(t)([Ras]_t - x_1)}{(1 + (x_4/K_I^{ERK})^{3.2})(K_M^u + [Ras]_t - x_1)} - \frac{V_{max}^{p_1}x_1}{K_M^{p_1} + x_1},$$

$$\frac{\mathrm{d}x_2}{\mathrm{d}t} = \frac{k_{cat}^{Ras}([Raf]_t - x_2)x_1}{K_M^{Ras} + [Raf]_t - x_2} - \frac{V_{max}^{p_2}x_2}{K_M^{p_2} + x_2},$$

$$\frac{\mathrm{d}x_3}{\mathrm{d}t} = \frac{k_{cat}^{Raf}([MEK]_t - x_3)x_2}{K_M^{Raf}(1 + (([RKIP]_t - x_5)/K_I^{RKIP})^{2.3}) + [MEK]_t - x_3}$$

$$- \frac{V_{max}^{p_3}x_3}{K_M^{p_3} + x_3}, \tag{15}$$

$$\frac{\mathrm{d}x_4}{\mathrm{d}t} = \frac{k_{cat}^{MEK}([ERK]_t - x_4)x_3}{K_M^{MEK} + [ERK]_t - x_4} - \frac{V_{max}^{p_4}x_4}{K_M^{p_4} + x_4},$$

$$\frac{\mathrm{d}x_5}{\mathrm{d}t} = \frac{k_{cat}^{ERK}([RKIP]_t - x_5)x_4}{K_M^{ERK} + [RKIP]_t - x_5} - \frac{V_{max}^{p_5}x_5}{K_M^{p_5} + x_5},$$

where $x_i, i = 1,\ldots,5$, denote the phosphorylated form of Ras, Raf, MEK, ERK, and RKIP, respectively, $[x_i]_t$ indicates the total amount of x_i (phosphorylated and dephosphorylated), $u(t)$ is the activating stimulation, k_{cat}^y is the turnover number of the enzyme y, K_M^y and $K_M^{p_i}$ are the Michaelis

Table 2. The initial conditions for all variables and kinetic parameters for the rate equations in the model described by Eq. (15).

Species	Value	Parameter	Value	Parameter	Value
x_1	0	k_{cat}^{u}	144	V_{max}^{P1}	7.56
x_2	0	k_{cat}^{Ras}	144	V_{max}^{P2}	7.56
x_3	0	k_{cat}^{Raf}	144	V_{max}^{P3}	7.56
x_4	0	k_{cat}^{MEK}	53.52	V_{max}^{P4}	20.11
x_5	0	k_{cat}^{ERK}	144	V_{max}^{P5}	28.8
$[Ras]_t$	0.35	K_{M}^{u}	1	K_{M}^{P1}	125.64
$[Raf]_t$	0.12	K_{M}^{Ras}	90	K_{M}^{P2}	95
$[MEK]_t$	0.36	K_{M}^{Raf}	11	K_{M}^{P3}	95
$[ERK]_t$	0.7	K_{M}^{MEK}	136.88	K_{M}^{P4}	191.079
$[RKIP]_t$	0.24	K_{M}^{ERK}	90	K_{M}^{P5}	95
$u(t)$	1.0	K_{I}^{ERK}	0.004	K_{I}^{RKIP}	0.07

constants for the enzyme y and the phosphatase i, respectively, K_I^{ERK} and K_I^{RKIP} are the inhibitor constants for ERK and RKIP, respectively and $V_{max}^{P_i}$ is the maximum initial velocity for the phosphatase i. The total amount for each species $[x_i]_t$ in μM, the values for the Michaelis constants (K_M) and inhibitor constants (K_I) in μM, the turnover numbers (k_{cat}) in m^{-1} and the maximum initial velocities (V_{max}) in μM m^{-1} are obtained from Shin *et al.*[88] and listed in Table 2.

4.3. *Computational results*

The model equations for the MAPK pathway expressed in Eq. (15) are analyzed by the AMCM algorithm using the initial conditions as described in Table 2. All dynamic sensitivities with respect to 22 parameters and the solution of the model are computed simultaneously. The dynamic profiles of the model are shown in Fig. 5. The activities of MEK and ERK are unable to reach a steady state due to the combined functions of positive- and negative-feedback loops in the MAPK pathway. The oscillatory behavior of MEK and ERK is predicted theoretically and simulated in the computational model, as shown in Fig. 5. The dynamic sensitivities could be efficiently obtained using the AMCM algorithm. Such results are also verified by the finite difference approximation method.

45. Vilela M, Borges C, Vinga S *et al.* (2007) Automated smoother for the numerical decoupling of dynamics models. *BMC Bioinformatics* **8**(1): 305.
46. Liu PK, Wang FS. (2008) Inverse problems of biological systems using multi-objective optimization. *J Chin Inst Chem Eng* **39**(5): 399–406.
47. Bäck T, Fogel DB, Michalewicz Z. (1997) *Handbook of Evolutionary Computation.* IOP Publishing Ltd. and Oxford University Press.
48. Edwards K, Edgar TF, Manousiouthakis VI. (1998) Kinetic model reduction using genetic algorithms. *Comput Chem Eng.* **22**(1–2): 239–246.
49. McKay B, Willis M, Barton G. (1997) Steady-state modelling of chemical process systems using genetic programming. *Comput Chem Eng* **21**(9): 981–996.
50. Polisetty PK, Voit EO, GEP. (2005) Yield optimization of saccharomyces cerevisiae using a GMA model and a milp-based piecewise linear relaxation method. In: *Proc Found Syst Biol Eng*, Santa Barbara, CA.
51. Kutalik Z, Tucker W, Moulton V. (2007) S-system parameter estimation for noisy metabolic profiles using newton-flow analysis. *IET Syst Biol* **1**(3): 174–180.
52. Gennemark P, Wedelin D. (2007) Efficient algorithms for ordinary differential equation model identification of biological systems. *IET Syst Biol* **1**(2): 120–129.
53. Peifer M, Timmer J. (2007) Parameter estimation in ordinary differential euations for biochemical processes using the method of multiple shooting. *IET Syst Biol* **1**(2): 78–88.
54. Chou I-C, Martens H, Voit E. (2006) Parameter estimation in biochemical systems models with alternating regression. *Theor Biol Med Model* **3**(1): 25.
55. Moles CG, Mendes P, Banga JR. (2003) Parameter estimation in biochemical pathways: A comparison of global optimization methods. *Genome Res* **13**: 2467–2474.
56. Gonzalez OR, Kuper C, Jung K, Naval J, Prospero C, Mendoza E. (2007) Parameter estimation using simulated annealing for S-system models of biochemical networks. *Bioinformatics* **23**(4): 480–486.
57. Balsa-Canto E, Peifer M, Banga J, Timmer J, Fleck C. (2008) Hybrid optimization method with general switching strategy for parameter estimation. *BMC Syst Biol* **2**(1): 26.
58. Rodriguez-Fernandez M, Mendes P, Banga JR. (2006) A hybrid approach for efficient and robust parameter estimation in biochemical pathways. *Biosyst.* **83**(2–3): 248–265. *5th Int Conf Syst Biol — ICSB* 2004.
59. Mendes P. (2001) Modeling large biological systems from functional genomic data: Parameter estimation. In: *Foundations of Systems Biology*, pp. 163–186. MIT Press, Cambridge.
60. Wang F, Liu PK. (2010) Inverse problems of biochemical systems using hybrid differential evolution and data collocation. *Int J Sys Synthetic Biol* **1**(1): 21–38.
61. Heinrich R, Schuster S. (1996) *The Regulation of Cellular Systems.* Chapman & Hall, New York.

62. Atherton RW, Schainker RB, Ducot ER. (1975) Statistical sensitivity analysis of models for chemical kinetics. *AIChE J* **21**: 441–448.

63. Dougherty EP, Hwang J-T, Rabitz H. (1979) Further developments and applications of the Green's function method of sensitivity analysis in chemical kinetics. *J Chem Phys* **71**: 1794–1808.

64. Dougherty EP, Rabitz H. (1980) Computational kinetics and sensitivity analysis of hydrogen–oxygen combustion. *J Chem Phys* **72**: 6571–6586.

65. Kramer MA, Calo JM, Rabitz H. (1981) An improved computational method for sensitivity analysis: Green's function method with "AIM". *Appl Math Model* **5**: 432–441.

66. Dunker AM. (1984) The decoupled direct method for calculating sensitivities coefficients in chemical kinetics. *J Chem Phys* **81**: 2385–2393.

67. Acerenza L, Sauro HM, Kacser H. (1989) Control analysis of time-dependent metabolic systems. *J Theor Biol* **137**: 424–444.

68. Acerenza L, Ortega F. (2006) Metabolic control analysis for large changes: Extension to variable elasticity coefficients. *IEE Proc Syst Biol* **153**(5): 323–326.

69. Ingalls BP, Sauro HM. (2003) Sensitivity analysis of stoichiometric networks: An extension of metabolic control analysis to non-steady state trajectories. *J Theor Biol* **222**(1): 23–36.

70. Ingalls BP. (2004) Autonomously oscillating biochemical systems: Parametric sensitivity of extrema and period. *IET Syst Biol* **1**: 62–70.

71. Gunawan R, Doyle FJ. (2007) III Phase sensitivity analysis of circadian rhythm entrainment. *J Biol Rhythms* **22**(2): 180–194.

72. Demin OV, Westerhoff HV, Kholodenko BN. (1999) Control analysis of stationary fixed oscillations. *J Phys Chem B* **103**: 10695–10710.

73. Kholodenko BN, Demin OV, Westerhoff HV. (1997) Control analysis of periodic phenomena in biological systems. *J Phys Chem B* **101**: 2070–2081.

74. Bier M, Teusink B, Kholodenko BN, Westerhoff HV. (1996) Control analysis of glycolytic oscillations. *Biophys Chem* **62**: 15–24.

75. Buré EG, Rozenvasser EN. (1974) The study of the sensitivity of oscillatory systems. *Automat Rem Contr* **7**: 1045–1052.

76. Wu W, Wang F, Chang M. (2008) Dynamic sensitivity analysis of biological systems. *BMC Bioinformatics* **9**(Suppl 12): S17.

77. Kreyszig E. (1978) *Introductory Functional Analysis with Applications*. Wiley, New York.

78. Wong KK. (2009) Recent developments in anti-cancer agents targeting the Ras/Raf/MEK/ERK pathway. *Recent Pat Anticancer Drug Discov* **4**(1): 28–35.

79. McCubrey JA, Steelman LS, Abrams SL *et al.* (2006) Roles of the RAF/MEK/ERK and PI3K/PTEN/AKT pathways in malignant transformation and drug resistance. *Adv Enzyme Regul* **46**(1): 249–279.

80. McCubrey JA, Steelman LS, Chappell WH *et al.* (2007) Roles of the Raf/MEK/ERK pathway in cell growth, malignant transformation and drug resistance. *Biochimica et Biophysica Acta (BBA) — Mol Cell Res* **1773**(8): 1263–1284.

81. Thompson N, Lyons J. (2005) Recent progress in targeting the Raf/MEK/ ERK pathway with inhibitors in cancer drug discovery. *Curr Opin Pharmacol* **5**(4): 350–356.

82. Wellbrock C, Karasarides M, Marais R. (2004) The RAF proteins take centre stage. *Nat Rev Mol Cell Biol* **5**(11): 875–885.

83. Sridhar SS, Hedley D, Siu LL. (2005) Raf kinase as a target for anticancer therapeutics. *Mol Cancer Ther* **4**(4): 677–685.

84. Beeram M, Patnaik A, Rowinsky EK. (2005) Raf: A strategic target for therapeutic development against cancer. *J Clin Oncol* **23**(27): 6771–6790.

85. Schreck R, Rapp UR. (2006) Raf kinases: Oncogenesis and drug discovery. *Int J Cancer* **119**(10): 2261–2271.

86. Wang D, Boerner SA, Winkler JD, LoRusso PM. (2007) Clinical experience of MEK inhibitors in cancer therapy. *Biochimica et Biophysica Acta (BBA) — Mol Cell Res* **1773**(8): 1248–1255.

87. Friday BB, Adjei AA. (2008) Advances in targeting the Ras/Raf/MEK/Erk mitogen-activated protein kinase cascade with MEK Inhibitors for cancer therapy. *Clin Cancer Res* **14**(2): 342–346.

88. Shin S-Y, Rath O, Choo S-M *et al.* (2009) Positive- and negative-feedback regulations coordinate the dynamic behavior of the Ras-Raf-MEK-ERK signal transduction pathway. *J Cell Sci* **122**(3): 425–435.

89. Heinrich R, Neel BG, Rapoport TA. (2002) Mathematical models of protein kinase signal transduction. *Mol Cell* **9**(5): 957–970.

90. Hornberg JJ, Bruggeman FJ, Binder B *et al.* (2005) Principles behind the multifarious control of signal transduction. ERK phosphorylation and kinase/phosphatase control. *FEBS J* **272**(1): 244–258.

91. Asthagiri AR, Lauffenburger DA. (2001) A computational study of feedback effects on signal dynamics in a mitogen-activated protein kinase (MAPK) pathway model. *Biotechnol Prog* **17**(2): 227–239.

92. Schoeberl B, Eichler-Jonsson C, Gilles ED, Muller, G. (2002) Computational modeling of the dynamics of the MAP kinase cascade activated by surface and internalized EGF receptors. *Nat Biotech* **20**(4): 370–375.

93. Downward J. (2003) Targeting RAS signalling pathways in cancer therapy. *Nat Rev Cancer* **3**(1): 11–22.

Part II

High-Throughput Omics Data and Analysis

Chapter 6

ChIP-Seq Analytics: Methods and Systems to Improve ChIP-Seq Peak Identification

Stuart Brown[*], D. Frank Hsu[†], Christina Schweikert[†] and Zuojian Tang[*]

1. Introduction

1.1. *ChIP-seq framework*

Systems Biology is a data-hungry field. In order to accurately model the metabolic and developmental pathways within a cell or an organism, it is necessary to compile a complete list of the molecules that participate in these activities and dynamically quantify their interactions and changes in abundance in response to an experimental stimulus or developmental program. High-throughput or Next-Generation DNA Sequencing (NGS) is a rich source of data that is particularly useful for this type of modeling approach. The Sanger method for the sequencing of DNA[1] was one of the most important biotechnologies developed in the 20th century (Sanger was awarded the Nobel prize in 1980). NGS technologies, recently developed by commercial vendors, including Roche-454, Illumina, and Life Technologies, have increased DNA sequencing capacity by three orders of magnitude over a period of about five years, making entirely new applications possible.[2,3] NGS technologies have the potential to transform biomedical research, allowing investigators to sequence the complete genomes of multiple strains

[*]New York University Langone Medical Center, New York, NY 10016, USA. {Stuart. Brown, Zuojian.Tang}@nyumc.org

[†]Department of Computer and Information Science, Fordham University, New York, NY 10023, USA. {cschweikert, hsu}@cis.fordham.edu

of pathogens to gain a deeper understanding of gene regulation and chromatin structure, to study entire populations of microorganisms in both environmental and medical contexts, to follow the evolution of viral and bacterial resistance in real time, to uncover the huge diversity of novel genes and alternative transcription of currently annotated gene loci, to sequence cancer genomes, and to make personal genomics practical and affordable.

Chromatin Immunoprecipitation Sequencing (ChIP-seq) is a method of using NGS to interrogate protein–DNA interactions, especially those interactions that are important for the regulation of gene expression. The ChIP-seq method is particularly important as a source of basic data for reverse engineering of regulatory pathways and as a form of empirical data to validate pathway predictions and predictions of transcription factor binding sites. Chromatin Immunoprecipitation (ChIP) involves crosslinking proteins to DNA with a reagent such as formaldehyde, randomly shearing the DNA into small fragments (200–500 base pairs), then using an antibody specific for a known DNA-interacting protein to isolate DNA fragments bound to the target protein.[4] A method known as ChIP-on-chip[5] or ChIP-chip[6] was developed to identify the DNA fragments isolated by ChIP using a microarray containing large numbers of probes of known genomic sequences. A recent modification of this method, known as ChIP-seq,[7] was developed using NGS to identify the DNA fragments isolated by ChIP (Fig. 1).

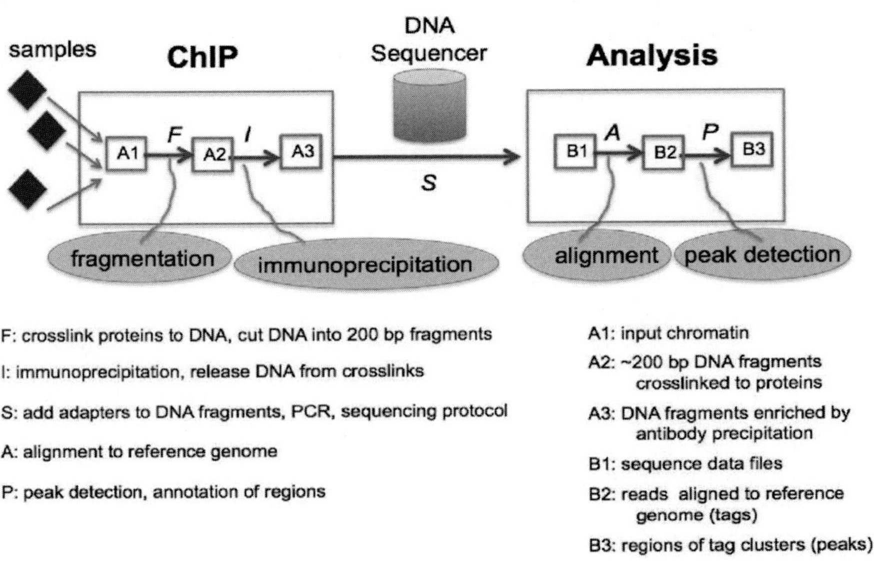

Fig. 1. A framework for the ChIP-seq experimental and analytic workflow.

The resulting sequence data is then mapped back onto the reference genome of the experimental organism. Clusters of reads mapped to a genomic locus (an enriched region) indicate a site of DNA binding by the protein targeted by the ChIP. ChIP-seq has the advantage of neither being limited by prior knowledge of potential binding sites nor by the inability of a microarray platform to hold a sufficient number of probes to span the entire genome at high resolution. Therefore ChIP-seq is able to discover active sites on DNA, including transcription factor and RNA polymerase binding sites, chromatin structure (histone modifications, DNAse hypersenstive sites), and patterns of DNA methylation, in an unbiased fashion across the entire genome. This is particularly important for the study of trans-acting transcriptional enhancers.[8] ChIP-seq is also relatively unaffected by base composition, cross-hybridization among similar sequences, and secondary structure, all of which may interfere with DNA hybridization to a microarray.

1.2. *Current computational methods and systems*

The ChIP-seq method relies on NGS machines such as the Illumina Genome Analyzer (GA), which are capable of simultaneously determining the sequences of millions of DNA fragments in a single sample with greater than 99% base calling accuracy. The sequence reads obtained from ends of ChIP-selected DNA fragments are typically 25–50 base pairs long. These short reads can then be mapped to a reference genome by a stringent DNA sequence alignment algorithm such as ELAND (Illumina Inc.), bwa,[9] or MAQ.[10] Sequence reads that do not map to a unique position on the genome (with two or fewer mismatches) are generally discarded. The final product of such a mapping procedure is a set of positions on the reference genome (known as tags) indicating the start and end of each short sequence read.

Once the reads are mapped to the genome, the tag positions must then be analyzed for clusters or "peaks," which indicate protein-binding (or histone modification) positions enriched by the ChIP. Simple algorithms can identify peaks as windows of the genome sequence that contain counts of mapped tags greater than a threshold, but the reproducibility of peak predictions across replicates and the quantification and comparison of peaks across samples from different experimental conditions is a challenging computational problem. Background and noise exist in sets of mapped sequence tags that may be caused by co-precipitation of unbound DNA fragments, nonspecific interactions of the ChIP target protein with DNA, and various types of contamination and amplification artifacts (see Sec. 2.2 of this chapter). Since the genomes of complex eukaryotes are billions of bases long, accurately and reproducibly identifying and quantifying the heights and/or areas of these peaks require a sophisticated computational solution. Our initial study of ChIP-seq peak-detection software relied on correlation of reported peaks with external data sources such as RefSeq gene annotations and correlation with experimental data from ChIP-chip experiments using the same experimental system. However, neither of these external data sets provides an adequate "gold standard" against which to measure the specificity and sensitivity of peak-detection results. Published results from other ChIP-seq experiments show only 30%–40% correlation of peak detection with ChIP-chip binding site predictions[11,12] We have subsequently studied ChIP-seq peak-detection software based on the reproducibility of peak detection across technical replicates. Even this approach does not provide a true measure of accuracy for peak-detection methods, since we cannot empirically determine if a peak-detection disagreement between replicates is the result of a false positive or a false negative.

Many computational methods have recently been developed for the analysis of ChIP-seq data. The proliferation of software is an indication that no well-validated general method exists that performs well in the hands of most investigators. We have studied the following methods and software for ChIP-seq peak-detection analysis: Genome Studio (Illumina Inc.), FindPeaks,[13] peak detection based on overlaps of extended sequence reads,[12] ChIPSeq Peak Finder/E-RANGE,[7,14] SISSRs,[15] CisGenome,[11] QuEST,[16] USeq,[17] and MACS.[18]

The commercial GenomeStudio^{TM} application (Illumina, Inc.) uses a simple window/threshold method to detect peaks in the sequence reads mapped to a reference genome by the ELAND alignment method. In our evaluation, windows of 400 bp and tag thresholds of 5–10 tags/window were optimal to identify both transcription factor binding and histone modification sites. This method does not provide for any form of background (control sample) correction nor does it provide a false discovery rate (FDR) or p-value calculation to support predicted peaks. Reproducibility of peak region detection across replicates varied from 38% to 90% depending on depth of sequencing (tag density) and the type of antibody target used for the ChIP (see Table 1).

FindPeaks[13] is an open source program that directly counts overlapping sequence reads (reads may be extended to the length of the original DNA

Table 1. Peak detection reproducibility

Experiment	Tag Correlation	PeakFinder	SISSRS	GenomeStudio
E2F4	0.86	17.2%	23.4%	38.0%
Sin3b	0.76	20.7%	16.0%	56.1%
H3K4	0.98	66.4%	25.6%	90.0%

fragments isolated by ChIP, 200 bp in our data). It calculates peak height as the maximum number of overlapping tags at any one base position within an enriched region (similar to Robertson *et al.*[12]). Optionally, it requires that the peak height be greater than a false discovery threshold value set by a Monte Carlo simulation of sequence reads randomly distributed over the genome. We have observed that a Poisson or Monte Carlo model does not accurately represent the background distribution of sequence reads mapped on the genome in data files from input DNA (non-immunoprecipitated genomic DNA) or ChIP-seq using an anti-IGG antibody. Nix *et al.*[17] have made detailed observations of background peak distributions and false discovery rates that confirm the nonrandom distribution of sequence tags in ChIP-seq control lanes. FindPeaks does not provide a method to subtract background or to compare experimental and control samples.

ChipSeq Peak Finder[7] clusters the reads and uses the ratio of the counts in the immunoprecipitated and the control sample to call peaks. Site identification from short sequence reads (SISSR) estimates high read counts using Poisson probabilities and calls regions where the peaks shift from the forward to the reverse strand.[15] The SISSRS method is attractive because it explicitly makes use of information from the orientation of tags around a protein-binding site — where it is expected that Forward strand tags will be found upstream of the true binding site and Reverse strand tags downstream. This allows for very precise prediction of the actual binding site. However, this method does not perform well for regions of low-tag density or for histone methylation ChIP, where tags are not neatly oriented. It tends to create many different peaks across enriched regions, which are not reproducible across replicates.

CisGenome[11] uses a two-pass algorithm for peak detection to ensure adjustment for DNA fragmentation length. It can analyze both ChIP-seq and ChIP-chip data, or combine the two. In order to correct many types of systemic bias created by sample preparation, amplification, sequencing (or hybridization), and alignment, it uses both a ChIP sample and a negative control sample (input DNA or mock-ChIP with IGG) to compute

FDR at each specific location. It also provides methods to detect binding regions, peak localization, and filtering. However, using FDR calculation to obtain a threshold may not be the best way to get an optimal value.

QuEST[16] provides a data-driven statistical analysis model to generate peak calls by leveraging the key attributes of the sequenced and aligned DNA reads, such as directionality (strand orientation) and the original size of ChIP-isolated DNA fragments. The statistical framework used is the kernel density probability estimation approach, which facilitates the aggregation of signals originated from densely packed sequence reads at protein interaction sites. Statistical estimation of FDR in QuEST requires two samples of control data for each experimental sample.

The USeq software package[17] uses control data to calculate window summary statistics with four different methods (window counts, window counts with subtracted control counts, normalized differences, p-value based on random binomial distribution). Pre-filtering is applied to remove all regions of the genome with significantly elevated p-values in control data (background peaks). Despite careful attention to FDR, window thresholds may be obtained by the application of inappropriate p-values. True peak located in the same genomic regions as a background peaks will not be detected.

MACS (Model-based Analysis of ChIP-Seq)[18] empirically models the shift size of ChIP-seq tags to enhance peak identification by taking advantage of the bimodal pattern of forward and reverse tags. MACS also utilizes a dynamic Poisson distribution to identify local biases in the genome.

2. Data Acquisition and Management

2.1. *Preliminary data*

We have observed that the peak-detection abilities of existing ChIP-seq software packages vary greatly depending on the type of protein that is targeted by the antibody used in the ChIP. Our preliminary work has identified three different types of protein–DNA interactions that yield very different results when ChIP-seq data is analyzed with a single peak-detection method. Transcription factors, such as E2F4, bind strongly to a single isolated DNA sequence (a motif) near the transcription start site of a gene, producing distinct ChIP-seq peaks $\sim$400 bases wide (twice the

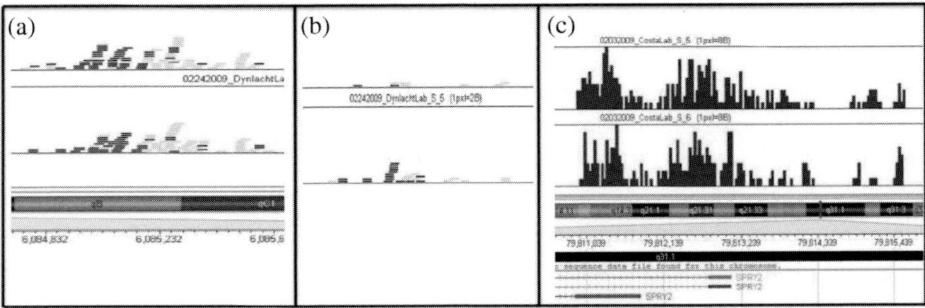

Fig. 2. (a) A single peak from E2F4 ChIP-seq showing oriented tags. (b) A peak region from Sin3b ChIP-seq showing weaker more dispersed binding. (c) A peak region from H3K4-3met histone ChIP-seq showing a much larger span of enriched tags extending downstream beyond the first exon of gene SPRY2.

size of ChIP DNA fragments), with oriented tags that approximately follow a normal distribution (Fig. 2(a)). A second pattern is observed with transcription factors, such as Sin3a, that bind weakly to DNA together with co-factors, yielding wider ChIP-seq peaks (800–1600 bases) with a flat distribution of lower tag density, and unoriented tags (Fig. 2(b)) and greater variability among replicates. Modified histone proteins, such as H3K4-3met, are a third type of ChIP-seq target that produce much wider peaks (∼4000 bases) and unoriented tags (Fig. 2(c)).

Our evaluation of several existing ChIP-seq peak-finding programs (PeakFinder, CisGenome, SISSRS, and GenomeStudio) with our preliminary data has shown poor reproducibility of peak detection across technical replicate data sets. The reproducibility of peak calls is particularly important to many investigators who wish to use the ChIP-seq technique to investigate biological differences — to discover how transcription factor binding or histone methylation changes in response to various developmental, environmental, genetic, or other factors. In order to detect meaningful biological differences, the peak-finding method must be both sensitive and robust (able to produce consistent results across replicates). A peak-detection technique that only identifies the strongest signals from binding sites that are well supported by many overlapping sequence tags will not be adequate to dissect subtle genome-wide changes.

As preliminary data, we have investigated ChIP-seq on an Illumina Genome Analyzer II for pairs of technical replicates for each of these three types of ChIP targets (E2F4, Sin3a, and H3K4-3met). Our initial exploration of the distribution of control sequence reads (input DNA) on the reference genome indicates that the tags are highly dispersed, and clusters of more than 5 tags per 500 bases (in a window moving along the genome) are rare.

We have developed methods for evaluating the reproducibility across replicates of peak detection by a specific peak-finding program at a fixed set of parameters and also for calculating the underlying similarity of the tag distribution in those replicates. We calculate reproducibility of peak detection between technical replicates (samples A and B) as follows: set all peak predictions to a width of 500 bases, count all peaks that overlap (by even a single base) between samples A and B as "shared peaks," count nonoverlapping peaks in samples A and B as "unique peaks," calculate reproducibility as #shared/(#shared + #unique). We calculate the similarity of the tag density of the underlying data sets (sample A versus B) as follows: count the number of tags in A and B in every region of the genome where tag density in A or B is greater than a threshold of 5 tags per 500 bp, rank the number of tags in each over-threshold region of A as compared to all windows in A, rank regions in B versus B, calculate the Spearman–Rho correlation of ranks for A versus B. This rank correlation of tag density across all genomic regions is a property of the two data sets and does not vary with different peak-detection software. For our preliminary data sets we find tag density correlations of E2F4 = 0.86, Sin3b = 0.76, and H3K4 = 0.98, but reproducibility of peak detection across replicates is much worse for most of the peak-finding programs that we have tested (see Table 1).

2.2. *Background and PCR artifacts*

In our initial study of sequence read distribution across the genome in ChIP-seq data from the Illumina GAII machine, we have noticed many instances of duplicate reads — many copies of identical sequences. In many of our ChIP-seq data sets, 50% or more of all tags are duplicates (i.e. sequence tags mapped to start positions at a distance of zero bases from at least one other tag, see Table 2). We believe that these duplicates are PCR artifacts, which are produced by biased amplification when the ChIP yields a small amount of DNA with dramatically reduced complexity compared to the original genome. These duplicated sequences frequently occur as stacks of Forward–Reverse pairs that are spaced 200–250 bases apart on the genome (Fig. 3(b)). A single DNA fragment can be amplified to high abundance in the sample by polymerase chain reaction (PCR) and copies are sequenced many times on the Illumina flowcell. Since the Illumina process is unoriented, $3'$ and $5'$ sequence reads are collected from these

Table 2. Duplicate reads.

	total # reads	Unique	# dups	% dups
Sin3a	3068290	1602570	1465720	47.77
Sin3a	3190943	1542869	1648074	51.65
Sin3b	3778662	1872330	1906332	50.45
Sin3b	3782435	1880457	1901978	50.28
Sin3b	1560551	599052	961499	61.61
input DNA	5867770	5020223	847547	14.44
IGG	2799222	1343425	1455797	52.01
E2F3	4430663	1981291	2449372	55.28
E2F3	2604726	1631194	973532	37.38
E2F4	2114152	1287288	826864	39.11
E2F4	3522269	1713440	1808829	51.35
E2F4	4420211	2989270	1430941	32.37
input DNA	6256743	5578686	678057	10.84
IGG	4417808	2191617	2226191	50.39
H3K4	3065816	2742533	323283	10.54
H3K4	2876048	2563675	312373	10.86

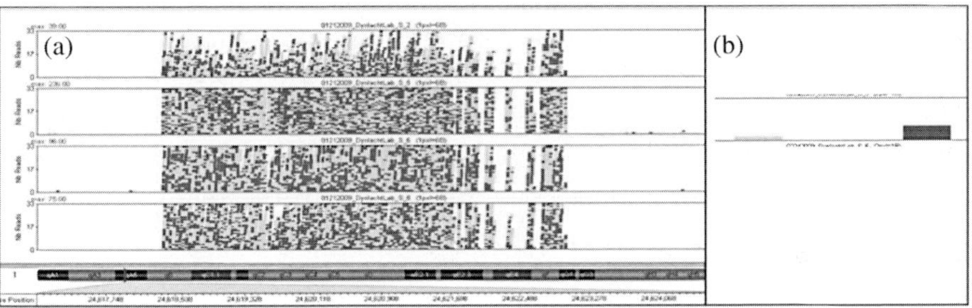

Fig. 3. Some clusters of nonspecific sequence tags in ChIP-seq data. (a) A collection of thousands of mapped sequence reads found on chromosome 1 in both ChIP and control samples. It does not resemble a protein interaction site enriched by ChIP. (b) This is an example of a set of perfectly duplicated sequence reads produced by biased PCR. Two stacks of sequence tags, on Forward and Reverse strands, are spaced ~200 bp apart, corresponding to the ends of an amplified PCR product.

duplicate DNA fragments. It is necessary to remove these duplicates prior to applying any peak-detection method so that they do not contribute to false-positive peaks.

We have also observed some tag clusters that do not appear to be related to actual protein binding or histone methylation sites. In one set of experiments, we found a cluster spanning approximately 4,000 bases on chromosome 1 that contained many thousands of perfectly aligned tags in all replicates of control and experimental samples (Fig. 3(a)). This is a clear example that illustrates that subtracting a control sample is necessary in any peak-detection algorithm. This example supports the general observation that Poisson or Monte Carlo distributions of tags across the genome do not adequately model the ChIP-seq background. We plan on extending our study of the background distribution of sequence reads produced from unselected genomic fragments (input DNA) and mock ChIP-seq conducted with the nonspecific IGG antibody.

3. ChIP-seq Analytics Methods Using Combinatorial Fusion Analysis

3.1. *Multiple scoring systems for ChIP-seq binding site detection using rank-score function*

In order to evaluate the performance of ChIP-seq peak-detection software, we propose that the peak detection for each of the binding sites be viewed as a scoring system on the set of all possible binding site regions. Different scoring systems for peak detection can represent different features/cues or different algorithms/methods. They can also represent different technical replicates or different biological replicates using each of the same set of features or cues or the same algorithm or method. By using multiple scoring systems defined on the set of possible binding site regions to detect peaks for each of the binding sites, we can study the reproducibility of peak calls among different replicates. We will draw from recent research in combinatorial fusion analysis (CFA).[19,20] Using a rank-score characteristic graph to measure the scoring diversity, combinatorial fusion analysis has been an active area of research in the past 10 years in a variety of application domains such as microarray gene expression analysis,[21] motif finding,[22] protein structure prediction,[23] virtual screening,[24] information retrieval,[25,26] and target tracking.[27]

Let $\mathbf{D} = \{d_1, d_2, \ldots, d_n\}$ be the set of possible regions. Let $\mathbf{s}$ be a function from $\mathbf{D}$ to $\mathbf{R}$, the set of real numbers. Let $\mathbf{r}$ be the function from $\mathbf{D}$ to $\mathbf{N} = [1, n] = \{1, 2, \ldots, n\}$ by sorting the values in $\mathbf{s(D)}$ into descending order and converting the function $\mathbf{s(D)}$ into the function $\mathbf{r(D)}$ using the rank as its function value. $\mathbf{s(D)}$ and $\mathbf{r(D)}$ are called the score function and rank function defined on the set $\mathbf{D}$, respectively. Given the function $\mathbf{s}$ and the function $\mathbf{r}$, the rank-score-function $\boldsymbol{f}$ is computed:

$$\boldsymbol{f} : \mathrm{N} = [1, \mathrm{n}] \to \mathrm{R} \quad \text{as } \boldsymbol{f}(\mathrm{i}) = (\mathrm{s} \circ \mathbf{r}^{-1})(\mathrm{i}) = \mathrm{s}(\mathrm{r}^{-1}(\mathrm{i})).$$

The rank-score-characteristic (RSC) graph is the graph of the rank-score function f on two-dimensional coordinate systems. Let $A_1, \ldots, A_t$ be t scoring systems on the set of regions $\mathbf{D}$. These t systems can occur in a variety of different scenarios:

(1) They are the scoring systems obtained from different technical replicates.
(2) They are the scoring systems obtained from different biological replicates.
(3) They represent different features/parameters/attributes/cues which are used to store the set of possible scenarios $\mathbf{D}$.
(4) They represent different algorithms/procedures/methods which are used to score the set of possible regions $\mathbf{D}$.

One of the main goals in scenarios (1) and (2) is to examine the reproducibility of the ChIP-seq technology and methodology on technical replicates and on biological replicates. The goal in scenarios (3) and (4) is to obtain a final scoring system that is robust and reliable. "Robust" means adaptation (to different experiments, different organisms, and different species) and scalability (to different sizes). "Reliable" means efficiency (fast and optimal) and effectiveness (low false-positive and false-negative rates). In both cases ((1), (2) and (3), (4)), the rank-score functions f_A and f_B and their corresponding RSC graphs are used to measure the diversity between scoring systems A and B (in the former case of (1) and (2)) or to determine the likelihood of success by combining A and B (in the latter case of (3) and (4)).

When two scoring systems A and B represent a sample and its replicate, the Pearson's $\mathbf{z}$ (score) and Spearman's **rho** (rank) are often used to calculate the correlation between A and B. In addition, we propose to also use the variation $\mathbf{d}(f_A, f_B)$ between rank-score function of A, f_A, and that of B, f_B, to measure the diversity $\mathbf{d}(A,B)$ between scoring systems A and B. Since the rank-score functions f_A and f_B and their corresponding RSC graphs are independent of the scores of each individual binding site region in $\mathbf{D}$, this variation $\mathbf{d}(f_A, f_B)$ reflects vividly the diversity between different scoring (and ranking) characteristics and behaviors of A and B. Moreover, these different scoring behaviors can be visualized and compared clearly by drawing the two RSC graphs on the same diagram with both score function values normalized to values in $[0, 1]$. For the H3K4 data set, the RSC graphs for the two replicates are produced in Fig. 4, which shows how little diversity (and hence much correlation), to validate its reproducibility.

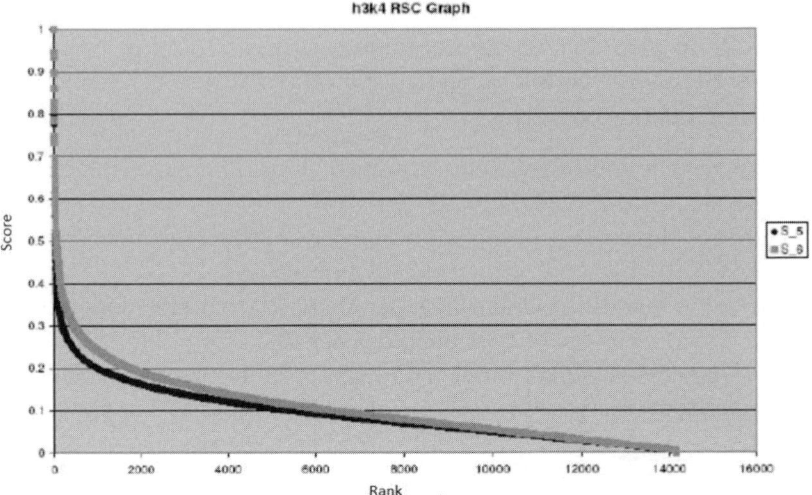

Fig. 4. RSC Graphs for H3K4 replicates.

3.2. *Variation of multiple scoring systems*

One of the most challenging problems in ChIP-seq binding site detection is the relatively high false-positive rate resulting from each specific scoring and tag counting method and the improperly chosen threshold to obtain the final "positive" list of binding site regions. If the threshold is chosen too high or too low, then the false-negative rate or the false-positive rate is increased, respectively. We illustrate two scenarios by two scoring systems A and B, where A has less and B has more noise, respectively, due to different counting methods on different biological or technical replicates. Figure 5 shows example window/threshold score distributions on 15,000 regions obtained from H3K4 ChIP-seq data and the same number of regions of synthetic data. The peak-detecting scenario A produces a much greater variance in scores. Less variance is observed for the uniform scoring scenario because the scoring system has difficulty distinguishing good or bad binding sites and the choice of a correct region of binding site is not a clear-cut situation. The frequency of scores for A and B are shown in Fig. 6 as $F_A(s)$ and $F_B(s)$, respectively. Let s_0 be the value of score s, when $F_A(s)$ and $F_B(s)$ intersect. The frequency graph $F_A(s)$ shows that there are fewer binding site regions with high scores (to the right of s_0) than other regions with clearly low scores (to the left of s_0). On the other hand, $F_B(s)$ has similar number of binding site regions with high and low scores.

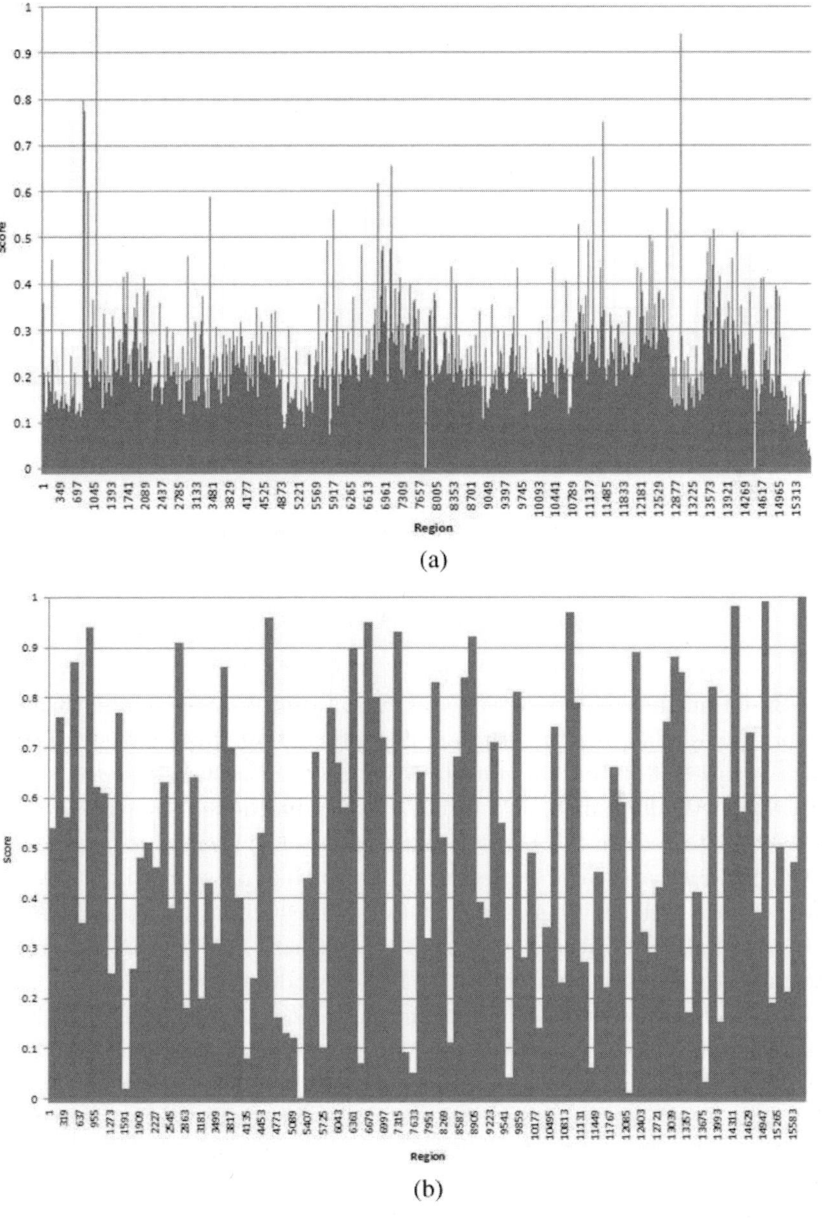

Fig. 5. Example score distributions s_A and s_B for A and B on 15,000 regions $\{d_1, \ldots, d_{15,000}\}$. (a) Scoring system A ($H_3K_4$), (b) Scoring system B (synthetic).

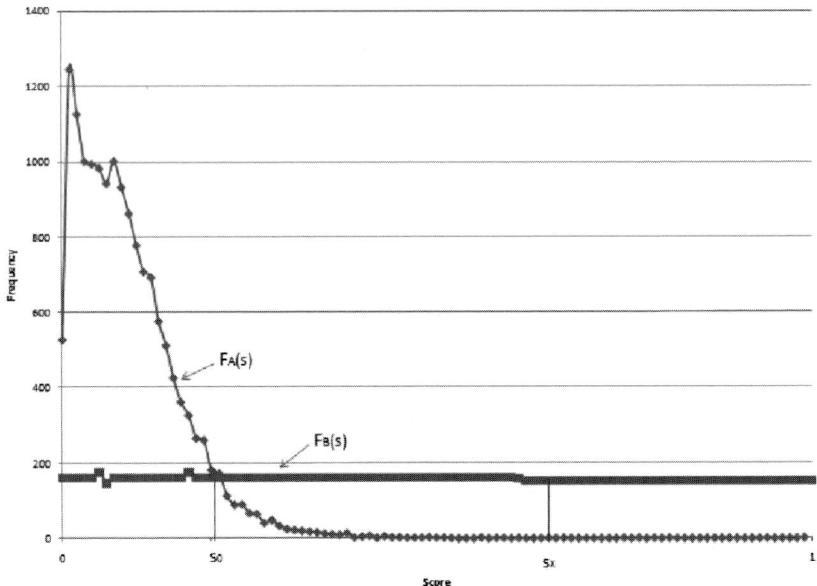

Fig. 6. Frequency of scores for A and B.

Based on these discussions, as in Figs. 6 and 7, we conclude that using a score threshold to reduce the number of binding site regions in D has a much greater effect on the variation in ranks in scoring scenario B than in A. In particular, since most of the existing methods use a threshold (i.e. a score) s_x to reduce the binding site region pool, then as long as $s_x > s_0$, it will produce a greater variation in ranks than in scores in scenarios A and B. These results illustrate that for scoring scenarios with much noise and greater false-positive rate such as the scoring scenario B, why using a threshold to produce a reduced final list of binding site regions can lead to an over-fit and under-fit situation and hence has higher false-positive rate and false-negative rate, respectively. However, this is often the case for the various existing algorithms, methods, and technical/biological replicates in detecting binding site regions in the ChIP-seq peak finding systems. It also illustrates that, under these conditions, using both rank function and score function has distinctive benefit (robustness and reliability) in scoring the binding site regions in the spectrum of whole genome-wide study. Since the rank-score function and its corresponding RSC graph is readily obtained from the score function and rank function, using the variation of the rank-score functions between two scoring systems to measure diversity enables

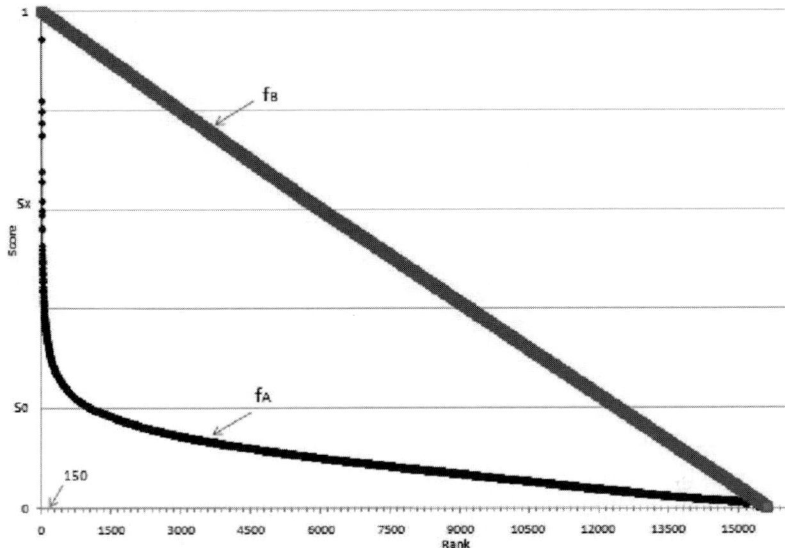

Fig. 7. RSC graphs f_A and f_B derived from h_A and h_B.

us to have a different perspective for documenting the reproducibility of different peak calls (in the cases of scenarios (1) and (2)). They can also guide us to determine the likelihood of success when combining A and B to obtain better algorithms/methods for identifying binding site regions (in the case of scenarios (3) and (4)).

Combinatorial fusion methods can be used to design, develop, and validate an improved algorithm for identifying the protein-binding site. In this context, multiple scoring systems can represent different features/cues in a method, which are then fused to produce the best performing result (feature fusion for scenario (3)). They can also represent different proposed methods/algorithms that can be fused to produce a better method/algorithms (decision fusion for scenario (4)). In both cases, system selection and combination are the main tasks. Given a list of multiple scoring systems $A_1, A_2, \ldots, A_t$ on the set D of possible binding site regions $d_1, d_2, \ldots, d_n$, it is possible to identify which systems should be fused to produce a better new system. It has been demonstrated[25] that combining multiple scoring functions improve the enrichment of true positive only if (a) each of the individual scoring functions has relatively high performance and (b) the individual scoring functions are diverse. These two prediction criteria are previously established for the performance of fusion approaches with respect to either score or rank combination.[23,24,26,27] We will use the rank-score functions $\boldsymbol{f}_A$ and $\boldsymbol{f}_B$ and their difference $\mathbf{d}(\boldsymbol{f}_A, \boldsymbol{f}_B)$ to measure the diversity $d(A,B)$ between two scoring systems A and B. Since the rank-score functions $\boldsymbol{f}_A, \boldsymbol{f}_B$ and their corresponding RSC graphs are independent of the performance of the individual scoring system, we can use $\mathbf{d}(\boldsymbol{f}_A, \boldsymbol{f}_B)$ to determine the likely success of combining these two scoring systems A and B. In the case of the ChIP-seq binding site identification, these two different and diverse features/cues are combined to design and validate the best algorithm (scenario (3)). In scenario (4), two different and diverse methods/algorithms are combined to obtain a method/algorithm, which is better than each of the individual systems.

4. Evaluation of ChIP-Seq Peak-Detection Systems

As ChIP-seq techniques and systems become popular, they lead to the development of many different analytical methods and systems. So far, there are over 30 open source programs.[28] Wilbanks and Facciotti[28] compared the performance of 11 peak calling programs on common empirical, transcription factor datasets, and measured their sensitivity, accuracy, and usability. A recent study by C. Schweikert *et al.*[29] evaluates six open source methods based on three attributes and their combinations. They also use combinatorial fusion analysis[20] to explore the space of all combinations among these six methods.

4.1. *Evaluation of algorithm performance*

Wilbanks and Facciotti[28] benchmarked 11 peak-calling algorithms against three transcription factor ChIP-seq datasets with control: human neuron-restrictive silencer factor (NRSF),[7] growth-associated binding protein (GABP),[16] and hepatocyte nuclear factor 3 α (Fox A1).[18] The 11 ChIP-seq peak calling programs selected for evaluation are as follows: CisGenome,[11] Minimal ChipSeq Peak Finder,[7] E-RANGE,[14] MACS,[18] QuEST,[16] HPeak,[30] Sole-Search,[31] PeakSeq,[32] SISSRs,[15] and SPP Package (wtd and mtc).[33]

For each of the three datasets, all peak callers reported a different number of peaks. The variation in the quantity of indentified peaks shows that default stringency levels are tuned differently among programs (Fig. 8). They conducted a series of pair-wise comparisons between the peak lists from each method to determine which peaks were shared. These results are presented as Fig. 9. They also explore further by examining PCR-validated

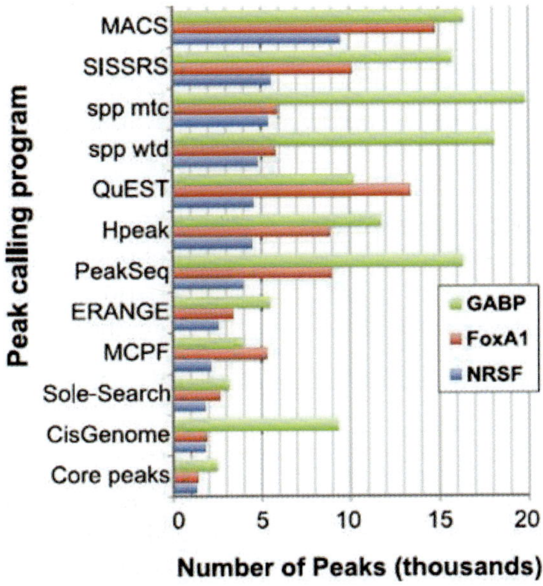

Fig. 8.   Quantity of peaks identified by systems.[1]

true positive sites available for NRSF and GABP. The sensitivity of these 11 methods was assessed by calculating the percentage of these true positives found by each program and validated by qPCR, which is shown in Fig. 10.

4.2. *Combining multiple ChIP-seq detection systems*

Schweikert *et al.*[29] evaluated six open source methods based on three attributes: tag count, p-value, and fold-change and their combinations. These six methods are: CisGenome,[11] MACS,[18] PeakSeq,[32] QuEST,[16] SISSRs,[15] and TRLocator.[29] For each of the six systems, they studied its performance on the three features tag count (T), p-value (P), and fold-change (F) and on the four systems resulting from the combinations TP, TF, PF, and TPF. Performance evaluation by average precision is based on (a) region overlap with a transcription start site (TSS), and (b) region overlap within 1,000 bases of a TSS (Tables 3(a) and (b)). Since the feature tag count (T) dominates the other two features, p-value (P) and fold enrichment (F), they use the tag count feature to compare all 2-combinations of the six systems. They perform two kinds of combinations: intersection and union. The union of two systems, x and y, is the set of all regions

NRSF

	CisGenome	Sole-Search	WOLD	ERANGE	PeakSeq	Hpeak	QuEST	wtd	mtc	SISSRS	MACS
CisGenome	X	80	76	64	44	40	36	37	33	31	19
Sole-Search	82	X	81	68	45	40	36	38	34	37	19
MCPF	91	95	X	81	53	48	42	47	41	48	22
ERANGE	91	93	94	X	61	54	47	52	46	49	26
PeakSeq	98	99	100	100	X	85	66	78	69	78	43
Hpeak	98	99	100	100	91	X	69	83	74	80	43
QuEST	91	92	91	89	76	74	X	74	68	76	44
spp wtd	98	99	99	97	87	85	72	X	84	76	45
spp mtc	98	98	99	96	87	86	75	94	X	77	47
SISSRS	97	98	100	99	89	86	75	88	79	X	46
MACS	100	99	100	100	97	94	87	93	88	93	X

GABP

	Sole-Search	MCPF	ERANGE	CisGenome	QuEST	Hpeak	SISSRS	PeakSeq	MACS	wtd	mtc
Sole-Search	X	63	50	30	28	26	22	18	18	16	15
MCPF	81	X	72	42	37	34	40	24	24	23	21
ERANGE	89	99	X	58	50	47	49	33	33	30	27
CisGenome	90	100	98	X	73	76	53	55	53	45	41
QuEST	96	97	96	82	X	81	81	61	59	56	52
Hpeak	100	100	100	96	89	X	83	68	66	60	55
SISSRS	97	100	100	84	92	86	X	70	67	60	55
PeakSeq	100	100	100	100	97	100	96	X	86	82	76
MACS	100	100	100	97	95	97	94	87	X	78	73
wtd	93	90	90	86	88	87	72	81	77	X	89
mtc	93	90	90	86	89	87	72	82	79	97	X

FoxA1

	CisGenome	Sole-Search	ERANCE	MCPF	wtd	mtc	Hpeak	PeakSeq	SISSRS	QuEST	MACS
CisGenome	X	58	52	33	30	30	21	21	18	13	12
Sole-Search	82	X	67	47	44	44	30	29	27	18	18
ERANGE	96	86	X	58	56	55	38	38	34	22	23
MCPF	93	95	90	X	74	72	56	57	52	31	35
wtd	93	98	95	81	X	93	65	63	55	36	39
mtc	93	97	94	80	95	X	66	63	56	37	39
Hpeak	100	100	100	94	99	99	X	86	79	51	55
PeakSeq	100	100	100	96	98	96	88	X	80	50	58
SISSRS	96	100	98	97	96	96	88	88	X	54	61
QuEST	94	91	89	78	84	85	77	73	73	X	60
MACS	99	100	100	99	99	99	92	96	91	67	X

Fig. 9. Pair-wise comparison of shared peaks.[1]

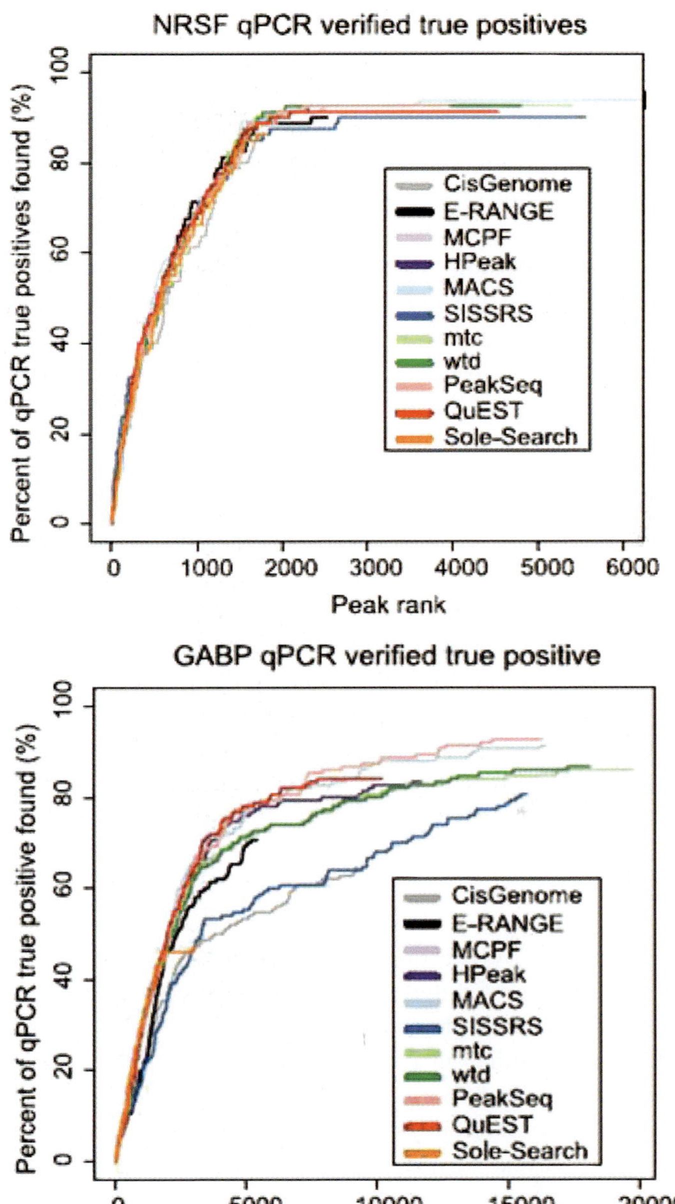

Fig. 10. Sensitivity assessment: True positives of NRSF and GABP Datasets.[1]

identified by x or y, where overlapping regions between the two methods are merged together to form new merged regions; the intersection of two systems consists of only the merged regions formed by overlapping regions between the two methods. When evaluating each system or combined system, average precision and coverage are used, where coverage refers to the number of unique TSS that overlap with regions identified by the system.[29] These results are listed in Fig. 11(a)–(d).

Several observations have been noted. First, there is no single answer for the selection of available methods and systems for ChIP-seq peak detection. It depends on the criteria for features and performance evaluations. For example, each of the six systems (A: CisGenome, B: MACS, C: PeakSeq, D: QuEST, E: SISSRs, F: TRLocator) is ranked as (4, 3, 5, 6, 2, 1) and (3, 6, 2, 4.5, 1, 4.5), when evaluating performance with average precision according to overlap with and distance within 1,000 bp of transcription start sites, respectively.[29] Second, combination of different methods or systems does improve results in many cases. For example, all 2-Combinations using intersection and evaluation by average precision, and all 2-Combinations using union and evaluation by coverage are positive cases where the combined system outperforms both of the individual systems (see Fig. 11(a) and (d)). In addition, the use of a rank function, other than the score function, in the evaluation of multiple detection systems provides a generic framework to study the preference and relative preference for the methods (or systems) selection process.

4.3. *Overall evaluation of ChIP-seq analytic methods*

The current utility of the ChIP-seq experimental method for biomedical researchers is limited by the quality of available data analysis software. There is an abundance of different software packages available, which are implementations of many different algorithms. These methods and systems perform differently on different types of ChIP-seq data, which have different types of inherent data distributions (i.e. transcription factor versus histone modification). Several different reviews and software benchmarking studies have noted that analysis with default parameters using different software packages will produce dramatically different results, allowing arbitrary software choices to bias the biological conclusions that researchers draw from their data. Our own work shows that even with a single software package, analysis of technical replicates shows poor reproducibility of peak-detection. Strong enrichment events (binding sites) can be detected by many different data processing methods, but weaker events are much

Table 3. (a) Average Precision* of the six methods A, B, C, D, E, F according to T, P, F and their combinations.[2] T: tag count, P: p-value, F: fold/enrichment (log ratio for TRLocator), A: CisGenome, B: MACS, C: PeakSeq, D: QuEST, E: SISSRs, F: TRLocator.

x	T $s_\mathrm{T}(\mathrm{x})$; $r_\mathrm{T}(\mathrm{x})$	P $s_\mathrm{P}(\mathrm{x})$; $r_\mathrm{P}(x)$	F $s_\mathrm{F}(\mathrm{x})$; $r_\mathrm{F}(\mathrm{x})$	S(TP) $s_{\mathrm{S(TP)}}(\mathrm{x})$; $r_*(\mathrm{x})$	S(TF) $s_{\mathrm{S(TF)}}(\mathrm{x})$; $r_*(\mathrm{x})$	S(PF) $s_{\mathrm{S(PF)}}(\mathrm{x})$; $r_*(\mathrm{x})$	S(TPF) $s_{\mathrm{S(TPF)}}(\mathrm{x})$; $r_*(\mathrm{x})$	R(TP) $s_{\mathrm{R(TP)}}(\mathrm{x})$; $r_*(\mathrm{x})$	R(TF) $s_{\mathrm{R(TF)}}(\mathrm{x})$; $r_*(\mathrm{x})$	R(PF) $s_{\mathrm{R(PF)}}(\mathrm{x})$; $r_*(\mathrm{x})$	R(TPF) $s_{\mathrm{R(TPF)}}(\mathrm{x})$; $r_*(\mathrm{x})$
A	0.827727; 5	0.802579; 3	0.805079; 5	0.810301; 4	0.814170; 5	0.802654; 3	0.808434; 3	0.820356; 2	0.820209; 4	0.802673; 3	0.815494; 4
B	0.902319; 2	0.640694; 5	0.839573; 2	0.642255; 6	0.879249; 2	0.625049; 6	0.625966; 6	0.792034; 4	0.890088; 2	0.677927; 5	0.868649; 2
C	0.863441; 3	0.710995; 4	0.814438; 3	0.794711; 5	0.863746; 3	0.711968; 4	0.797201; 4	0.786969; 5	0.856481; 3	0.740196; 4	0.834999; 3
D	0.828119; 4	0.606828; 6	0.741773; 6	0.823837; 2	0.790938; 6	0.695407; 5	0.767535; 5	0.743815; 6	0.810023; 6	0.632052; 6	0.758494; 6
E	0.821222; 6	0.812032; 2	0.809098; 4	0.816196; 3	0.819450; 4	0.809098; 2	0.814192; 2	0.817271; 3	0.817103; 5	0.809098; 2	0.814165; 5
F	0.921679; 1	0.927470; 1	0.841910; 1	0.921653; 1	0.887119; 1	0.857774; 1	0.893524; 1	0.921608; 1	0.909710; 1	0.910864; 1	0.917984; 1

*Average Precision is based on region overlap with a TSS.

Table 3. (b) Average Precision[†] of the six methods A, B, C, D, E, F according to T, P, F and their combinations.[2] T: tag count, P: p-value, F: fold/enrichment (log ratio for TRLocator), A: CisGenome, B: MACS, C: PeakSeq, D: QuEST, E: SISSRs, F: TRLocator.

x	T $s_T(x)$; $r_T(x)$	P $s_P(x)$; $r_P(x)$	F $s_F(x)$; $r_F(x)$	S(TP) $s_{S(TP)}(x)$; $r_*(x)$	S(TF) $s_{S(TF)}(x)$; $r_*(x)$	S(PF) $s_{S(PF)}(x)$; $r_*(x)$	S(TPF) $s_{S(TPF)}(x)$; $r_*(x)$	R(TP) $s_{R(TP)}(x)$; $r_*(x)$	R(TF) $s_{R(TF)}(x)$; $r_*(x)$	R(PF) $s_{R(PF)}(x)$; $r_*(x)$	R(TPF) $s_{R(TPF)}(x)$; $r_*(x)$
A	0.932456; 3	0.930686; 4	0.931956; 3	0.931538; 2	0.931938; 3	0.930688; 3	0.931344; 3	0.933201; 3	0.933185; 3	0.930685; 3	0.932667; 3
B	0.925842; 5	0.926661; 5	0.889186; 5	0.924610; 6	0.912956; 5	0.920649; 4	0.922047; 4	0.925479; 5	0.919436; 5	0.915393; 6	0.922146; 6
C	0.937024; 2	0.939849; 2	0.935452; 2	0.938810; 4	0.937594; 2	0.940150; 2	0.939087; 2	0.938834; 2	0.941785; 2	0.940967; 2	0.941488; 2
D	0.924747; 6	0.925191; 6	0.899949; 4	0.924937; 5	0.916143; 4	0.910957; 5	0.919508; 5	0.925056; 6	0.922330; 4	0.921773; 4	0.925563; 4
E	0.948529; 1	0.945481; 1	0.944346; 1	0.946616; 1	0.948093; 1	0.944346; 1	0.946056; 1	0.947331; 1	0.947281; 1	0.944346; 1	0.946255; 1
F	0.927621; 4	0.932859; 3	0.863957; 6	0.927604; 3	0.900902; 6	0.877136; 6	0.906088; 6	0.927564; 4	0.918757; 6	0.919628; 5	0.925234; 5

† Average Precision is based on region overlap within 1,000 bases of a TSS.

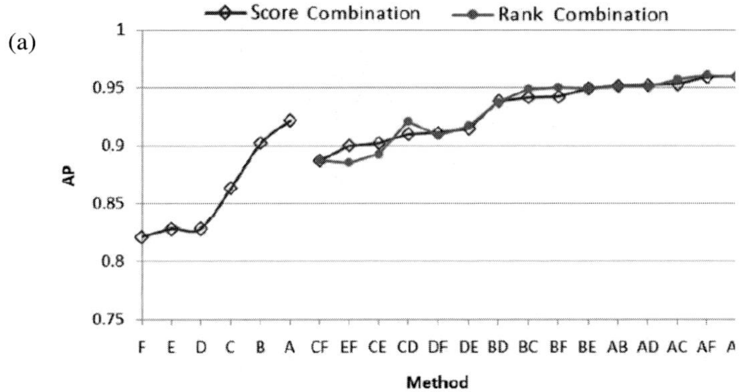

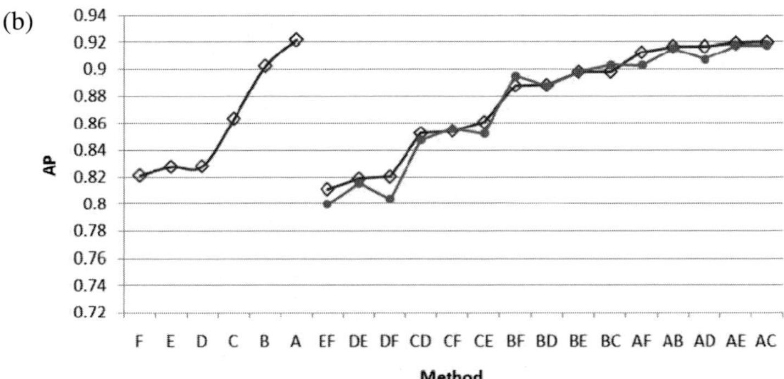

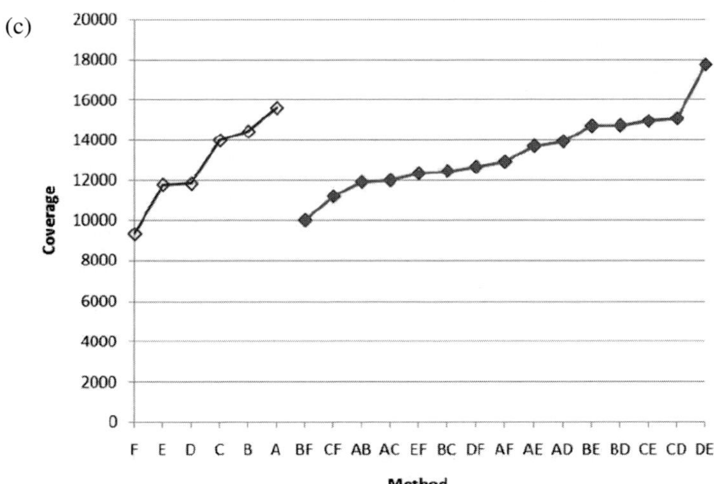

Fig. 11. (*Continued*)

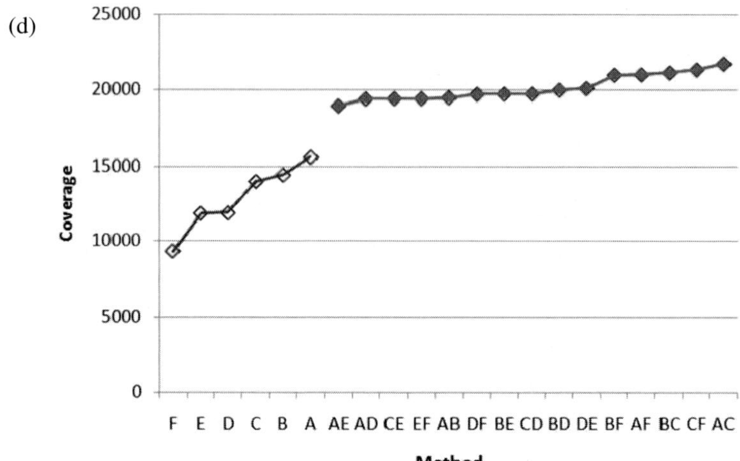

A = TRLocator, B = MACS, C = PeakSeq, D = QuEST, E = CisGenome, F = SISSRs

Fig. 11. (a) Average precision for 2-combinations by intersection[2]; (b) Average precision for 2-combinations by union[2]; (c) Coverage for 2-combinations by intersection[2]; (d) Coverage for 2-combinations by union.[2]

less reproducible. The use of control data as background (input DNA or IGG control) allows for improved results with a reduction in false-positive peak-detections. Improved software is required in order to allow the use of the ChIP-seq method for quantitiatve studies that compare the impact of different biological conditions on genome wide transcription factor binding or epigenetic modifications. Our preliminary work indicates that information fusion methods, in particular, Combinatorial Fusion Analysis (CFA), can be used to improve the quality of ChIP-seq peak-detection by combining algorithms or software with diverse characteristics. Following our work reviewed in Section 4.2,[29] we will use rank-score functions and their corresponding rank-score characteristic (RSC) graphs[34] to investigate variations in each of (and among) the four scenarios described in Section 4.3.[34-37] Future work may also be directed to create software that can evaluate the data distribution of a data set and optimize its own parameters, such as machine learning methods.

References

1. Sanger F, Nicklen S, Coulson AR. (1977) DNA sequencing with chain-terminating inhibitors. *Proc Natl Acad Sci USA* **74**(12): 5463–5467.
2. Mardis ER. (2008) The impact of next-generation sequencing technology on genetics. *Trends Genet* **24**(3): 133–141.

3. Shendure J, Ji H. (2008) Next-generation DNA sequencing. *Nat Biotechnol* **26**(10): 1135–1145.

4. Solomon MJ, Larsen PL,Varshavsky A. (1988) Mapping protein–DNA interactions *in vivo* with formaldehyde: Evidence that histone H4 is retained on a highly transcribed gene. *Cell* **53**(6): 937–947.

5. Huebert DJ, Kamal M, O'Donovan A, Bernstein BE. (2006) Genome-wide analysis of histone modifications by ChIP-on-chip. *Methods* **40**(4): 365–369.

6. Kim TH, Barrera LO, Ren B. (2007) ChIP-chip for genome-wide analysis of protein binding in mammalian cells. *Curr Protoc Mol Biol* Chapter 21: Unit 21.13.

7. Johnson DS, Mortazavi A, Myers RM, Wold B. (2007) Genome-wide mapping of *in vivo* protein–DNA interactions. *Science* **316**(5830): 1497–1502.

8. Visel A, Blow MJ, Li Z, Zhang T, Akiyama JA *et al.* (2009) ChIP-seq accurately predicts tissue-specific activity of enhancers. *Nature* **457**(7231): 854–858.

9. Li H, Durbin R. (2009) Fast and accurate short read alignment with Burrows–Wheeler transform. *Bioinformatics* **25**(14): 1754–1760.

10. Li H, Ruan J, Durbin R. (2008) Mapping short DNA sequencing reads and calling variants using mapping quality scores. *Genome Res* **18**(11): 1851–1858.

11. Ji H, Jiang H, Ma W, Johnson DS, Myers RM *et al.* (2008) An integrated software system for analyzing ChIP-chip and ChIP-seq data. *Nat Biotechnol* **26**(11): 1293–1300.

12. Robertson G, Hirst M, Bainbridge M, Bilenky M, Zhao Y *et al.* (2007) Genome-wide profiles of STAT1 DNA association using chromatin immuno-precipitation and massively parallel sequencing. *Nat Methods.* **4**(8): 651–657.

13. Fejes AP, Robertson G, Bilenky M, Varhol R, Bainbridge M *et al.* (2008) FindPeaks 3.1: A tool for identifying areas of enrichment from massively parallel short-read sequencing technology. *Bioinformatics* **24**(15): 1729–1730.

14. Mortazavi A, Williams BA, McCue K, Schaeffer L, Wold B. (2008) Mapping and quantifying mammalian transcriptomes by RNA-seq. *Nat Methods* **5**(7): 621–628.

15. Jothi R, Cuddapah S, Barski A, Cui K, Zhao K. (2008) Genome-wide identification of *in vivo* protein–DNA binding sites from ChIP-seq data. *Nucleic Acids Res* **36**(16): 5221–5231.

16. Valouev A, Johnson DS, Sundquist A, Medina C, Anton E *et al.* (2008) Genome-wide analysis of transcription factor binding sites based on ChIP-Seq data. *Nat Methods* **5**(9): 829–834.

17. Nix DA, Courdy SJ, Boucher KM. (2008) Empirical methods for controlling false positives and estimating confidence in ChIP-Seq peaks. *BMC Bioinformatics* **9**: 523.

18. Zhang Y, Liu T, Meyer CA, Eeckhoute J, Johnson DS *et al.* (2008) Model-based analysis of ChIP-Seq (MACS). *Genome Biol* **9**(9): R137.

19. Ho TK, Hull JJ, Srihari SN. (1994) Decision combination in multiple classifier systems. *IEEE Trans Pattern Anal Mach Intell* **16**(1): 66–75.

20. Hsu DF, Chung Y-S, Kristal BS. (2006) Combinatorial fusion analysis: Methods and practice of combining multiple scoring systems. In: *Advanced Data Mining Technologies in Bioinformatics*, pp. 32–36. Idea Group Inc., Hershey, Philadelphia.

21. Chuang H-Y, Liu H, Brown S, McMunn-Coffran C, Kao C-Y *et al.* (2004)
    Identifying significant genes from microarray data. *In Procs. IEEE Bioinfor-
    matics and Bioengineering (BIBE'04)* IEEE Comput Soc, 358–365.
22. Peng CH, Hsu JT, Chung YS, Lin YJ, Chow WY *et al.* (2006) Identification
    of degenerate motifs using position restricted selection and hybrid ranking
    combination. *Nucleic Acids Res* **34**(22): 6379–6391.
23. Lin KL, Lin CY, Huang CD, Chang HM, Yang CY *et al.* (2007) Feature
    selection and combination criteria for improving accuracy in protein structure
    prediction. *IEEE Trans Nanobioscience* **6**(2): 186–196.
24. Yang JM, Chen YF, Shen TW, Kristal BS, Hsu DF. (2005) Consensus scoring
    criteria for improving enrichment in virtual screening. *J Chem Inf Model*
    45(4): 1134–1146.
25. Ng KB, Kantor PB. (2000) Predicting the effectiveness of naive data fusion
    on the basis of system characteristics. *J Am Soc Inf Sci* **51**(13): 1177–1189.
26. Hsu DF, Taksa I. (2005) Comparing rank and score combination methods for
    data fusion in information retrieval. *Inf Retrieval* **8**(3): 449–480.
27. Lyons DM, Hsu DF. (2009) Combining multiple scoring systems for target
    tracking using rank-score characteristics. *Inf Fusion* **10**(2): 124–136.
28. Wilbanks EG, Facciotti MT. (2010) Evaluation of algorithm performance in
    ChIP-seq peak detection. *PLoS One* **5**(7): e11471.
29. Schweikert C, Liu J, An W, Brown S, Smith PR *et al.* (2010) Combining
    Multiple Detection Systems for Improving ChIP-seq Peak Identification of
    Protein Binding Sites. *Department of Computer and Information Science
    Technical Report 11-2010, Fordham University, New York, NY.*
30. Qin ZS, Yu J, Shen J, Maher CA, Hu M *et al.* (2010) HPeak: An HMM-
    based algorithm for defining read-enriched regions in ChIP-seq data. *BMC
    Bioinformatics* **11**: 369.
31. Blahnik KR, Dou L, O'Geen H, McPhillips T, Xu X *et al.* (2010) Sole-Search:
    An integrated analysis program for peak detection and functional annotation
    using ChIP-seq data. *Nucleic Acids Res* **38**(3): e13.
32. Rozowsky J, Euskirchen G, Auerbach RK, Zhang ZD, Gibson T *et al.*
    (2009) PeakSeq enables systematic scoring of ChIP-seq experiments relative
    to controls. *Nat Biotechnol* **27**(1): 66–75.
33. Kharchenko PV, Tolstorukov MY, Park PJ. (2008) Design and analysis of ChIP-
    seq experiments for DNA-binding proteins. *Nat Biotechnol* **26**(12): 1351–1359.
34. Hsu DF, Taksa I. (2010) Combinatorial fusion analysis for meta search
    information retrieval. Chapter 8 in *Web-Based Support Systems*, Jing Tao
    Yao (Ed.), Springer, pp. 147–165.
35. Hsu DF, Kristal BS, Schweikert C. (2010) Rank-score characteristic function
    and cognitive diversity, in *Brain Informatics*, Yiyu Yao *et al.* (Eds.), LNAI
    6334, Springer, pp. 42–54.
36. Li Y, Hsu DF, Chung SM. (2009) Combining multiple feature selection
    methods for text categorization by using rank-score characteristics. *21st IEEE
    International Conference on Tools with Artificial Intelligence*, pp. 508–517.
37. Mesterharm C, Hsu DF. (2008) Combinatorial fusion with on-line learn-
    ing algorithms. *The 11th International Conference on Information Fusion*,
    pp. 1117–1124.

Chapter 7

Discovery of Transcription Factor Binding Sites and Its Applications in Cancer Study

Chien-Yu Chen[*]

1. Introduction

Transcription factors (TFs) play an important role in gene regulation. TFs recognize in genomes specific nucleotide sequences called transcription factor-binding sites (TFBSs) to activate or repress gene expression. Identification of TF-binding motifs and their genomic locations underlies the understanding of gene regulation.[1,2] Many computational methods have been developed to discover automatically over-represented subsequences from a set of related sequences.[3] Applying such methods to data generated by recent high-throughput technologies such as ChIP-chip or ChIP-seq further helps to connect the discovered motifs with the TFs of interest. Alternatively, this problem has been tackled from the viewpoint of computational chemistry. Employing quantitative structure-activity relationship (QSAR) models on existing protein–DNA co-crystallized structure also facilitates identification of position specificity in protein–DNA interactions.[4]

A TF usually binds to a variety of similar sequences. The TFBSs bound by the same TF can be summarized as a motif using either a consensus or a profile. By analyzing sequence entropy of aligned binding sites, it is possible to distinguish the degenerated positions from the conserved ones. A motif that contains one or more successive highly degenerated positions is called a "gapped" motif. Since allowing presence of degenerated positions

[*]Department of Bio-Industrial Mechatronics Engineering, National Taiwan University, Taipei 106, Taiwan. cychen@mars.csie.ntu.edu.tw

113

usually greatly increases computational cost, discovery of gapped motifs is a challenging task. Most sequence-based computational approaches request the users to set in advance the shape of a motif, such as the length of the motifs, the type of motifs (gapped or non-gapped), the number of wildcards (degenerated positions) in motifs, and so forth. However, it is difficult to determine *a priori* about the shape of the motifs before conducting biological experiments.

For sequence-based computational approaches, identifying a potential sequence set for motif finding is the first step for TFBS discovery. This task can usually be achieved by looking for co-expressed genes from microarray data.[5] However, it remains challenging to distinguish co-expressed genes affected by a single regulator from co-expressed genes influenced by multiple regulators. That is, it is usually the case that only a considerably small proportion of the co-expressed genes share a similar binding site. This fact complicates the procedure of motif discovery. Another difficulty is that for a gene, a TFBS may be located far from the transcription start site (TSS). In this regard, even if the list of target genes is reliable, a computational method may still fail to find the correct motifs if the search regions are not broad enough. Indeed, if the search is restricted to a short region in the promoter (e.g. from -500 bp to the TSS), it may happen that only a small proportion (e.g. $<15\%$) of the input sequences contain the binding site. Of course, it is possible to looking for TFBSs in a broader region (e.g. from -5000 bp to the TSS or even the downstream of the TSS). However, enlarging the search region usually complicates the task and exponentially expands the search time.

Due to the advance of chromatin immunoprecipitation (ChIP) related techniques such as ChIP-chip and ChIP-seq, the last decade has witnessed great progress in solving this problem.[6-8] Experimental data derived from ChIP-chip helps researchers to compile a list of potential target genes or sequences, which can then be used to determine the regulatory motif and the location of TFBSs in the genome for the TF of interest. Recently, the technique of chromatin immunoprecipitation coupled with massively parallel segueing (ChIP-seq)[7] has become the main focus of research interest in uncovering protein–DNA interactions. ChIP-seq is considered to provide more precise (narrower regions) and more accurate (with more true positives) information than ChIP-chip.[9] In addition, ChIP-seq technique, when compared with ChIP-chip, is not biased by priori design of probe sequences. That is, TFBS discovery based on ChIP-seq data potentially spans all the regions of the genome.

In addition to sequence-based approaches, an alterative way to predict binding targets of regulatory proteins is exploiting co-crystallized structures of proteins and DNA. To predict the specificity of a particular DNA sequence, empirical or physics-based energy functions can be applied to structure models. Kono and Sarai developed a statistical energy function obtained from existing protein–DNA complexes to estimate binding specificity by the calculated Z-scores.[10] While knowledge-based potentials for direct readouts are helpful in predicting protein–DNA recognition, some physics-based energy functions designed in particular for measuring influence from water molecules and DNA bending also contribute to enhancing predictive power.[11] Different from sequence-based methods that require a sufficient quantity of potential target sequences for motif discovery, structure-based approaches can make reliable prediction simply based on a single protein–DNA co-crystallized structure. Before real structure models can be determined experimentally using X-ray or Nuclear Magnetic Resonance (NMR) techniques, computational approaches, including homology modeling and molecular docking, can also be employed to enlarge the pool of structure models for motif profile prediction. Furthermore, a recent study demonstrated that sequence-based approaches incorporated with ChIP data and structure-based approaches can be combined to complement each other.[12]

Many cancer studies have shown that TFs are related to the development of cancer cells. For example, defects in some important TFs such as RUNX1 greatly increase the risks of certain types of leukemia. Identifying the relationships between TFs and a particular type of cancer facilitates early diagnosis and drug treatment. Microarray data is now widely used in studying differential genes to compare patient samples against normal tissues. It is generally believed that many of these differential genes are due to differences in the activities of TFs. Researchers also use time-course array data derived from cancer cell lines to discover target genes of an interested TF. For example, array experiments after treatment of estrogen have been used to identify direct or indirect target genes of estrogen receptor (ER).[13] Afterward, motif discovery can be applied on the derived gene sets to identify TFBSs. The derived motifs are then compared with known matrixes in motif database or further verified by biological experiments. Alternatively, known motifs in motif databases such as TRANSFAC and JASPAR can be used to scan promoter regions of the differential gene sets to further explain the mechanism of cancer development. However, in human and other mammals, we are still far from the objective of determining the TFBSs for most human TFs.

This review will first briefly describe both the sequence- and structure-based computational approaches and also illustrate the concept of how structure information can be incorporated in sequence-based approaches to enhance predictive accuracy. Section 3 discusses how motif discovery algorithms and existing motif matrixes can facilitate TF-related cancer studies. The final section provides concluding remarks and suggestions for future directions.

2. Computational Methods for TFBS Discovery

A TFBS is usually 5–15 bp long and may contain a number of highly degenerate positions (known as wildcards in pattern recognition). There are variations between several binding sites to which a TF can bind. The most popular way to present a set of TFBSs bound by a TF is using position weight matrixes (PWMs), which has the advantage of quantitatively modeling the variability of a TFBS when compared to a regular expression form (consensus).

There are two major categories of computational methods to derive binding profiles (motifs) of a TF: sequence-based approaches and structure-based approaches. As illustrated in Fig. 1, a sequence-based computational method first collects a set of DNA sequences from the promoter regions of related or co-expressed genes and conducts pattern mining or sequence alignment for detecting over-represented subsequences. A structure-based approach starts with a protein–DNA co-crystallized structure and performs position-wise specificity analysis in accordance with an energetic scoring function. In this section, we first review the basic ideas and some of existing motif-finding packages for the sequence-based computational methods. Next, we introduce an efficient web server eTFBS as an example designed for a general purpose of motif discovery. Then we further explain the basic ideas behind the structure-based computational methods and review a hybrid approach to enhance the accuracy of predicting TFBSs.

2.1. *Sequence-based approaches*

Many sequence-based computational methods have been proposed to discover TFBSs,[14–18] most of which involve the identification of overrepresented subsequences among a set of related sequences. Computational methods can be categorized based on the method of representing a motif, oligo- or profile-based, corresponding to the consensus and PWM presentation, respectively.

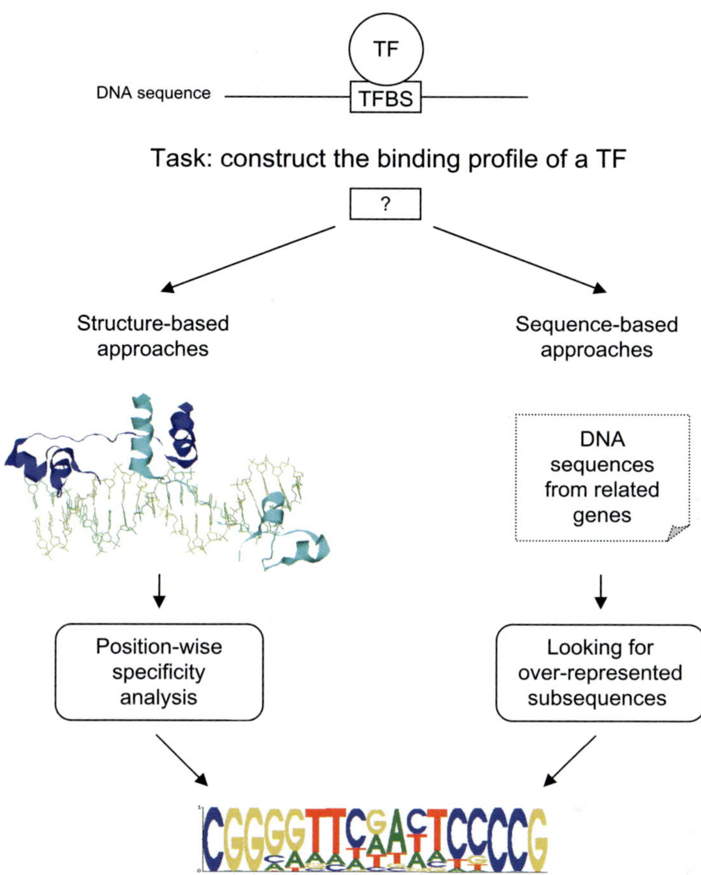

Fig. 1. TFBS prediction based on sequence- or structure-based computational methods.

An oligo-based representation is a consensus including ambiguous positions represented by IUPAC (International Union of Pure and Applied Chemistry) codes. Though 15 letters are employed to express the diversity of positions, they are still less flexible than using PWMs to represent a TFBS.

Sequence-based methods compare a set of DNA sequences that putatively contain binding sites for a particular TF (a positive set) with either a set of sequences that are not bound by the TF or a general background set (a negative set). A search is then conducted for motifs that are over-represented in the positive set and considered candidates for the TF-binding motif. In 2005, Tompa *et al.* conducted performance evaluation on

13 motif-discovery tools. They selected these programs because they satisfy the criterion that no auxiliary information, such as gene expression or phylogenetic trees, is required as input. Hu *et al.* also compared five stand-alone programs on prokaryotic data.[19] They concluded that currently available methods are good at discovering TFBSs at the motif level but not at the site or position level. In addition, several algorithms, including RSAT, BioProspector, BIPAD, SeSiMCMC, SPACER, SPACE,[20–25] and one of our previous studies,[26] were especially designed for discovering gapped motifs. BioProspector allows gaps with varied lengths in a TFBS, while SPACE considers more than two subunits in a motif.

2.1.1. *Discovering TFBSs based on ChIP-chip data*

In recent years, studies have used various types of data in addition to genomic sequences to identify TFBSs, including ChIP-chip data,[5,27–30] microarray gene expression data,[5,31–35] and evolutionary conservation of motif sequences.[5,36–39] These studies demonstrate that refining the list of potential target genes consistently improves the predictive accuracy of computational methods. Toward this end, we previously developed a method for discovering yeast TFBSs based on ChIP-chip data, especially gapped motifs, by gradually growing patterns from scratch.[26] The major advantage of using pattern growth (data-driven approaches) over examining all possible k-mer motifs (pattern-driven approaches) is that it is not necessary to specify the type of the motif (gapped or ungapped, long or short) in advance.

Central to our method is a novel ranking scheme that favors candidate motifs that are evolutionarily conserved, have positions that are repeatedly matched by frequently observed patterns, and are found in promoters that are bound by the TF with a high probability in ChIP-chip experiments. Our ranking scheme favors patterns that match popular positions in input sequences — positions that are matched many times by similar patterns. Instead of merging overlapped similar motifs, our method uses position concurrence as a source of signal intensity on the true binding site. The idea behind this ranking scheme is that a true motif might evolve into different forms during evolution. Note further that accounting for the TF-binding strength from ChIP-chip data improves the reliability of inferred motifs.[29] Thus our method weights the contribution of each putative promoter sequence according to the probability in the ChIP-chip data that a TF binds to that promoter.

Empirical tests on 32 known yeast TFBSs show that the new method is highly accurate in identifying gapped motifs, outperforming current methods, and it also works well on ungapped motifs. Predictions on additional 54 yeast TFs successfully discovered 11 gapped and 38 ungapped motifs supported by the literature. Our method achieves high sensitivity and specificity for predicting experimentally verified TFBSs. However, for 203 TFs with ChIP-chip data published by Harbison *et al.*[40] the binding motifs of more than half TFs still remain unknown or unclear.[41]

2.1.2. *eTFBS: A general-purpose web server for discovering motifs*

The high success rate of our method in discovering yeast TFBSs prompted us to develop the web server eTFBS (http://biominer.bime.ntu.edu.tw/etfbs/) without relying on auxiliary information such as conservation. In order to adopt eTFBS to general applications, we implemented the kernel of the mining algorithm in our previous study[26] and simply used the two-sample proportion test to rank over-represented motifs. Thus we can use eTFBS as a mining tool for generating a list of potential motif patterns for further analysis or experimental validation.

Both the ungapped and gapped patterns are collected together and then ranked according to the preferential occurrence of a pattern in the positive group (G^α) relative to the negative group ($G^{-\alpha}$), denoted by S_d, which is calculated by one-tailed two-sample proportion test. Patterns with a z score (i.e. S_d) smaller than $z_{1-0.01}$ are treated as nonsignificant and are removed before the ranking process.

A segment in a promoter sequence is said to be matched by a pattern (e.g. consensus "ACnCGT") if it can be aligned with the consensus without any mismatches (except wildcards), insertions, or deletions. A position in a sequence may be matched by several similar patterns. For example, the segment "ACGCGT" can be simultaneously matched by patterns "AnGCGT," "ACnCGT," "ACGnGT," and so on. In this way, the occurrences of a pattern in the positive and negative sets are calculated respectively.

Users define positive and negative sequence sets. A positive set contains the DNA sequences in FASTA format on which the motif discovery is performed. Three parameters can be set from the web user interface: (1) support (the minimum proportion of input sequences that contain the discovered [putative] motifs); (2) gap length (the maximum length of the gap in between the two pattern blocks of the discovered motifs); and (3) background GC content. The sequences that match a pattern in the consensus form are called the "supporting sequences of a pattern," with "support" defined as the ratio of the number of supporting sequences to the total number of input sequences. It is usually the case that not all the sequences in the input set match the

However, Morozov and Siggia showed in their recent study that half of the predictions from structural models are not consistent with the previously predicted TFBSs from sequence approaches. This suggests the need to develop computational approaches that aim at integrating both structure and sequence information as well as the state-of-art machine learning algorithms for predicting protein–DNA interactions.

3. Applications in Cancer Study

Many differential genes discovered in cancer-related microarray studies are caused by abnormal activities of TFs. Some TFs play an important role in the establishment, progression, or treatment of human cancer. Defects on such TFs result in diseases and considerably affect patient outcome. For example, in array data of breast cancer patients, the expression of many genes is largely correlated to the status of the enhancer estrogen receptor alpha (ERα). In other words, the differential genes are mainly due to the presence or absence of the ERα protein. The expression level of a TF will affect the expression of its target genes, some of which are also TFs. The change on such TFs will further affect more downstream genes and result in dramatic differences in patient outcome. When a microarray dataset with two categories is prepared, biologists can conduct fold-change analysis or statistical tests such as a t-test to find differential genes. With the gene set, the next step is to find the key regulators in order to construct potential transcriptional regulatory networks. The key procedures are mining the promoter regions of such gene sets for over-represented motifs and recognizing which TFs might bind to the discovered motifs. However, usually the TFs might recruit other co-regulatory proteins to achieve the desired regulation. In other words, usually such differential genes contain more than one TFBS in their promoters. Motif discovery can help to discover automatically these over-represented subsequences and unravel the complicated network structures. In this section, we provide two examples of using gene expression data of breast cancer patients or cell lines to exemplify how motif discovery can be employed to find potential regulators from a set of differential or co-expressed genes.

3.1. *Motif discovery based on co-expressed genes*

In the first example, we illustrate how motif discovery can be integrated with time-course data analysis to identify potential regulators. The first step involves identifying co-expressed genes based on a set of experiments after estrogen treatment. The experiments used Affymatrix microarrays. After estrogen treatment, many genes are activated or repressed to adapt to this external stimulus due to the activation of ERα. Previous studies have discussed that upon binding of estrogen, ERα might use a classical pathway to modulate its target genes through the estrogen response element (ERE).[58] Alternatively, ERα invokes the activities of other TFs to affect downstream genes. As discussed in many previous studies, the E2F family is very important in the related pathways.[59]

First, we used time-course gene expression profiles on MCF-7 upon estrogen treatment[60] to compile a list of 302 potential estrogen responsive genes through trajectory clustering.[61] We downloaded the MCF-7 time course expression array (Affymetrix human genome u133 plus 2 arrays) from supplementary data in online publication of Carroll *et al.*[60] The data contain gene expression profiles of MCF-7 on estrogen treatment for four, eight, and twelve hours, respectively. After gene filtering and statistical test (ANOVA), there were 1,438 genes left. We then fed those genes into the trajectory clustering algorithm.[61] After trajectory clustering, 302 genes in total were classified as continuously upregulated estrogen responsive genes (the III pattern). Next, we used the web server eTFBS to perform motif discovery, with eTFBS accepting a list of human gene names as the query. eTFBS used the gene names to collect corresponding promoter regions defined by [−500, TSS]. Meanwhile, for conducting over-representation tests, we prepared a negative set containing promoter regions of equal length retrieved by 1,000 randomly selected genes. After the motif discovery process terminated, we compared the top ten motifs with known matrixes in the TRANSFAC database and observed that about 33% of the genes in the positive set contained the binding sites of TFs belonging to the E2F family. This suggests the important role of the E2F family in related pathways. In addition, eTFBS found two motifs that are similar to the motifs discovered in a previous study by comparison of several mammals.[62] This indicates that the motifs discovered by eTFBS are conserved across different species and deserve further studies. The results are summarized in Table 1.

3.2. *Motif discovery based on differential genes*

Another mining result based on a gene set of potential ER secondary target genes also reveals that the E2F family plays an important role in gene expression after estrogen treatment. We referred to a recent study

Table 1. Mining results by eTFBS for the first example.

	Top-1 motif	Top-2 motif	Top-7 motif
E2F1DP1 (TRASFAC matrix: M00736)			
A motif discovery by a recent study[62]			
A motif discovery by a recent study[62]			

for collecting 104 potential secondary target genes of ER.[13] These genes have differential expression after estrogen treatment, but this expression is absent if cycloheximide (CHX) is supplied. Among the top ten motifs generated by eTFBS, the second motif is similar to E2F-related matrixes. In addition, the fourth motif is similar to the annotated matrix of CKROX.

4. Conclusion

Computational identification of sequence motifs is important to the study of regulatory networks. It helps to discover TFBSs from a list of related genes or sequences. Both sequence- and structure-based approaches are able to detect binding elements effectively. This review in particular introduces the web server eTFBS and illustrates how it can be used in cancer-related research. Examples of mining co-expressed genes and differential genes successfully discover the binding sites of potential regulators.

Two directions are suggested for the future development of computational approaches. Using auxiliary information to refine the gene list for motif discovery might greatly enhance the predictive accuracy, and combination of sequence- and structure-based approaches can provide more detailed information about specificity of binding positions.

References

1. Pilpel Y, Sudarsanam P, Church GM. (2001) Identifying regulatory networks by combinatorial analysis of promoter elements. *Nat Genet* **29**(2): 153–159.
2. Lee TI, Rinaldi NJ, Robert F *et al.* (2002) Transcriptional regulatory networks in *Saccharomyces cerevisiae*. *Science* **298**(5594): 799–804.
3. Sandve GK, Drablos F. (2006) A survey of motif discovery methods in an integrated framework. *Biology Direct* **1**: 11.
4. Hansch C, Hoekman D, Leo A *et al.* (2002) Chem-bioinformatics: Comparative QSAR at the interface between chemistry and biology. *Chem Rev* **102**(3): 783–812.
5. Tsai HK, Huang GT, Chou MY *et al.* (2006) Method for identifying transcription factor binding sites in yeast. *Bioinformatics* **22**(14): 1675–1681.
6. Euskirchen GM, Rozowsky JS, Wei CL *et al.* (2007) Mapping of transcription factor binding regions in mammalian cells by ChIP: Comparison of array- and sequencing-based technologies. *Genome Res* **17**(6): 898–909.
7. Johnson DS, Mortazavi A, Myers RM *et al.* (2007) Genome-wide mapping of *in vivo* protein–DNA interactions. *Science* **316**(5830): 1497–1502.
8. Mardis ER. (2007) ChIP-seq: Welcome to the new frontier. *Nat Methods* **4**(8): 613–614.
9. Ji HK, Jiang H, Ma WX *et al.* (2008) An integrated software system for analyzing ChIP-chip and ChIP-seq data. *Nat Biotech* **26**(11): 1293–1300.
10. Kono H, Sarai A. (1999) Structure-based prediction of DNA target sites by regulatory proteins. *Proteins-Struct Funct Genet* **35**(1): 114–131.
11. Olson WK, Gorin AA, Lu XJ *et al.* (1998) DNA sequence-dependent deformability deduced from protein–DNA crystal complexes. *Proc Nat Acad Sci USA* **95**(19): 11163–11168.
12. Morozov AV, Siggia ED. (2007) Connecting protein structure with predictions of regulatory sites. *Proc Nat Acad Sci USA* **104**(17): 7068–7073.
13. Bourdeau V, Deschenes J, Laperriere D *et al.* (2008) Mechanisms of primary and secondary estrogen target gene regulation in breast cancer cells. *Nucleic Acids Res* **36**(1): 76–93.
14. Bulyk ML. (2003) Computational prediction of transcription-factor binding site locations. *Genome Biol* **5**(1): 201.
15. Wasserman WW, Sandelin A. (2004) Applied bioinformatics for the identification of regulatory elements. *Nat Rev Genet* **5**(4): 276–287.
16. Zhang C, Liu S, Zhu QQ *et al.* (2005) A knowledge-based energy function for protein–ligand, protein–protein, and protein–DNA complexes. *J Med Chem* **48**(7): 2325–2335.
17. Stormo GD. (2000) DNA binding sites: Representation and discovery. *Bioinformatics* **16**(1): 16–23.
18. Tompa M, Li N, Bailey TL *et al.* (2005) Assessing computational tools for the discovery of transcription factor binding sites. *Nat Biotechnol* **23**(1): 137–144.
19. Hu J, Li B, Kihara D. (2005) Limitations and potentials of current motif discovery algorithms. *Nucleic Acids Res* **33**(15): 4899–4913.

20. van Helden J, Rios AF, Collado-Vides J. (2000) Discovering regulatory elements in non-coding sequences by analysis of spaced dyads. *Nucleic Acids Res* **28**(8): 1808–1818.
21. Liu X, Brutlag DL, Liu JS. (2001) BioProspector: Discovering conserved DNA motifs in upstream regulatory regions of co-expressed genes. *Pac Symp Biocomput*: 127–138.
22. Bi C, Rogan PK. (2004) Bipartite pattern discovery by entropy minimization-based multiple local alignment. *Nucleic Acids Res* **32**(17): 4979–4991.
23. Favorov AV, Gelfand MS, Gerasimova AV *et al.* (2005) A Gibbs sampler for identification of symmetrically structured, spaced DNA motifs with improved estimation of the signal length. *Bioinformatics* **21**(10): 2240–2245.
24. Chakravarty A, Carlson JM, Khetani RS *et al.* (2007) SPACER: Identification of cis-regulatory elements with non-contiguous critical residues. *Bioinformatics* **23**(8): 1029–1031.
25. Wijaya E, Rajaraman K, Yiu SM *et al.* (2007) Detection of generic spaced motifs using submotif pattern mining. *Bioinformatics* **23**(12): 1476–1485.
26. Chen CY, Tsai HK, Hsu CM *et al.* (2008) Discovering gapped binding sites of yeast transcription factors. *Proc Nat Acad Sci USA* **105**(7): 2527–2532.
27. Smith AD, Sumazin P, Das D *et al.* (2005) Mining ChIP-chip data for transcription factor and cofactor binding sites. *Bioinformatics* **21**(Suppl 1): i403–i412.
28. Hong P, Liu XS, Zhou Q *et al.* (2005) A boosting approach for motif modeling using ChIP-chip data. *Bioinformatics* **21**(11): 2636–2643.
29. Eden E, Lipson D, Yogev S *et al.* (2007) Discovering motifs in ranked lists of DNA sequences. *PLoS Comput Biol* **3**(3): e39.
30. Liu XS, Brutlag DL, Liu JS. (2002) An algorithm for finding protein–DNA binding sites with applications to chromatin-immunoprecipitation microarray experiments. *Nat Biotech* **20**(8): 835–839.
31. Bannai H, Inenaga S, Shinohara A *et al.* (2004) Efficiently finding regulatory elements using correlation with gene expression. *J Bioinform Comput Biol* **2**(2): 273–288.
32. Sinha S, Tompa M. (2000) A statistical method for finding transcription factor binding sites. *Proc Int Conf Intell Syst Mol Biol* **8**: 344–354.
33. Bussemaker HJ, Li H, Siggia ED. (2001) Regulatory element detection using correlation with expression. *Nat Genet* **27**(2): 167–171.
34. Jensen LJ, Knudsen S. (2000) Automatic discovery of regulatory patterns in promoter regions based on whole cell expression data and functional annotation. *Bioinformatics* **16**(4): 326–333.
35. Roth FP, Hughes JD, Estep PW *et al.* (1998) Finding DNA regulatory motifs within unaligned noncoding sequences clustered by whole-genome mRNA quantitation. *Nat Biotech* **16**(10): 939–945.
36. Doniger SW, Huh J, Fay JC. (2005) Identification of functional transcription factor binding sites using closely related Saccharomyces species. *Genome Res* **15**(5): 701–709.

37. Elemento O, Tavazoie S. (2005) Fast and systematic genome-wide discovery of conserved regulatory elements using a non-alignment based approach. *Genome Biol* **6**(2): R18.

38. Emberly E, Rajewsky N, Siggia ED. (2003) Conservation of regulatory elements between two species of Drosophila. *BMC Bioinformatics* **4**: 57.

39. Kellis M, Patterson N, Endrizzi M *et al.* (2003) Sequencing and comparison of yeast species to identify genes and regulatory elements. *Nature* **423**(6937): 241–254.

40. Harbison CT, Gordon DB, Lee TI *et al.* (2004) Transcriptional regulatory code of a eukaryotic genome. *Nature* **431**(7004): 99–104.

41. Zhu C, Byers K, McCord R *et al.* (2009) High-resolution DNA binding specificity analysis of yeast transcription factors. *Genome Res* **19**(4): 556–566.

42. Rhead B, Karolchik D, Kuhn RM *et al.* (2010) The UCSC genome browser database: Update 2010 *Nucleic Acids Res* **38**(Database issue): D613–D619.

43. Gromiha MM, Siebers JG, Selvaraj S *et al.* (2004) Intermolecular and intramolecular readout mechanisms in protein–DNA recognition. *J Mol Biol* **337**(2): 285–294.

44. Morozov AV, Havranek JJ, Baker D *et al.* (2005) Protein–DNA binding specificity predictions with structural models. *Nucleic Acids Res* **33**(18): 5781–5798.

45. Ahmad S, Kono H, Arauzo-Bravo MJ *et al.* (2006) ReadOut: Structure-based calculation of direct and indirect readout energies and specificities for protein–DNA recognition. *Nucleic Acids Res* **34**: W124–W127.

46. Endres RG, Wingreen NS. (2006) Weight matrices for protein–DNA binding sites from a single co-crystal structure. *Phys Rev E* **73**(6): 061921.

47. Jamal Rahi S, Virnau P, Mirny LA *et al.* (2008) Predicting transcription factor specificity with all-atom models. *Nucleic Acids Res* **36**(19): 6209–6217.

48. Angarica VE, Perez AG, Vasconcelos AT *et al.* (2008) Prediction of TF target sites based on atomistic models of protein–DNA complexes. *BMC Bioinformatics* **9**: 436.

49. Moroni E, Caselle M, Fogolari F. (2007) Identification of DNA-binding protein target sequences by physical effective energy functions: Free energy analysis of lambda repressor-DNA complexes. *BMC Struct Biol* **7**: 61.

50. Stormo GD. (1998) Information content and free energy in DNA–protein interactions. *J Theor Biol* **195**(1): 135–137.

51. Donald JE, Chen WW, Shakhnovich EI. (2007) Energetics of protein–DNA interactions. *Nucleic Acids Res* **35**(4): 1039–1047.

52. Endres RG, Schulthess TC, Wingreen NS. (2004) Toward an atomistic model for predicting transcription-factor binding sites. *Proteins-Struct Funct Bioinformatics* **57**(2): 262–268.

53. Liu ZJ, Mao FL, Guo JT *et al.* (2005) Quantitative evaluation of protein–DNA interactions using an optimized knowledge-based potential. *Nucleic Acids Res* **33**(2): 546–558.

54. Xu BS, Yang YD, Liang HJ *et al.* (2009) An all-atom knowledge-based energy function for protein–DNA threading, docking decoy discrimination, and prediction of transcription-factor binding profiles. *Proteins-Struc Funct Bioinformatics* **76**(3): 718–730.

55. van Dijk M, van Dijk ADJ, Hsu V *et al.* (2006) Information-driven protein–DNA docking using HADDOCK: It is a matter of flexibility. *Nucleic Acids Res* **34**(11): 3317–3325.

56. Liu ZJ, Guo JT, Li T *et al.* (2008) Structure-based prediction of transcription factor binding sites using a protein–DNA docking approach. *Proteins-Struct Funct Bioinformatics* **72**(4): 1114–1124.

57. Gao MSkolnick J. (2008) DBD-Hunter: A knowledge-based method for the prediction of DNA–protein interactions. *Nucleic Acids Res* **36**(12): 3978–3992.

58. Stender JD, Frasor J, Komm B *et al.* (2007) Estrogen-regulated gene networks in human breast cancer cells: Involvement of E2F1 in the regulation of cell proliferation. *Mol Endocrinol* **21**(9): 2112–2123.

59. Jin VX, Leu YW, Liyanarachchi S *et al.* (2004) Identifying estrogen receptor alpha target genes using integrated computational genomics and chromatin immunoprecipitation microarray. *Nucleic Acids Res* **32**(22): 6627–6635.

60. Carroll JS, Meyer CA, Song J *et al.* (2006) Genome-wide analysis of estrogen receptor binding sites. *Nat Genet* **38**(11): 1289–1297.

61. Phang TL, Neville MC, Rudolph M *et al.* (2003) Trajectory clustering: A non-parametric method for grouping gene expression time courses, with applications to mammary development. *Pac Sym Biocomput* **8**: 351–362.

62. Xie XH, Lu J, Kulbokas EJ *et al.* (2005) Systematic discovery of regulatory motifs in human promoters and 3′ UTRs by comparison of several mammals. *Nature* **434**(7031): 338–345.

Chapter 8

Cancer Epigenomics

Jer-Wei Chang[*] and Yi-Ching Wang[*]

1. Introduction

Cancer is characterized by abnormal growth of cells that have uncontrolled proliferation and in most cases can invade or metastasize to other tissues and organs. In genomics study, cancer has been well understood not only as a genetic but also as an epigenetic disease (Fig. 1). Genetic events include alterations in DNA sequence that may result in changes in the structure and function of a gene product. Epigenetic event does not involve changes in nucleotide sequences but is known to produce alterations in gene expression, regulation, reprogramming, and chromatin structure. Epigenetic modifications include two main categories: DNA methylation and histone modifications. DNA methylation at the carbon-5 position of cytosine is catalyzed by DNA 5′-cytosine-methyltransferases (DNMTs) (Fig. 2(a))[1] and usually occurs at CpG islands (CpG dinucleotide rich regions) located at the promoter and first exon of genes. Subsequently, DNA methylation is associated with transcriptional silencing of imprinted genes in normal cells or inactivated X-chromosomes of females.[2,3] In addition, DNA methylation is associated with stable gene silencing and formation of heterochromatin structures. Furthermore, DNA methylation may directly or indirectly inhibit gene expression and maintain key repressive elements of the histone code at a hypermethylated gene promoter in cancers.[4−6] In neoplastic cells, CpG island hypermethylation is now accepted as a key pathway leading to transcriptional silencing of tumor suppressor genes (TSGs) (Fig. 2(b)).[7,8]

[*]Department of Pharmacology and Institute of Basic Medical Science, National Cheng Kung University, Tainan 701, Taiwan.

129

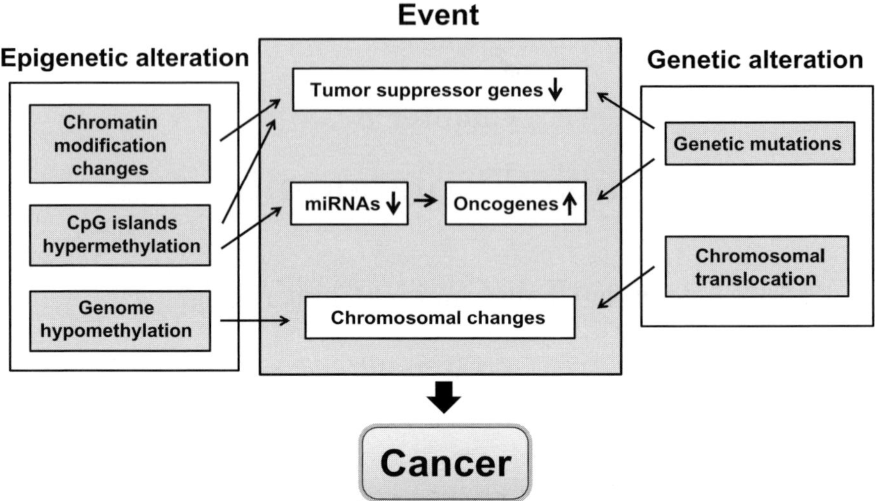

Fig. 1. Cancer is the result of both genetic and epigenetic alteration of key genes involved in tumorigenesis. Chromatin modification changes and hypermethylation of the promoter CpG island in tumor suppressor genes (TSGs) result in stable inactivation. This mechanism also inactivates microRNAs (miRNAs), which, in turn, regulate oncogenes. Global hypomethylation has been associated with chromosomal instability. Genetic mutations account for both inactivation of TSGs and activation of oncogenes. Chromosomal translocations also have an impact on many different tumorigenic processes.

Chromatin is the protein–nucleic acid complex in the genome. The basic unit of chromatin is the nucleosome, consisting of an octamer of two copies of each of the 4-core histone proteins (H2A, H2B, H3, and H4). The histones are not only DNA-packaging proteins but also molecular structures that participate in the regulation of gene expression. They store epigenetic information through a complex set of post-translational modifications. Most of these modifications (such as lysine acetylation, arginine and lysine methylation, or serine phosphorylation) occur at amino acid residues located in protruding N-terminal tails. Modifications to histone tails are clearly associated with transcriptional activation and silencing. For example, acetylation of histone lysines (K) is generally associated with transcriptional activation.[9,10] The methylation of histones depends on the type of residue, e.g. methylation of histone H3 at K4 is closely linked to transcriptional activation, whereas methylation of H3 at K9 or K27 and of H4 at K20 is associated with transcriptional repression.[11] In cancer studies, trimethylation of K20-H4 is commonly reduced in cancer cells.[12–14] One study found that low levels of acetylation at K9 and K18

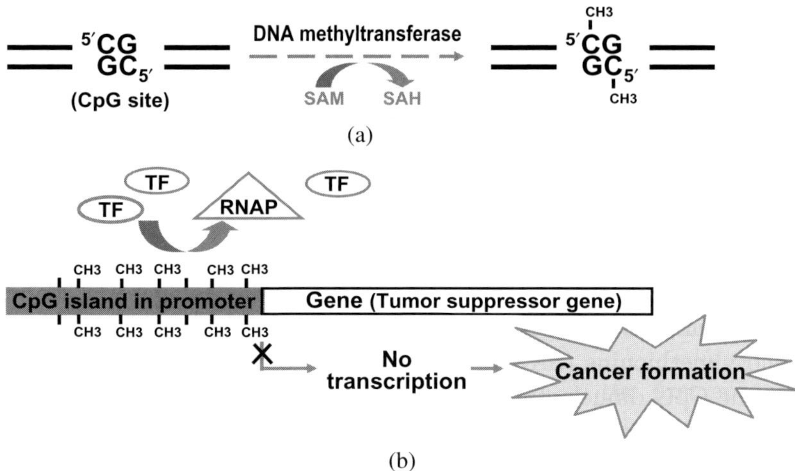

Fig. 2. DNA methylation in CpG sites. (a) DNA methyltransferase catalyzes DNA methylation by transferring the methyl group from the donor S-adenosylmethionine (SAM) to the carbon-5 position of cytosine within the CpG dinucleotide; (b) Hypermethylation of the promoter with CpG island in tumor suppressor gene (TSG) results in loss of binding of transcription factor (TF) and mRNA polymerase (RNAP) and leads to gene silencing and cancer formation.

of histone H3 increase the recurrence level of prostate cancer.[15] In another study, H3 acetylation has been found to be reduced in human primary tumors and cell lines of colon cancer.[16,17]

Recently, more evidence demonstrated the multiple connections between DNA methylation and histone modifications. Genes that are methylated are usually related to deacetylated and inactive chromatin, whereas unmethylated promoters and active genes are associated with an open euchromatin.[18] DNA methylation determines histone acetylation status. The methylated DNA-binding protein (MBD) recruits histone methyltransferase (HMT) and histone deacetylase (HDAC) activity to methylation-regulated genes.[19] However, the other studies seem to be a bidirectional control: chromatin inactivation recruits DNMTs to regulatory regions of genes.[20,21]

As mentioned above changes in chromatin structure are affected by DNA methylation and histone modifications and play a key role in genes transcriptional activity. Therefore, epigenetic modifications are essential for defining the cellular transcriptome at several levels. Aberrant changes in the pattern of epigenetic modifications would result in altered nuclear activity, and thereby an altered transcriptome, which would transform the cell. In

example, silencing of TSGs mediated by DNA methylation has been shown for *glutathione S-transferase 1, O^6-methylguanine-DNA methyltransferase (MGMT), p15^{INK4b}, RB, E-cadherin (CDH1), H-cadherin (CDH13)*, and *death-associated protein kinase 1 (DAPK1)*.[55] In addition, in our previous publications, we showed that silencing of TSGs is mediated by DNA methylation and is an important player in lung tumorigenesis. These genes include *hMLH1*,[56] *hMSH2*,[57] *FHIT*,[58] *BRCA1*,[59] *BRCA2*,[59] *XRCC5*,[59] *p16INK4a*,[60] *p14ARF*,[61] *RARβ*,[62] *SLIT2*,[63] *TIMP3*,[63] *RASSF1A*,[64] and *HIC1*.[65] The CpG island hypermethylation and subsequent inactivation of genes, such as TSGs and cell cycle regulated genes, can provide growth advantage to cancer cells.[44] There is substantial evidence that CpG island hypermethylation of genes occurs early in the neoplastic process. In colorectal cancer studies, the CpG island hypermethylation has been shown to start in normal mucosa, early- and pre-neoplasia. The genes with higher levels of methylation show to increase the risk of colon cancer.[66,67] In addition, hypermethylation of the *p16^{INK4a}* gene is frequently seen during metaplastic progression in Barrett's esophagus,[68] lung,[69] cervical,[69] and gastric carcinoma.[70] Hypermethylation of *14-3-3 sigma* is an early event in breast cancer.[71] Because of their heritable nature, hypermethylated CpG islands leave molecular footprints from which the event of epigenetic progression can be reconstructed during tumorigenesis (Fig. 3). In addition, aberrant methylation in the promoter regions of multiple genes is now known to exist in both early and advanced stages of cancer. Release of cells or free DNA containing these aberrantly methylated genes into surrounding luminal fluids or blood might thus permit the early detection of cancer or the identification of individuals at high risk of developing cancer. Therefore, DNA methylation is a useful biomarker for molecular classification of different tumor types (Table 1).

2.3. *DNA methylation in relation to transcription repression*

Previous studies have shown that specific sequences in the promoter region may serve as docking sites to attract repression complexes, including HDACs, MBDs, DNMTs, and others so that the sequences can be hypermethylated and their histones become hypoacetylated and methylated.[19,44,72,73] Sequences within and/or around CpG dinucleotides may also play a role in the association of proteins to methylated cytosines.[4,74,75] The interference of these proteins in the binding of

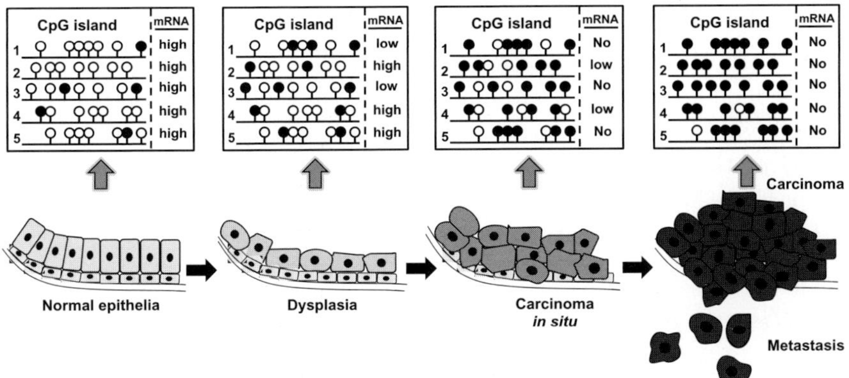

Fig. 3. An epigenetic model for cancer initiation and progression. Five promoter CpG islands of genes and their mRNA expression critical to tumorigenesis are depicted here. In the initial neoplastic step, *de novo* methylation occurs at the flanking CpG sites and progressively spreads into the core of a CpG island, resulting in silencing of the corresponding gene. This methylation spread may occur later in some other loci important for certain stages of neoplasm. In general, the density of methylated CpG sites within a locus as well as the number of methylated loci increase in more advanced stages of cancer. Black circle: methylated CpG dinucleotide; white circle: unmethylated CpG dinucleotide; mRNA expression status: "high" means mRNA normal expression, "low" means mRNA lower expression than normal cell, and "No" means loss of mRNA expression.

Table 1. Biological fluids and tissues in which cancer-derived methylated genes have been shown to have potential diagnostic value.

Fluid/tissue	Cancer type	Methylated genes
Sputum	Lung	$p16^{INK4A}$, RASSF1, MGMT
Ejaculate	Prostate	GSTP1
Urine	Prostate	GSTP1, RASSF1A, RARβ2, APC
Biopsy	Prostate	GSTP1, RARβ2, APC, TIG1
Nipple fluid	Breast	RASSF1
Serum	Breast	RASSF1, APC, DAPK
Serum	Ovary	BRCA1, RASSF1
Peritoneal	Ovary	BRCA1, RASSF1
Stools	Colorectum	SFRP2, CDKN2A, hMLH1
Serum	Colorectum	SEPT9, TMEEF2, NGFR
Urine	Bladder	RASSF1, APC, $p14^{ARF}$, DAPK, BCL2, TERT

transcription factors was recently reported,[76] suggesting that transcription factor abundance may compete with methylation machinery and this may also be relevant to the initiation of CpG island hypermethylation in cancer.

3. Histone Modification

DNA methylation and histone modification can influence each other. Histone modifications play an important role in gene-expression regulation by modifying the tertiary structure to an open or accessible status (euchromatin) or to closed and inaccessible configuration (heterochromatin). In this part, we summarize epigenetic alterations in the histone modification pattern in the process of tumorigenesis.

3.1. *Histone post translational modification*

The basic unit of chromatin includes two copies of each of the four core histones H2A, H2B, H3, and H4 wrapped by 146 bp of DNA. The histone tails can have various modifications, including acetylation, methylation, phosphorylation, poly-ADP ribosylation, ubiquitination, and glycosylation.[77] The chromatin structure and gene expression can be regulated by modifying the histone code. Most modifications have been found to be dynamic, and enzymes that remove the modification have been identified. Histone acetyltransferases (HATs) have usually been linked to transcriptional activation, whereas histone deacetylases (HDACs) play a major role in transcriptional repression. Histone methylation is another well-studied histone modification. It is linked to both transcriptional activation and repression.[77] Methylation of lysine 9 (K9) of histone H3 by the histone methyltransferase (HMT) is linked to heterochromatin formation and gene silencing, whereas methylation of lysine 4 (K4) is linked to transcriptional activity.[78,79] In addition, the histone methylation level depends on the precise balance between the HMTs and histone demethylases. In some cases, the specificity of enzymes that modify histones can be influenced by other factors, such as proteins that are associated with the enzyme affecting its selection of residue to modify or the degree of methylation (mono-, di-, or tri-) at a specific site.[80,81] In budding yeast study, dimethylation of K4 is linked to potential transcriptional activity, whereas trimethylation of the same residue occurs when the gene is actually actively transcribed.[82] The histone modifications that have been classified into repressing and activating are listed in Table 2.

3.2. *Histone modification and DNA methylation*

Recently, chromatin remodeling and DNA methylation in cancer have been linked. The histone modifications and DNA methylation appear to be functionally coupled in fungi, plants, and mammalian systems.[74,83,84]

Table 2. Chromatin modifications.

Mark	Transcriptionally relevant sites[a]	Transcriptional role[b]
Acetylated lysine (Kac)	H3 (9, 14, 18, 56), H4 (5, 8, 13, 16), H2A, H2B	Activation
Phosphorylated serine/threonine (S/Tph)	H3 (3, 10, 28), H2A, H2B	Activation
Methylated arginine (Rme)	H3 (17, 23), H4 (3)	Activation
Methylated lysine (Kme)	H3 (4, 36, 79), H3 (9, 27), H4 (20)	Activation Repression
Ubiquitylated lysine (Kub)	H2B (123[c]/120[d]), H2A (119[d])	Activation Repression
Sumoylated lysine (Ksu)	H2B (6/7), H2A (126)	Repression
Isomerized proline (Pisom)	H3 (30–38)	Activation/ Repression

[a]The number in parentheses refers to amino acid residue.
[b]The role that the epigenetic mark is associated with activation or repression.
[c]Yeast (*Saccharomyces cerevisiae*).
[d]Mammals.

MBDs and DNMTs can recruit HDACs to methylated promoters, and co-work to deacetylate histones, to maintain chromatin in the repressed state.[74,85] Together with the observations that MBD-containing corepressor complexes are associated with methylated CpG islands,[86] it has been suggested that histone modifications are secondary to DNA methylation. However, mutations in a putative methyltransferase specific for histone H3 lysine 9 result in loss of cytosine methylation in fungus,[83] providing evidence that histone methylation can initiate DNA methylation. Furthermore, recently study has shown that CpG island methylation depends on histone H3 methyltransferase in plants,[84] suggesting that histone methylation can actually direct DNA methylation. Finally, a specific "histone code" has been proposed to target methylation to DNA.[72] Modifications of histones, such as by deacetylation, methylation, or phosphorylation, may be somehow required to initiate or maintain DNA methylation, perhaps by recruiting proteins, such as MBDs, transcriptional activators, or repressors, and these interactions then determine which genes are expressed or silenced.

3.3. *Histone modification in cancer*

Hypermethylation of the CpG islands in the promoter regions of TSGs in cancer cells is associated with a particular combination of histone markers: deacetylation of histones H3 and H4, loss of H3K4 trimethylation, and gain of H3K9 methylation and H3K27 trimethylation.[6,87] For example, $P21^{WAF1}$ gene is silent at transcriptional level with hypermethylation of CpG island when hypoacetylated and hypermethylated histones H3 and H4 are present.[88]

In terms of histone modifying enzymes, changes in the expression of these proteins have important and tumor-specific roles in cancer development. The acetylation of H4K16 is catalyzed by specific HATs, such as MORF, MOZ, MOF, TIP60, and HBO1. In leukemias and uterine myomas, chromosomal translocations that involve *HATs* genes, such as CREB-binding protein (CBP–MOZ and CBP–MORF fusion proteins) disrupt the acetylation of H4K16.[89,90] For the trimethyl-H4K20 mark, the observed loss in cancer cells[12] and the demonstration that knockout mice for the HMT SUV39H are prone to developing cancer,[91] indicate that HMTs for H4K20 could function as TSGs. In addition, HDAC2 is another histone deacetylase that is targeted for mutational inactivation in colorectal cancer, and the cancer shows an altered acetylation profile and is associated with microsatellite instability.[92]

4. Methods for Analysis of the Cancer Methylome

In recent years, a range of approaches have been developed for DNA methylation analysis in a gene-specific or genome-wide manner, and described in a number of recent reviews.[93–95] Several techniques of analyzing DNA methylation, such as bisulfite conversion, methylation-sensitive enzyme restriction, and affinity purification of methylation DNA, have been adapted.[95] We focus on the three major approaches: (1) the isoschizomers and gel-based methylation analysis, (2) the microarray-based methylation analysis, and (3) the sequencing-based methylation analysis.

4.1. *Three major methods for analysis of the cancer methylome*

4.1.1. *The isoschizomers and gel-based methylation analysis*

In the year 2000, restriction landmark genomic scanning (RLGS) was one of the earliest methods, which was based on the differential cleavage of isoschizomers with distinct methylation sensitivity, to be adapted for genome-wide methylation analysis in primary human tumors.[53] With this technique, DNA from normal and tumor samples is cleaved by methylation-sensitive endonuclease, such as *NotI* restriction enzyme. DNA is radioactively labeled at cleaved *NotI* sites. Following a second methylation-sensitive endonuclease digestion, the digested DNA is separated in first-dimension agarose gel and *in situ* digested with *HinfI* endonuclease. Finally, the digestion fragments are separated by second-dimension acrylamide gel. When tumor and normal samples are compared, more than 1,000 CpG islands methylation pattern can be analyzed. However, the limitations of RLGS include its reliance on specific digestion sites that are not present in all CpG islands and the fact that not all resulting fragments can be resolved in the two-dimensional electrophoresis steps.

4.1.2. *The microarray-based methylation analysis*

Undoubtedly, microarray technologies are one of the most efficient means of studying CpG-island methylation on a genome-wide scale. These array-based methods can interrogate much larger number of CpG islands than the other nonarray-based approaches, and provide DNA methylation platform for uncovering the biological role of DNA methylation in both normal and tumor cells. In addition, there are three major types of microarray-based methods for cancer methylation analysis. The first method uses methylation-sensitive restriction enzymes to digest DNA, followed by ligation-mediated PCR amplification and hybridization to CpG island array.[96] The second method is the methyl-DNA immunoprecipitation assay,[97] which utilizes an antibody specific for methylated cytosine to immunocapture methylated genomic fragments and the captured methylated fragments are hybridized with a differentially labeled total DNA control. The third method, which combines bisulfite conversion DNA and bead-array system, can detect DNA methylation profiling, such as Illumina GoldenGate methylation analysis system.[98] These methods have provided useful and high-throughput tools in studying the phenomenon of DNA methylation.

4.1.2.1. The isoschizomers-based microarray analysis

Array-based DNA methylation profiling in cancer genome analysis was first described in 1999.[96] The method, called differential methylation hybridization (DMH), identifies hypermethylated sequences in tumor cells by screening many CpG island loci derived from a CpG genomic library.[99] (Fig. 4(a)). The genomic DNA from the normal and tumor samples are

human contains a much larger genome (~3000 megabase) than does *Arabidopsis* (~120 megabase). Therefore, the interested genome regions are selected for methylation analysis by bisulfite sequencing in human cancers. Taylor *et al.* had shown that Roche GS FL-X sequencer technology can be used in the high-throughput sequencing of bisulfite PCR amplicons in human.[131] They analyzed 25 gene-related CpG islands from normal lymphocytes, acute lymphoblastic leukemia (ALL), chronic lymphocytic leukemia (CLL), follicular lymphoma (FL), and mantle cell lymphoma (MCL). More than 100 PCR amplicons per run were sequenced and a total of 294,631 sequences were generated with short-read sequencing. A significant increase in methylation was detected in ALL and FL samples in comparison with CLL and MCL. Another group used the GS FL-X platform to explore the complexity of tumor-specific methylation patterns by bisulfite pyrosequencing of more than 700,000 DNA fragments derived from 50 breast cancer tissues and cancer-free serum samples.[132] After computational BLAST and statistical analysis, tumor samples displayed more variation in methylation level than normal samples. In 2008, Meissner *et al.* reported the reduced representation bisulfite sequencing (RRBS), which provided DNA methylation profiles at nucleotide resolution in mouse and perhaps in other species with large genomes.[133] This approach involves digesting DNA with the methylation-insensitive restriction enzyme *Msp*I and size selection for 40–220 bp fragments, followed by linker ligation, bisulfite conversion, PCR amplification, and high-throughput sequencing. Using an Illumina 1G genome analyzer, the RRBS could reveal sequences from 90% of all CpG islands in mouse cells.

4.2. *Novel methylation spectrum analysis for candidate gene*

4.2.1. *Methylation-specific oligonucleotide microarray*

In addition to genome-wide methylation analysis, single gene array-based DNA methylation profiling in cancer was also described in 2002.[134] This methylation-specific oligonucleotide (MSO) microarray can be applied to map methylation CpG sites within the CpG island of a single gene in normal or tumor DNA. Essentially, this technique involves sodium bisulfite conversion of genomic DNA and PCR amplification of regions of interest. The fluorescently labeled PCR products are hybridized to arrayed oligonucleotides that can discriminate between methylated and unmethylated alleles in interest region and provide readout of the original methylation state at that CpG site. The MSO microarray has been used for mapping methylation changes in CpG island loci of *ERα*, *RASSF1A*, *p16^{INK4a}*, and *hMLH1* genes in different cancers.[134–137]

4.2.2. *Mass array based technology for methylation spectrum*

EpiTYPER method adapts MassARRAY system (www.sequenom.com) for quantitative discrimination between methylated and nonmethylated DNA samples. This approach uses bisulfite-converted genomic DNA for PCR amplification. The PCR primers bind to both methylated and nonmethylated template, and a T7-promoter tag is annealed to the 3′ region of the reverse primer. After the PCR amplification, *in vitro* RNA transcription is performed on the reverse strand, followed by RNase A base-specific cleavage. MALDI-TOF MS analyzes the cleavage products, and a distinct signal pair pattern results from the methylated and nonmethylated template DNA by EpiTYPER software. The application of EpiTYPER was used to characterize DNA methylation of a total of 1,425 CpG sites in 47 genes of 48 nonsmall cell lung cancer patients.[138] EpiTYPER method provides several benefits, such as: (1) quantitative assessment of the degree of DNA methylation, (2) detection of methylation levels as low as 5% in sample mixtures, and (3) gene expression, SNP discovery, genotyping, and methylation studies running on the same platform. However, EpiTYPER platform cannot be high-throughput compared with microarray. This method may provide DNA methylation status of multiple CpG sites in candidate genes of interest or validate CpG sites methylation status selected by other genome-wide methods.[139,140]

4.3. *Applications for genomic methylation analysis*

4.3.1. *Pharmacological unmasking analysis*

Array-based method can be applied to compare mRNA re-expression from cancer cell lines before and after treatment with demethylating agents, and to identify novel cancer-specific hypermethylated genes.[141–143] (Fig. 5(a)). In 2008, Hoque *et al.* used 20 different human cancer cell lines of 5 major cancer types and treated cells with 5-aza-2′-deoxycytidine (5-aza-dC) by comparing the re-expression microarray data of cells treated with or without 5-aza-dC[143]; they identified 175 novel cancer-specific methylation genes, which were confirmed with bisulfite DNA sequencing

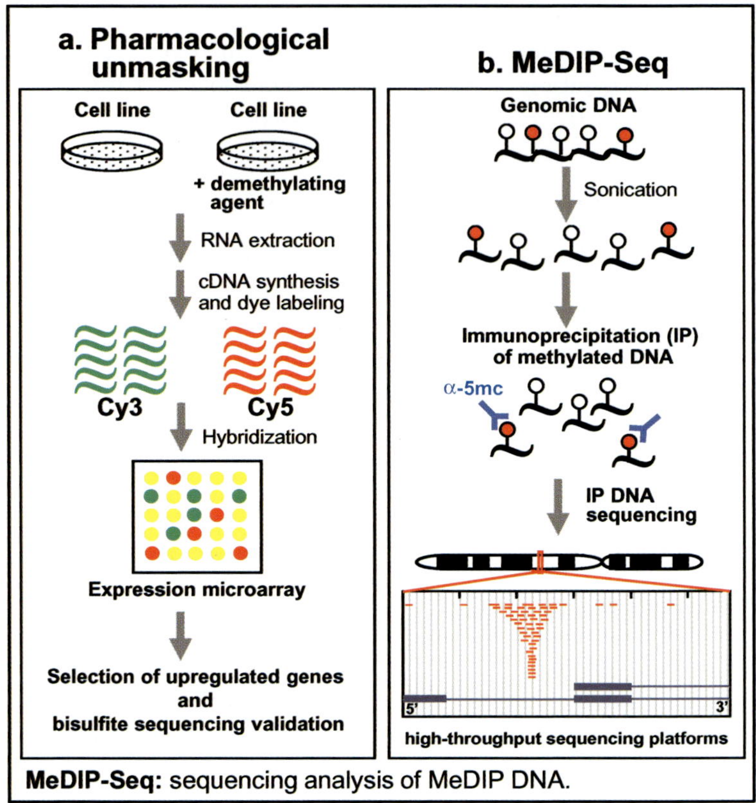

Fig. 5.   Application for cancer methylome analysis.

and methylation-specific PCR (Fig. 5(a)). This approach is a major advance over previous empirical techniques that require too much experimental works and yield only a few cancer-specific methylated genes. Therefore, a combination of re-expression arrays and promoter bisulfite sequence pattern provides a higher yield of genes harboring promoter methylation in cancer.

4.3.2. *Direct sequencing analysis of ChIP DNA samples*

Affinity-based purification of DNA sequences can also be applied to high-throughput sequencing. The approach focused on histone methylation profiling and used ChIP to pull down the histone methylation region DNA. Direct sequencing analysis of ChIP DNA samples (termed ChIP-seq) was conducted using Illumina 1G genome analyzer.[125] After data analysis, genome-wide distributions of histone modifications and chromatin protein target sites could be revealed. Recently, NimbleGen (http://www. nimblegen.com/) designed a tool to capture DNA sequences from microarray chip, and then the captured DNA fragments are amplified for high-throughput short-read sequencing. Using this technique, only the interested genome regions are hybridized, eluted, and sequenced. Recaptured DNA from ChIP-microarray and high-throughput sequencing may provide a highly efficient and cost-effective cancer methylome approach. In addition, MeDIP-Seq assay in combination with high-throughput sequencing platforms may be another alternative for methylome analysis in the future (Fig. 5(b)).

5. Conclusions

Cancer is a genetic and epigenetic disease. Methylation at DNA CpG dinucleotide is the major epigenetic modification in mammalian genomes and is known to regulate gene expression. DNA methylation profiling technologies promise to enable the characterization of distinct methylation signatures from cancers and normal tissues with diagnostic implications. The DNA methylation profile is now considered a potential biomarker in cancer detection.[144,145] Methylation profiling of cancer treated with 5-aza-dC provides a focus on reactivation of methylation-silenced genes as therapeutic targets and forms a rational basis for designing future treatment strategies in cancer alteration processes.

6. Future Perspective

Systematic whole-genome epigenomic studies are needed for gaining insights into mechanisms involved in various diseases including cancer. Recent advances in epigenetic research have provided new high-throughput technologies for the analyses of DNA methylation patterns and histone modification marks that give structure and function to the human genome. Recently, several research centers have projects for the exploration of methylomes and epigenomes in health and diseases including cancers.[95] For example, Human Epigenome Project (HEP) (http://www.epigenome.org/) aims to identify, catalog, and interpret genome-wide DNA methylation profiles of all human genes in all major tissues.[146] The methylation variable positions provided by HEP are thought to reflect gene activity, tissue type specificity, and disease state and to provide useful epigenetic markers of the genome. In cancer epigenetic research, Epigenetic Treatment of Neoplastic Disease

(EPITRON) (http://www.epitron.eu/) project aims to define and validate epigenetic cancer treatment. The major aims include defining epigenetic alterations in cancer, identifying therapeutic targets, and developing epi-drugs. Although most of the current methods are not ready to provide a complete picture of the human methylome, continuous rapid improvements in technology will make the study of cancer methylome more accessible and exciting. In the next few years, genome-wide, high-throughput methylation analysis will advance further and is expected to be widely available in cancer research.

References

1. Chiang PK, Gordon RK, Tal J, Zeng GC, Doctor BP, Pardhasaradhi K, McCann PP. (1996) S-Adenosylmethionine and methylation. *Faseb J* **10**: 471–480.
2. Li E, Beard C, Jaenisch R. (1993) Role for DNA methylation in genomic imprinting. *Nature* **366**: 362–365.
3. Singer-Sam J, Riggs AD. (1993) X chromosome inactivation and DNA methylation. *Exs* **64**: 358–384.
4. Nguyen CT, Gonzales FA, Jones PA. (2001) Altered chromatin structure associated with methylation-induced gene silencing in cancer cells: Correlation of accessibility, methylation, MeCP2 binding and acetylation. *Nucleic Acids Res* **29**: 4598–4606.
5. Fahrner JA, Eguchi S, Herman JG, Baylin SB. (2002) Dependence of histone modifications and gene expression on DNA hypermethylation in cancer. *Cancer Res* **62**: 7213–7218.
6. Ballestar E, Paz MF, Valle L, Wei S, Fraga MF, Espada J, Cigudosa JC, Huang TH, Esteller M. (2003) Methyl-CpG binding proteins identify novel sites of epigenetic inactivation in human cancer. *The EMBO J* **22**: 6335–6345.
7. Jones PA. (1996) DNA methylation errors and cancer. *Cancer Res* **56**: 2463–2467.
8. Baylin SB, Herman JG. (2000) DNA hypermethylation in tumorigenesis: Epigenetics joins genetics. *Trends Genet* **16**: 168–174.
9. Mack GS. (2006) Epigenetic cancer therapy makes headway. *J Natl Cancer Inst* **98**: 1443–1444.
10. Bernstein BE, Meissner A, Lander ES. (2007) The mammalian epigenome. *Cell* **128**: 669–681.
11. Karpf AR, Matsui S. (2005) Genetic disruption of cytosine DNA methyltransferase enzymes induces chromosomal instability in human cancer cells. *Cancer Res* **65**: 8635–8639.
12. Fraga MF, Ballestar E, Villar-Garea A, Boix-Chornet M, Espada J, Schotta G, Bonaldi T, Haydon C, Ropero S, Petrie K, Iyer NG, Perez-Rosado A *et al.* (2005) Loss of acetylation at Lys16 and trimethylation at Lys20 of histone H4 is a common hallmark of human cancer. *Nat Genet* **37**: 391–400.

13. Pogribny IP, Ross SA, Tryndyak VP, Pogribna M, Poirier LA, Karpinets TV. (2006) Histone H3 lysine 9 and H4 lysine 20 trimethylation and the expression of Suv4-20h2 and Suv-39h1 histone methyltransferases in hepatocarcinogenesis induced by methyl deficiency in rats. *Carcinogenesis* **27**: 1180–1186.

14. Tryndyak VP, Kovalchuk O, Pogribny IP. (2006) Loss of DNA methylation and histone H4 lysine 20 trimethylation in human breast cancer cells is associated with aberrant expression of DNA methyltransferase 1, Suv4-20h2 histone methyltransferase and methyl-binding proteins. *Cancer Biol Ther* **5**: 65–70.

15. Seligson DB, Horvath S, Shi T, Yu H, Tze S, Grunstein M, Kurdistani SK. (2005) Global histone modification patterns predict risk of prostate cancer recurrence. *Nature* **435**: 1262–1266.

16. Chen YX, Fang JY, Lu R, Qiu DK. (2007) Expression of p21(WAF1) is related to acetylation of histone H3 in total chromatin in human colorectal cancer. *World J Gastroenterol* **13**: 2209–2213.

17. Chen YX, Fang JY, Zhu HY, Lu R, Cheng ZH, Qiu DK. (2004) Histone acetylation regulates p21WAF1 expression in human colon cancer cell lines. *World J Gastroenterol* **10**: 2643–2646.

18. Razin A. (1998) CpG methylation, chromatin structure and gene silencing-a three-way connection. *EMBO J* **17**: 4905–4908.

19. Ballestar E, Wolffe AP. (2001) Methyl–CpG-binding proteins. Targeting specific gene repression. *Eur J Biochem* **268**: 1–6.

20. Cervoni N, Szyf M. (2001) Demethylase activity is directed by histone acetylation. *J Biol Chem* **276**: 40778–40787.

21. Cervoni N, Detich N, Seo SB, Chakravarti D, Szyf M. (2002) The oncoprotein Set/TAF-1beta, an inhibitor of histone acetyltransferase, inhibits active demethylation of DNA, integrating DNA methylation and transcriptional silencing. *J Biol Chem* **277**: 25026–25031.

22. Pradhan S, Talbot D, Sha M, Benner J, Hornstra L, Li E, Jaenisch R, Roberts RJ. (1997) Baculovirus-mediated expression and characterization of the full-length murine DNA methyltransferase. *Nucleic Acids Res* **25**: 4666–4673.

23. Bestor TH. (1988) Cloning of a mammalian DNA methyltransferase. *Gene* **74**: 9–12.

24. Chuang LS, Ian HI, Koh TW, Ng HH, Xu G, Li BF. (1997) Human DNA-(cytosine-5) methyltransferase-PCNA complex as a target for p21WAF1. *Science* **277**: 1996–2000.

25. Beard C, Li E, Jaenisch R. (1995) Loss of methylation activates Xist in somatic but not in embryonic cells. *Genes Dev* **9**: 2325–2334.

26. Li E, Bestor TH, Jaenisch R. (1992) Targeted mutation of the DNA methyltransferase gene results in embryonic lethality. *Cell* **69**: 915–926.

27. Kimura H, Shiota K. (2003) Methyl–CpG-binding protein, MeCP2, is a target molecule for maintenance DNA methyltransferase, DNMT1. *J Biol Chem* **278**: 4806–4812.

28. Lin X, Nelson WG. (2003) Methyl-CpG-binding domain protein-2 mediates transcriptional repression associated with hypermethylated GSTP1 CpG islands in MCF-7 breast cancer cells. *Cancer Res* **63**: 498–504.

61. Wang YC, Lin RK, Tan YH, Chen JT, Chen CY. (2005) Wild-type p53 overexpression and its correlation with MDM2 and p14ARF alterations: An alternative pathway to non-small-cell lung cancer. *J Clin Oncol* **23**: 154–164.

62. Lin RK, Hsu HS, Chang JW, Chen CY, Chen JT, Wang YC. (2007) Alteration of DNA methyltransferases contributes to 5′CpG methylation and poor prognosis in lung cancer. *Lung Cancer* **55**: 205–213.

63. Lin RK, Hsu CH, Wang YC. (2007) Mithramycin A inhibits DNA methyltransferase and metastasis potential of lung cancer cells. *Anticancer Drugs* **18**: 1157–1164.

64. Agathanggelou A, Cooper WN, Latif F. (2005) Role of the Ras-association domain family 1 tumor suppressor gene in human cancers. *Cancer Res* **65**: 3497–3508.

65. Tseng RC, Lee CC, Hsu HS, Tzao C, Wang YC. (2009) Distinct HIC1-SIRT1-p53 loop deregulation in lung squamous carcinoma and adenocarcinoma patients. *Neoplasia* **11**: 763–770.

66. Toyota M, Ahuja N, Ohe-Toyota M, Herman JG, Baylin SB, Issa JP. (1999) CpG island methylator phenotype in colorectal cancer. *Proc Natl Acad Sci USA* **96**: 8681–8686.

67. Chan AO, Issa JP, Morris JS, Hamilton SR, Rashid A. (2002) Concordant CpG island methylation in hyperplastic polyposis. *Am J Pathol* **160**: 529–536.

68. Wong DJ, Paulson TG, Prevo LJ, Galipeau PC, Longton G, Blount PL, Reid BJ. (2001) p16(INK4a) lesions are common, early abnormalities that undergo clonal expansion in Barrett's metaplastic epithelium. *Cancer Res* **61**: 8284–8289.

69. Nuovo GJ, Plaia TW, Belinsky SA, Baylin SB, Herman JG. (1999) *In situ* detection of the hypermethylation-induced inactivation of the p16 gene as an early event in oncogenesis. *Proc Natl Acad Sci USA* **96**: 12754–12759.

70. Kang GH, Shim YH, Jung HY, Kim WH, Ro JY, Rhyu MG. (2001) CpG island methylation in premalignant stages of gastric carcinoma. *Cancer Res* **61**: 2847–2851.

71. Umbricht CB, Evron E, Gabrielson E, Ferguson A, Marks J, Sukumar S. (2001) Hypermethylation of 14-3-3 sigma (stratifin) is an early event in breast cancer. *Oncogene* **20**: 3348–3353.

72. Jenuwein T, Allis CD. (2001) Translating the histone code. *Science* **293**: 1074–1080.

73. Graff JR, Gabrielson E, Fujii H, Baylin SB, Herman JG. (2000) Methylation patterns of the E-cadherin 5′ CpG island are unstable and reflect the dynamic, heterogeneous loss of E-cadherin expression during metastatic progression. *J Biol Chem* **275**: 2727–2732.

74. Ballestar E, Esteller M. (2002) The impact of chromatin in human cancer: Linking DNA methylation to gene silencing. *Carcinogenesis* **23**: 1103–1109.

75. Magdinier F, Wolffe AP. (2001) Selective association of the methyl-CpG binding protein MBD2 with the silent p14/p16 locus in human neoplasia. *Proc Natl Acad Sci USA* **98**: 4990–4995.

76. Curradi M, Izzo A, Badaracco G, Landsberger N. (2002) Molecular mechanisms of gene silencing mediated by DNA methylation. *Mol Cell Biol* **22**: 3157–3173.

77. Kouzarides T. (2007) Chromatin modifications and their function. *Cell* **128**: 693–705.

78. Noma K, Allis CD, Grewal SI. (2001) Transitions in distinct histone H3 methylation patterns at the heterochromatin domain boundaries. *Science* **293**: 1150–1155.

79. Grewal SI, Elgin SC. (2002) Heterochromatin: New possibilities for the inheritance of structure. *Curr Opin Genet Dev* **12**: 178–187.

80. Metzger E, Wissmann M, Yin N, Muller JM, Schneider R, Peters AH, Gunther T, Buettner R, Schule R. (2005) LSD1 demethylates repressive histone marks to promote androgen-receptor-dependent transcription. *Nature* **437**: 436–439.

81. Steward MM, Lee JS, O'Donovan A, Wyatt M, Bernstein BE, Shilatifard A. (2006) Molecular regulation of H3K4 trimethylation by ASH2L, a shared subunit of MLL complexes. *Nat Struct Mol Biol* **13**: 852–854.

82. Santos-Rosa H, Schneider R, Bannister AJ, Sherriff J, Bernstein BE, Emre NC, Schreiber SL, Mellor J, Kouzarides T. (2002) Active genes are trimethylated at K4 of histone H3. *Nature* **419**: 407–411.

83. Tamaru H, Selker EU. (2001) A histone H3 methyltransferase controls DNA methylation in Neurospora crassa. *Nature* **414**: 277–283.

84. Jackson JP, Lindroth AM, Cao X, Jacobsen SE. (2002) Control of CpNpG DNA methylation by the KRYPTONITE histone H3 methyltransferase. *Nature* **416**: 556–560.

85. Fuks F, Burgers WA, Brehm A, Hughes-Davies L, Kouzarides T. (2000) DNA methyltransferase Dnmt1 associates with histone deacetylase activity. *Nat Genet* **24**: 88–91.

86. Nan X, Ng HH, Johnson CA, Laherty CD, Turner BM, Eisenman RN, Bird A. (1998) Transcriptional repression by the methyl–CpG-binding protein MeCP2 involves a histone deacetylase complex. *Nature* **393**: 386–389.

87. Jones PA, Baylin SB. (2007) The epigenomics of cancer. *Cell* **128**: 683–692.

88. Richon VM, Sandhoff TW, Rifkind RA, Marks PA. (2000) Histone deacetylase inhibitor selectively induces p21WAF1 expression and gene-associated histone acetylation. *Proc Natl Acad Sci USA* **97**: 10014–10019.

89. Yang XJ. (2004) The diverse superfamily of lysine acetyltransferases and their roles in leukemia and other diseases. *Nucleic Acids Res* **32**: 959–976.

90. Moore SD, Herrick SR, Ince TA, Kleinman MS, Dal Cin P, Morton CC, Quade BJ. (2004) Uterine leiomyomata with t(10;17) disrupt the histone acetyltransferase MORF. *Cancer Res* **64**: 5570–5577.

91. Peters AH, O'Carroll D, Scherthan H, Mechtler K, Sauer S, Schofer C, Weipoltshammer K, Pagani M, Lachner M, Kohlmaier A, Opravil S, Doyle M *et al.* (2001) Loss of the Suv39h histone methyltransferases impairs mammalian heterochromatin and genome stability. *Cell* **107**: 323–337.

92. Ropero S, Fraga MF, Ballestar E, Hamelin R, Yamamoto H, Boix-Chornet M, Caballero R, Alaminos M, Setien F, Paz MF, Herranz M, Palacios J *et al.* (2006) A truncating mutation of HDAC2 in human cancers confers resistance to histone deacetylase inhibition. *Nat Genet* **38**: 566–569.

93. Esteller M. (2007) Cancer epigenomics: DNA methylomes and histone-modification maps. *Nature Rev* **8**: 286–298.

94. Zilberman D, Henikoff S. (2007) Genome-wide analysis of DNA methylation patterns. *Dev* (Cambridge, England) **134**: 3959–3965.

95. Beck S, Rakyan VK. (2008) The methylome: Approaches for global DNA methylation profiling. *Trends Genet* **24**: 231–237.

96. Huang TH, Perry MR, Laux DE. (1999) Methylation profiling of CpG islands in human breast cancer cells. *Hum Mol Genet* **8**: 459–470.

97. Weber M, Davies JJ, Wittig D, Oakeley EJ, Haase M, Lam WL, Schubeler D. (2005) Chromosome-wide and promoter-specific analyses identify sites of differential DNA methylation in normal and transformed human cells. *Nat Genet* **37**: 853–862.

98. Bibikova M, Lin Z, Zhou L, Chudin E, Garcia EW, Wu B, Doucet D, Thomas NJ, Wang Y, Vollmer E, Goldmann T, Seifart C *et al.* (2006) High-throughput DNA methylation profiling using universal bead arrays. *Genome Res* **16**: 383–393.

99. Cross SH, Charlton JA, Nan X, Bird AP. (1994) Purification of CpG islands using a methylated DNA binding column. *Nat Genet* **6**: 236–244.

100. Yan PS, Efferth T, Chen HL, Lin J, Rodel F, Fuzesi L, Huang TH. (2002) Use of CpG island microarrays to identify colorectal tumors with a high degree of concurrent methylation. *Methods* **27**: 162–169.

101. Yan PS, Perry MR, Laux DE, Asare AL, Caldwell CW, Huang TH. (2000) CpG island arrays: An application toward deciphering epigenetic signatures of breast cancer. *Clin Cancer Res* **6**: 1432–1438.

102. Ahluwalia A, Yan P, Hurteau JA, Bigsby RM, Jung SH, Huang TH, Nephew KP. (2001) DNA methylation and ovarian cancer. I. Analysis of CpG island hypermethylation in human ovarian cancer using differential methylation hybridization. *Gynecol Oncol* **82**: 261–268.

103. Paz MF, Wei S, Cigudosa JC, Rodriguez-Perales S, Peinado MA, Huang TH, Esteller M. (2003) Genetic unmasking of epigenetically silenced tumor suppressor genes in colon cancer cells deficient in DNA methyltransferases. *Hum Mol Genet* **12**: 2209–2219.

104. Shi H, Yan PS, Chen CM, Rahmatpanah F, Lofton-Day C, Caldwell CW, Huang TH. (2002) Expressed CpG island sequence tag microarray for dual screening of DNA hypermethylation and gene silencing in cancer cells. *Cancer Res* **62**: 3214–3220.

105. Khulan B, Thompson RF, Ye K, Fazzari MJ, Suzuki M, Stasiek E, Figueroa ME, Glass JL, Chen Q, Montagna C, Hatchwell E, Selzer RR *et al.* (2006) Comparative isoschizomer profiling of cytosine methylation: The HELP assay. *Genome Res* **16**: 1046–1055.

106. Hatada I, Fukasawa M, Kimura M, Morita S, Yamada K, Yoshikawa T, Yamanaka S, Endo C, Sakurada A, Sato M, Kondo T, Horii A *et al.* (2006) Genome-wide profiling of promoter methylation in human. *Oncogene* **25**: 3059–3064.

107. Toyota M, Ho C, Ahuja N, Jair KW, Li Q, Ohe-Toyota M, Baylin SB, Issa JP. (1999) Identification of differentially methylated sequences in colorectal cancer by methylated CpG island amplification. *Cancer Res* **59**: 2307–2312.

108. Estecio MR, Yan PS, Ibrahim AE, Tellez CS, Shen L, Huang TH, Issa JP. (2007) High-throughput methylation profiling by MCA coupled to CpG island microarray. *Genome Res* **17**: 1529–1536.

109. Kuang SQ, Tong WG, Yang H, Lin W, Lee MK, Fang ZH, Wei Y, Jelinek J, Issa JP, Garcia-Manero G. (2008) Genome-wide identification of aberrantly methylated promoter associated CpG islands in acute lymphocytic leukemia. *Leukemia* **22**: 1529–1538.

110. Gao W, Kondo Y, Shen L, Shimizu Y, Sano T, Yamao K, Natsume A, Goto Y, Ito M, Murakami H, Osada H, Zhang J *et al.* (2008) Variable DNA methylation patterns associated with progression of disease in hepatocellular carcinomas. *Carcinogenesis* **29**: 1901–1910.

111. Nouzova M, Holtan N, Oshiro MM, Isett RB, Munoz-Rodriguez JL, List AF, Narro ML, Miller SJ, Merchant NC, Futscher BW. (2004) Epigenomic changes during leukemia cell differentiation: Analysis of histone acetylation and cytosine methylation using CpG island microarrays. *J Pharmacol Exp Ther* **311**: 968–981.

112. Ibrahim AE, Thorne NP, Baird K, Barbosa-Morais NL, Tavare S, Collins VP, Wyllie AH, Arends MJ, Brenton JD. (2006) MMASS: An optimized array-based method for assessing CpG island methylation. *Nucleic Acids Res* **34**: e136.

113. Irizarry RA, Ladd-Acosta C, Carvalho B, Wu H, Brandenburg SA, Jeddeloh JA, Wen B, Feinberg AP. (2008) Comprehensive high-throughput arrays for relative methylation (CHARM). *Genome Res* **18**: 780–790.

114. Lippman Z, Gendrel AV, Black M, Vaughn MW, Dedhia N, McCombie WR, Lavine K, Mittal V, May B, Kasschau KD, Carrington JC, Doerge RW *et al.* (2004) Role of transposable elements in heterochromatin and epigenetic control. *Nature* **430**: 471–476.

115. Lippman Z, Gendrel AV, Colot V, Martienssen R. (2005) Profiling DNA methylation patterns using genomic tiling microarrays. *Nat Methods* **2**: 219–224.

116. Ordway JM, Bedell JA, Citek RW, Nunberg A, Garrido A, Kendall R, Stevens JR, Cao D, Doerge RW, Korshunova Y, Holemon H, McPherson JD *et al.* (2006) Comprehensive DNA methylation profiling in a human cancer genome identifies novel epigenetic targets. *Carcinogenesis* **27**: 2409–2423.

117. Keshet I, Schlesinger Y, Farkash S, Rand E, Hecht M, Segal E, Pikarski E, Young RA, Niveleau A, Cedar H, Simon I. (2006) Evidence for an instructive mechanism of de novo methylation in cancer cells. *Nat Genet* **38**: 149–153.

118. Jacinto FV, Ballestar E, Ropero S, Esteller M. (2007) Discovery of epigenetically silenced genes by methylated DNA immunoprecipitation in colon cancer cells. *Cancer Res* **67**: 11481–11486.

119. Rauch T, Li H, Wu X, Pfeifer GP. (2006) MIRA-assisted microarray analysis, a new technology for the determination of DNA methylation patterns, identifies frequent methylation of homeodomain-containing genes in lung cancer cells. *Cancer Res* **66**: 7939–7947.

120. Lin Z, Thomas NJ, Bibikova M, Seifart C, Wang Y, Guo X, Wang G, Vollmer E, Goldmann T, Garcia EW, Zhou L, Fan JB *et al.* (2007) DNA methylation markers of surfactant proteins in lung cancer. *Int J Oncol* **31**: 181–191.

121. Bjornsson HT, Brown LJ, Fallin MD, Rongione MA, Bibikova M, Wickham E, Fan JB, Feinberg AP. (2007) Epigenetic specificity of loss of imprinting of the IGF2 gene in Wilms tumors. *J National Cancer Institute* **99**: 1270–1273.

122. Frommer M, McDonald LE, Millar DS, Collis CM, Watt F, Grigg GW, Molloy PL, Paul CL. (1992) A genomic sequencing protocol that yields a positive display of 5-methylcytosine residues in individual DNA strands. *Proc Nat Acad Sci USA* **89**: 1827–1831.

123. Lewin J, Schmitt AO, Adorjan P, Hildmann T, Piepenbrock C. (2004) Quantitative DNA methylation analysis based on four-dye trace data from direct sequencing of PCR amplificates. *Bioinformatics* (Oxford, England) **20**: 3005–3012.

124. Eckhardt F, Lewin J, Cortese R, Rakyan VK, Attwood J, Burger M, Burton J, Cox TV, Davies R, Down TA, Haefliger C, Horton R *et al.* (2006) DNA methylation profiling of human chromosomes 6, 20 and 22. *Nat Genet* **38**: 1378–1385.

125. Barski A, Cuddapah S, Cui K, Roh TY, Schones DE, Wang Z, Wei G, Chepelev I, Zhao K. (2007) High-resolution profiling of histone methylations in the human genome. *Cell* **129**: 823–837.

126. Bentley DR. (2006) Whole-genome re-sequencing. *Curr Opin Genet Dev* **16**: 545–552.

127. Margulies M, Egholm M, Altman WE, Attiya S, Bader JS, Bemben LA, Berka J, Braverman MS, Chen YJ, Chen Z, Dewell SB, Du L *et al.* (2005) Genome sequencing in microfabricated high-density picolitre reactors. *Nature* **437**: 376–380.

128. Kim J, Bhinge AA, Morgan XC, Iyer VR. (2005) Mapping DNA-protein interactions in large genomes by sequence tag analysis of genomic enrichment. *Nat Methods* **2**: 47–53.

129. Cokus SJ, Feng S, Zhang X, Chen Z, Merriman B, Haudenschild CD, Pradhan S, Nelson SF, Pellegrini M, Jacobsen SE. (2008) Shotgun bisulphite sequencing of the Arabidopsis genome reveals DNA methylation patterning. *Nature* **452**: 215–219.

130. Lister R, O'Malley RC, Tonti-Filippini J, Gregory BD, Berry CC, Millar AH, Ecker JR. (2008) Highly integrated single-base resolution maps of the epigenome in Arabidopsis. *Cell* **133**: 523–536.

131. Taylor KH, Kramer RS, Davis JW, Guo J, Duff DJ, Xu D, Caldwell CW, Shi H. (2007) Ultradeep bisulfite sequencing analysis of DNA methylation patterns in multiple gene promoters by 454 sequencing. *Cancer Res* **67**: 8511–8518.

132. Korshunova Y, Maloney RK, Lakey N, Citek RW, Bacher B, Budiman A, Ordway JM, McCombie WR, Leon J, Jeddeloh JA, McPherson JD. (2008) Massively parallel bisulphite pyrosequencing reveals the molecular complexity of breast cancer-associated cytosine-methylation patterns obtained from tissue and serum DNA. *Genome Res* **18**: 19–29.

133. Meissner A, Mikkelsen TS, Gu H, Wernig M, Hanna J, Sivachenko A, Zhang X, Bernstein BE, Nusbaum C, Jaffe DB, Gnirke A, Jaenisch R *et al.* (2008) Genome-scale DNA methylation maps of pluripotent and differentiated cells. *Nature* **454**: 766–770.

134. Gitan RS, Shi H, Chen CM, Yan PS, Huang TH. (2002) Methylation-specific oligonucleotide microarray: A new potential for high-throughput methylation analysis. *Genome Res* **12**: 158–164.
135. Yan PS, Shi H, Rahmatpanah F, Hsiau TH, Hsiau AH, Leu YW, Liu JC, Huang TH. (2003) Differential distribution of DNA methylation within the RASSF1A CpG island in breast cancer. *Cancer Res* **63**: 6178–6186.
136. Mund C, Beier V, Bewerunge P, Dahms M, Lyko F, Hoheisel JD. (2005) Array-based analysis of genomic DNA methylation patterns of the tumour suppressor gene p16INK4A promoter in colon carcinoma cell lines. *Nucleic Acids Res* **33**: e73.
137. Zhang D, Bai Y, Ge Q, Qiao Y, Wang Y, Chen Z, Lu Z. (2006) Microarray-based molecular margin methylation pattern analysis in colorectal carcinoma. *Anal Biochem* **355**: 117–124.
138. Ehrich M, Nelson MR, Stanssens P, Zabeau M, Liloglou T, Xinarianos G, Cantor CR, Field JK, van den Boom D. (2005) Quantitative high-throughput analysis of DNA methylation patterns by base-specific cleavage and mass spectrometry. *Proc Nat Acad Sci USA* **102**: 15785–15790.
139. Yamada N, Nishida Y, Tsutsumida H, Hamada T, Goto M, Higashi M, Nomoto M, Yonezawa S. (2008) MUC1 expression is regulated by DNA methylation and histone H3 lysine 9 modification in cancer cells. *Cancer Res* **68**: 2708–2716.
140. Smith JF, Mahmood S, Song F, Morrow A, Smiraglia D, Zhang X, Rajput A, Higgins MJ, Krumm A, Petrelli NJ, Costello JF, Nagase H *et al.* (2007) Identification of DNA methylation in 3′ genomic regions that are associated with upregulation of gene expression in colorectal cancer. *Epigenetics* **2**: 161–172.
141. Suzuki H, Gabrielson E, Chen W, Anbazhagan R, van Engeland M, Weijenberg MP, Herman JG, Baylin SB. (2002) A genomic screen for genes upregulated by demethylation and histone deacetylase inhibition in human *Colorectal Cancer. Nat Genet* **31**: 141–149.
142. Yamashita K, Upadhyay S, Osada M, Hoque MO, Xiao Y, Mori M, Sato F, Meltzer SJ, Sidransky D. (2002) Pharmacologic unmasking of epigenetically silenced tumor suppressor genes in esophageal squamous cell carcinoma. *Cancer Cell* **2**: 485–495.
143. Hoque MO, Kim MS, Ostrow KL, Liu J, Wisman GB, Park HL, Poeta ML, Jeronimo C, Henrique R, Lendvai A, Schuuring E, Begum S *et al.* (2008) Genome-wide promoter analysis uncovers portions of the cancer methylome. *Cancer Res* **68**: 2661–2670.
144. Esteller M, Herman JG. (2002) Cancer as an epigenetic disease: DNA methylation and chromatin alterations in human tumours. *J Pathol* **196**: 1–7.
145. Issa JP. (2004) CpG island methylator phenotype in cancer. *Nat Rev Cancer* **4**: 988–993.
146. Jones PA, Martienssen R. (2005) A blueprint for a Human Epigenome Project: The AACR Human Epigenome Workshop. *Cancer Res* **65**: 11241–11246.

cancer genes encoded targets, can be generated. The abnormal protein kinase activities with their corresponding changes in protein phosphorylation states have been implicated in the onset of tumor formation and cancer progression,[3,5] Moreover, protein kinases are found in the pathways of oncogenes or tumour suppressor genes and thus become attractive targets for the development of therapeutic agents against cancer.[6-8] Protein kinases are some of the best targets of small molecule inhibitors,[9] and novel cancer treatments.[10] Among the three amino acids, tyrosine phosphorylation plays prominent role in the cancer biology. Cellular protein phosphorylation is controlled by more than 500 members of the protein kinase superfamily — one of the largest enzyme families encoded by the human genome[11] — of which more than 100 kinases have been implicated in human cancers. Some of the malignancy and tumor formation are derived from unregulated tyrosine kinase activity that causes inappropriate proliferation and survival signals.[2] Thus, the study of tyrosine kinases and the pathways driving tumor growth provides great opportunity in the development of effective drug therapies.[12]

1.2. *Examples of phosphoproteins that regulate cancer progression*

Cancer-specific phosphosignatures can be used as diagnostic markers and/or the drug targets. Table 1 shows some US Food and Drug Administration (FDA)-approved anticancer drugs targeting protein kinases.[13] Antibodies, highly specific for their targeted antigens, are commonly used to inhibit the activity of these kinases. The mAb trastuzumab (Herceptin) is the first antikinase drug approved to treat advanced ErbB2 (Her2)-positive breast cancer. It was further approved in 2006 for the treatment of early-stage breast cancer after primary therapy. In addition, the overexpression of EGFR commonly occurs in a wide spectrum of cancer types.[14] Approved antibodies, Cetuximab (Erbitux) and Ppanitumumab (Vectibix), target EGFR for the treatment of metastatic colorectal carcinoma (CRC).

Small-molecule kinase inhibitors represent another class of antikinase drugs that have been shown to effectively treat different types of cancer such as chronic myelogenous leukemia (CML), gastrointestinal stromal tumor (GIST), non–small cell lung cancer (NSCLC), renal cell carcinoma (RCC), etc. Gefitinib (Iressa) and erlotinib (Tarceva) are examples of highly specific small-molecule kinase inhibitors that target EGFR for the treatment of advanced NSCLC.[15,16] In addition, many more kinase

Table 1. US Food and Drug Administration (FDA) approved anticancer drugs targeting protein kinases.[13]

Drug name	Commercial names	Drug target protein	Kinase types[a]	Cancer type[b]	Year approved
Small molecules					
Imatinib	Gleevec	BCR-ABL	NRTK	CML	2001
Imatinib	Gleevec	c-Kit, PDGFR	RTK	GIST	2002
Gefitinib	Iressa	EGFR	RTK	Metastatic NSCLC	2003
Erlotinib	Tarceva	EGFR	RTK	Metastatic NSCLC	2004
Sorafenib	Nexavar	VEGFR, Raf1, PDGFR, RET, c-Kit, FLT3	RTK, S/T Kinase	Advanced RCC	2005
Sunitinib	Sutent	PDGFR, VEGFR, c-Kit, FLT3	RTK	Advanced RCC, GIST	2006
Dasatinib	Sprycel	BCR-ABL, Src	NRTK	Gleevec-resistant CML	2006
Antibodies					
Bevacizumab	Avastin	VEGF	RTK	Metastatic CRC	2004
Cetuximab	Erbitux	EGFR	RTK	Metastatic CRC	2004
Trastuzumab	Herceptin	ERBB2	RTK	Breast cancer	1998, 2006
Pantumumab	Vectibix	EGFR	RTK	Metastatic CRC	2006

[a]NRTK, nonreceptor tyrosine kinase; RTK, receptor tyrosine kinase; S/T kinase, serine/threonine kinase.

[b]CML, chronic myelogenous leukemia; CRC, colorectal carcinoma; GIST, gastrointestinal stromal tumors; NSCLC, nonsmall-cell lung cancer; RCC, renal cell carcinoma.

inhibitors are currently under preclinical investigation and clinical trial.[17,18] Since most cancers are the result of numerous mutations and stimulation of various pathways,[19] one would expect that targeting multiple kinases in different pathways would potentially have better therapeutic efficacy than the therapy with single agents. Based on the scenario, tools for the identification of cancer-related kinase/phosphatase are urgently needed for cancer drug discovery.

2. Technologies for Identification and Quantitation of Phosphoproteome

To facilitate the discovery of protein targets and elucidate the molecular basis of cancer, it is crucial to identify the specific phosphorylation sites and quantify their temporal and dynamic changes associated with cancer progression. Recently, advances in phosphoproteomics and mass spectrometry (MS) have emerged as reliable and sensitive tools for the identification and quantitation of thousands of known and novel phosphorylation sites,[20–22] Despite the biological significance of protein phosphorylation and the advances in MS, characterization of site-specific phosphorylation is still challenged by their dynamic modification patterns, substoichiometric concentrations, heterogeneous forms of phosphoproteins, and low MS response. Methodologies that specifically enrich the transient phospho-subproteome in a robust, comprehensive manner are important to study phosphorylation-dependent cellular signaling.

2.1. *Phosphorylation site identification by mass spectrometry*

In MS-based approaches, identification of peptides and its modifications are implemented by the process of tandem mass spectrometry (MS/MS). MS/MS is performed to dissociate peptide into sequence-informative product ions, which can be deduced by database searching to acquire the peptide sequence and modification site. There are two primary MS/MS techniques in peptide fragmentation; collision-induced dissociation (CID) method and electron-derived dissociation methods, which comprise of electron-capture dissociation (ECD) and electro-transfer dissociation (ETD).

2.1.1. *Collision-induced dissociation*

Among the methods of peptide fragmentation, collision-induced dissociation (CID) is extensively used. When the collision process occurred, partial kinetic energy will be converted into internal energy of peptides. As the proton locating at energetically less favored sites such as N-terminal amine, oxygen and nitrogen in amide bond and side chain, protonation of amide nitrogen gives rise to weakening bond and the bond is subsequently dissociated when there is enough internal energy. Once dissociated at the amide bonds, fragments composed of the charge retaining on the N-terminus and C-terminus are indicated as b- and y-ions, respectively.

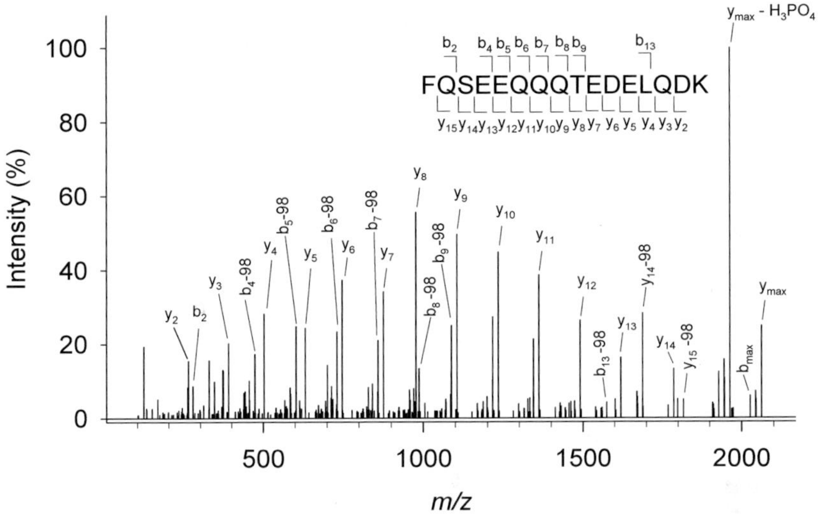

Fig. 1. A MS/MS mass spectrum of (FQpSEEQQQTEDELQDK, purified from β-casein) generated from CID fragmentation.

To demonstrate how to localize the phosphorylation site by MS/MS spectrum, a phosphorylated peptide (FQpSEEQQQTEDELQDK) from tryptic β-casein was used. The fragmentation mass spectrum of this peptide was shown in Fig. 1. We first examined N-terminal fragments. Fragment ions b_4 to b_8 ions have a neutral loss of H_3PO_4 ($-98\,Da$), revealing that these fragment ions are phosphorylation-containing fragments. The observation of b_2 and b_4 implies that Ser-3 is phosphorylated. To further confirm whether Ser-3 is a correct phosphorylation site, C-terminal fragment ions were examined. The spectrum showed that y_{14} and y_{15} ions contain phosphate group, suggesting that the phosphorylated site is located on Ser-3. During the process of fragmentation by CID, a dominant peak that yielded from loss of H_3PO_4 or HPO_3, was frequently observed in CID fragmentation (e.g. y_{max}- H_3PO_4 in Fig. 1). Sometimes this peak dominates the whole spectrum and hampers the observation of other informative product ions, which could result in the failure to sequence identification and phosphorylation sites determination during database matching.

2.1.2. *Electron-derived dissociation methods*

To circumvent the problem of neutral loss dominated in the spectrum, a complement technique of electron-derived dissociation has been introduced to not only dissociate peptides which are not fragment efficiently by CID but also conserve labile post-translational medications, including phosphate and glycan groups.[23] Electron capture dissociation (ECD) occurs in a Fourier transform ion cyclotron resonance (FT-ICR) mass spectrometer at which multiply protonated peptides coexist with near-thermal electrons ($<0.2\,eV$). After capturing an electron, hydrogen transferred from the basic residues amino group to carbonyl oxygen of amide bond leads to fragmentation of $N-C_\alpha$ bond and yield of c- and z-type product ions (Fig. 2).[24] However, ECD technique is limited in FT-ICR mass spectrometer because it can maintain thermal energy of electron to several milliseconds by its magnetic and electronic field rather than radiofrequency ion traps only to microseconds.[25]

In ion trap mass spectrometer, alternative electron-transferring approach, term ETD, is introduced by some anion compounds to release electrons to multiple-charged peptides, yielding ECD-like fragmentation.[25,26] ETD also has the same advantages in phosphopeptide fragmentation, that is, complete backbone fragmentation without loss of labile modifications. The comparison on the mass spectra of the same phosphopeptide subjected to CID or ETD fragmentation reported by

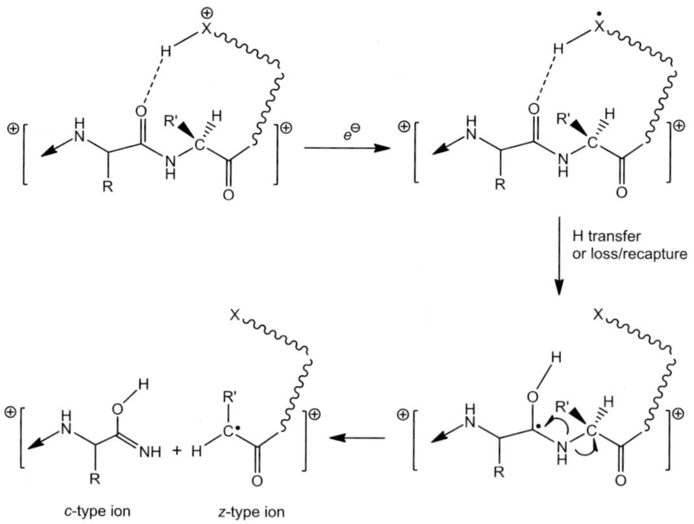

Fig. 2. The mechanism of phosphopeptide fragmentation by electron capture dissociation. In brief, an electron enters a charge-stabilized electronic state of delocalized amide group. This group functions as a superbase which abstracts a proton from a sterically proximate amino acid residue to form a labile aminoketyl radical that dissociated by $N-C_\alpha$ bond cleavage.[24]

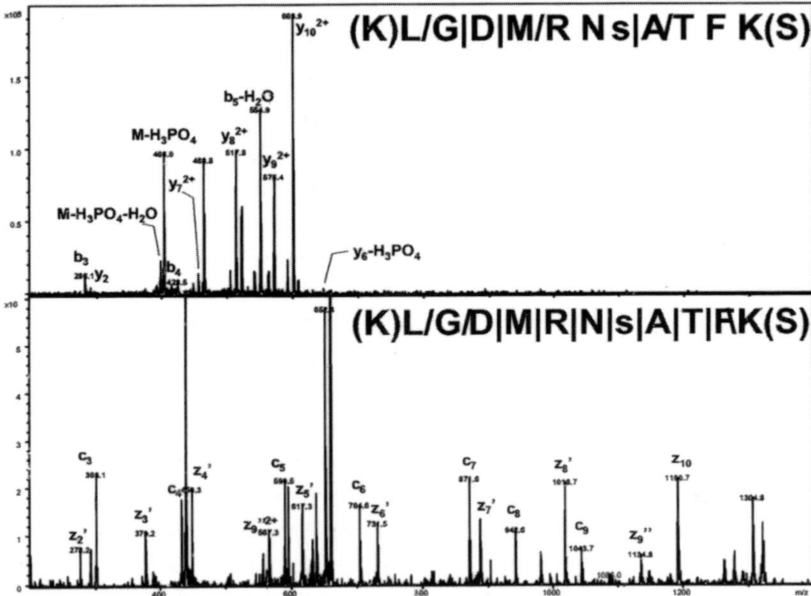

Fig. 3. Fragmentation pattern of phosphopeptide from CID (upper) and ETD (lower).[27]

Molina *et al.* was shown in Fig. 3.[27] A neutral loss peak was clearly shown in CID spectra (upper panel), which results in fewer b or y ions. As for ETD spectra, more c or z backbone fragment ions were presented without detectable neutral loss, facilitating the sequence identification and phosphorylation site localization.

2.2. *Enrichment of phosphoproteins/phosphopeptides*

Although MS has been employed in mapping phosphorylation sites on protein and quantification of their changes, some intrinsic characters of phosphopeptides have hindered their direct analysis in biological samples by mass spectrometry. First, the stoichiometric level of phosphoprotein may be very low. Second, phosphopeptides tend to have relatively low ionization efficiency; signal suppression effect often occurs especially in the presence of non-phosphorylated peptides. Third, phosphate groups on phosphoserine and phosphothreonine are labile, which could decompose during improper sample preparation or peptide fragmentation. Thus, enrichment steps prior to MS analysis are crucial to specifically isolate phosphoprotein/phosphopepitdes. So far, the most widely utilized approaches for comprehensive phosphopeptide purification are based on either antibody or metal ion affinity method; the latter includes Immobilized metal affinity chromatography (IMAC) and metal oxide affinity chromatography (MOAC).

2.2.1. *Immunoprecipitation with antibodies*

Antibodies raised against phosphopeptides can be utilized as prior enrichment (e.g. immunoprecipitation) of phosphorylated species from complex samples. A more generally way would be amino-specific antibodies being raised against a consensus sequence for a specific kinase motif (e.g. SXR, X is any amino acid).[28] There are several commercially available antibodies that are able to bind to phosphotyrosine, phosphoserine, and phosphothreonine. Among them, phosphotyrosine-specific antibodies are considered relatively efficient reagents to purify tyrosine phosphorylated proteins and has been used to profile the tyrosine phosphoproteome in cancer cells.[29] In the work published by Rush *et al.* a number of activated protein kinases and their phosphorylated substrates were identified, revealing the oncogenic signaling networks in comparison with the normal cells. Using similar reagents, Thelemann *et al.* purified tyrosine phosphorylated proteins prior to proteolysis and mass spectrometry analysis, which facilitated the characterization of phosphotyrosine signaling networks in epidermal growth factor receptor overexpressing squamous carcinoma cells.[30] On the other hand, the lack of excellent antiphosphoserine and antiphosphothreonine antibodies has hindered the application of this purification approach for phosphoserine and phosphothreonine analysis.

2.2.2. *Immobilized metal affinity chromatography*

Immobilized metal affinity chromatography (IMAC) relies on the electrostatic interactions and coordination bonding between the unoccupied orbital of metal ions and negatively charged phosphate groups. The concept of phosphoprotein enrichment by IMAC (Fe^{3+}) was first presented by Andersson *et al.* to isolate tryptic phosphopeptides.[31] Other metal ions, such as Ga^{3+}, Zn^{2+}, and Zr^{4+} that immobilized on a solid phase by metal-chelating ligands such as nitrilotriacetic acid (NTA) and iminodiacetic acid (IDA), were also proposed for phosphopeptide enrichment.[32–34] On the level of proteome-scale study, the performance of IMAC is deteriorated by co-purification of nonphosphorylated peptides such as acidic residues-containing peptides (e.g. glutamic acid and aspartic acids) or peptides that can bind to IMAC due to hydrophobic interaction. To improve the purification specificity, several approaches have been proposed. Kokubu *et al.* suggested the use of organic solvent, 40–60% of acetonitrile for effective prevention of the adsorption of hydrophobic nonphosphopeptides and thus improved the specificity of IMAC.[35] More recently, Tsai *et al.* proposed a simple and robust enrichment procedure to effectively inhibit nonspecific binding by controlling the pH value and the organic acid concentration. The protocol reported greater than 90% enrichment specificity and recovery.[36] Fractionation of complex sample prior to IMAC enrichment also reduce the interference of nonspecific binding and increase the number of phosphoprotein identification. Phosphopeptides was separated from nonphosphopeptides according to discrepancy of charge state and hydrophobisity by ion exchange chromatography,[33,37] hydrophilic interaction chromatography (HILIC)[38] and electrostatic repulsion hydrophilic interaction chromatography (ERLIC).[39]

2.2.3. *Metal oxide affinity chromatography*

In the past few years, metal oxide affinity chromatography (MOAC) has evolved rapidly as an efficient alternative to IMAC for phosphopeptide enrichment.[40] It is known that metal oxide have amphoteric properties, that is, it can play a role either as Lewis acid or base depending on pH value of aqueous solution.[41] In acidic conditions, metal oxide behaves as a Lewis acid with positively charged surface, which selectively purifies negatively charged phosphopeptides. Several different metal oxides materials such as titanium dioxide (TiO_2), zirconium dioxide (ZrO_2), aluminum oxide (Al_2O_3), and niobium oxide (Nb_2O_5) have been applied to purify phosphopeptides.[40,42–44] Ikeguchi *et al.* firstly showed that TiO_2 could selectively concentrate phospho-amino acids prior to the anion-exchange chromatography.[45] TiO_2 is then employed for enrichment of tryptic phosphopeptide from standard phosphoprotein.[46] Peptides with multi-acidic acid residues could also adsorb onto TiO_2 precolumnto cause co-purification of nonphosphopeptides. The enrichment specificity can be improved by the use of aliphatic hydroxy acid such as lactic and β-hydroxypropanoic acid.[47] Another challenge of using MOAC is that of the recovery of multiply phosphorylated peptides.[48] Because of strong binding between metal oxide and phosphate, multiply phosphorylated peptide is more difficult to be eluted and detected in MS analysis compared to monophosphopeptide. For the comprehensive analysis of phosphoproteome, SIMAC (sequential elution from IMAC) strategy was proposed to combine IMAC for purification of multiply phosphorylated peptides followed by TiO_2 purification to enrich monophosphopeptides. The number of multiply phosphorylated peptides purified by SIMAC is three times the number more than that enriched by TiO_2 only.[49]

2.2.4. *Nanomaterials for purification of phosphopeptides*

The use of nanomaterials ($<100\,nm$) has been demonstrated an efficient platform to enrich phosphopeptides due to its high surface-area-to-volume ratio and flexible surface functionalization.[50] The large surface-area-to-volume ratio of nanoparticle may potentially form higher amount of ligands on nanoparticles than on microscale materials, providing multivalent binding affinity to concentrate trace amount of phosphopeptide. Thus, surface functionalized nanoparticles with either metal-chelating compounds or metal oxides have been recently developed to purify phosphopeptides.

Chen *et al.* synthesized metal oxide–coated magnetic nanoparticles (MNPs) coated with TiO_2, ZrO_2, and Al_2O_3 to demonstrated concentration of phosphopeptides from low amount of digested standard phosphoprotein.[51–53] Instead of coating SiO_2 onto nanoparticle, Li *et al.* synthesized $Fe_3O_4@C$ particle to improve enrichment specificity by polymerization and carbonization of glucose, followed by addition of tetrabutyltitanate and conversion into titania by calcination.[54] In addition to $Fe_3O_4@TiO_2$ microsphere, $Fe_3O_4@ZrO_2$, $Fe_3O_4@Al_2O_3$ and $Fe_3O_4@Ga_2O_3$ core-shell microspheres have also been synthesized in similar ways and applied for standard phosphoprotein and phosphoproteome analysis.[55–57]

The other type of nanoscale material for enrichment of phosphopeptides is immobilized metal ion affinity chromatography (IMAC)-based magnetic nanoparticles. Li *et al.* first demonstrated the use of immobilized Zr^{4+} or Ga^{3+} ion on NTA-magnetic nanoparticles for phosphopeptide enrichment.[58] Pan *et al.* demonstrated that Fe^{3+}-IMAC nanoparticles has higher magnetic nanoparticles were used for analyzing mouse liver and Chang liver cell phosphoproteom.[59,60] Recently, Wu *et al.* demonstrated the complemtary enrichment of phosphopeptides between nano-size Ti^{4+}-NTA-PEG@MNP, chromatography-based Fe^{3+}-IMAC and TiO_2 methods. On the standard protein mixture test, nano-size Ti^{4+}-NTA-PEG@MNP detected more comprehensive coverage of both mono- and multiply phosphorylated peptides ($\alpha4$, $\alpha6$, $\alpha7$, $\alpha8$, $\alpha9$ and $\beta2$) (Fig. 4(b)), whereas chromatography-based Fe^{3+}-IMAC and TiO_2 methods showed only enrichment of monophosphorylated peptides ($\alpha1$, $\alpha2$, $\alpha5$ and $\beta1$) (Fig. 4(c)–4(d)).[61]

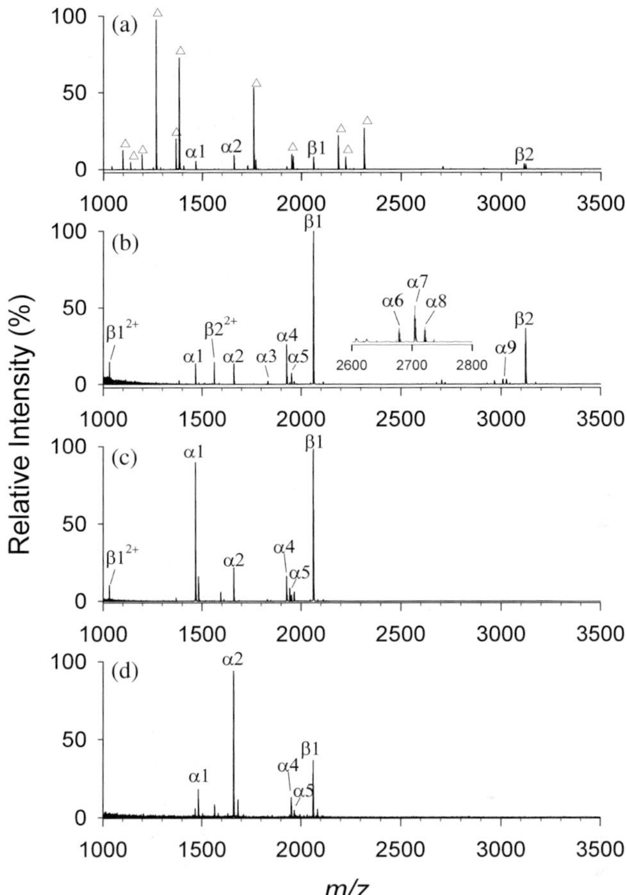

Fig. 4. MALDI-TOF mass spectrum of 20 pmole tryptic peptide mixture of α-casein and β-casein (a) before and (b) after phosphopeptide enrichment by Ti^{4+}-NTA-PEG@MNP. Peptide mixture (20 pmole of α-casein and β-casein) enriched by (c) Chromatography-based Fe^{3+}-IMAC and (d) TiO_2 method. Triangles represent nonphosphopeptides of β-casein.

Furthermore, in proteome-scale study, Ti^{4+}-NTA-PEG@MNP enriched two-fold and three-fold number of multiply phosphorylated peptide (195, 15%) compared to chromatography-based Fe^{3+}-IMAC (100, 5.2%) and TiO_2 method (64, 7.3%) (Fig. 5).[61] This may be attributable to the large surface area-to-volume ratio of nanoscale magnetic particle for loading more amount of Ti^{4+} ion on the surface. The high amount of Ti^{4+} may provide a stronger binding affinity with neighboring multiply phosphorylated residues through the well-documented multivalent effect of nanomaterials.[62]

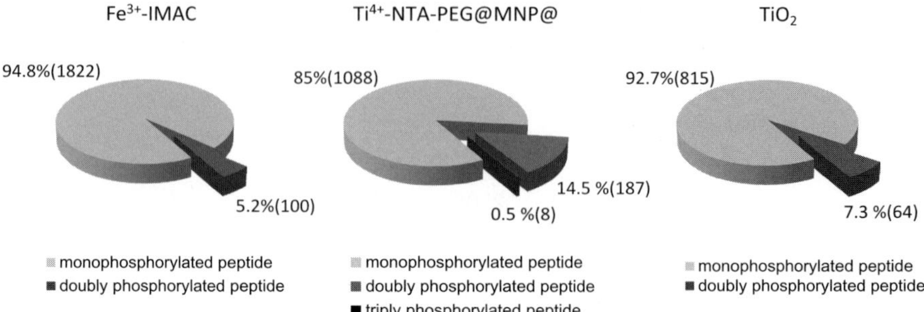

Fig. 5. Distribution of identified singly-, doubly-, and triply phosphorylated peptides from Raji B cell lysate (400 μg) by (a) chromatography-based Fe^{3+}-IMAC; (b) Ti^{4+}-NTA-PEG@MNP and (c) TiO_2 methods.[61]

Although most of the phosphopeptide purification approaches have not been demonstrated at proteomic scale, they provide promising tools for phosphoproteomics by the demonstrated high sensitivity and specificity achieved by the refinement of the nano-scale approaches.

2.3. *Quantitation method for phosphoproteomics*

Although large-scale identification of proteins and post-translational modification has enhanced the understanding of the cell process, quantitative analysis associated with dynamic changes of phosphorylation provides more direct information for the signal transduction of external stimuli. Moreover, to elucidate the molecular basis of diseases and its progressions, it is crucial to quantify their temporal and dynamic changes of specific phosphorylation sites. Such information may reveal the operation and connectivity of these pathways, which will further facilitate the identification of cancer-causing oncoprotein or tumor-suppressor proteins for cancer diagnosis and rational design for therapeutics.

Recently, the developments of quantitative proteomic strategies lead phosphoproteomic filed into a new perspective. In the past decade, stable isotope labeling is the most commonly adapted strategy for quantitation of phosphoproteome. This approach can be distinguished into three major categories, including chemical labeling (e.g. isobaric tags for relative and absolute quantification, iTRAQ), enzymatic labeling, e.g. ^{18}O-labeling, and metabolic labeling, e.g. SILAC (Stable isotope labeling of amino acid in cell culture). Among these strategies, metabolic labeling is the most commonly used method for quantitative analysis of protein phosphorylation. These three methods were described in the following paragraph.

2.3.1. *Chemical labeling*

Among the labeling methods, iTRAQ labeling has gained popularity for quantitative phosphoproteomics. The quantitation strategy for iTRAQ-labeling is based on a set of four isobaric reagents, each of which comprises three groups: reporter, balance, and reactive groups.[63] The multiplexing reagents label the N-terminus and the lysine side chains of peptides, more peptides per protein can be labeled. In contrast to other quantitative methods based on peptide intensity, the intensity of signature ions, 114, 115, 116, and 117, of the reporter group in an MS/MS spectrum is used to quantify protein expression levels. This quantitation technique has been apply to identify expression changes in novel tyrosine kinase substrates associated with breast cancer progression.[64]

2.3.2. *Enzymatic labeling*

Enzymatic labeling with ^{18}O water and trypsin is a simple and straight-forward method based on the corporation of isotopic oxygen at the C-terminus of peptides to generate mass difference of 4 amu.[65] After enzymatic labeling, the labeled and unlabeled peptide pairs are measured by high resolution LC-MS/MS. The peak intensity of the peptide pair can be used for quantification. To further improve the quantitation accuracy, Gygi introduced absolute quantification (AQUA) by spiking a synthetic analogous peptide incorporating stable isotope (^{13}C, ^{15}N) labeled amino acid into protein sample during proteolytic digestion to serve as the internal standard peptide.[66] They applied this approach for directly measuring the dynamic changes in cell cycle-dependent phosphorylation of Ser-1126 of human separase from Hela cell lysates.

2.3.3. *Metabolic labeling*

The most widely used metabolic labeling is stable isotope labeling by amino acid in cell culture (SILAC), which offers more options for selection of isotopically labeled amino acids such as arginine, leucine, lysine, serine, methionine and tyrosine. SILAC can effectively reduce systematic quantitation error because the different isotopic SILAC population cells

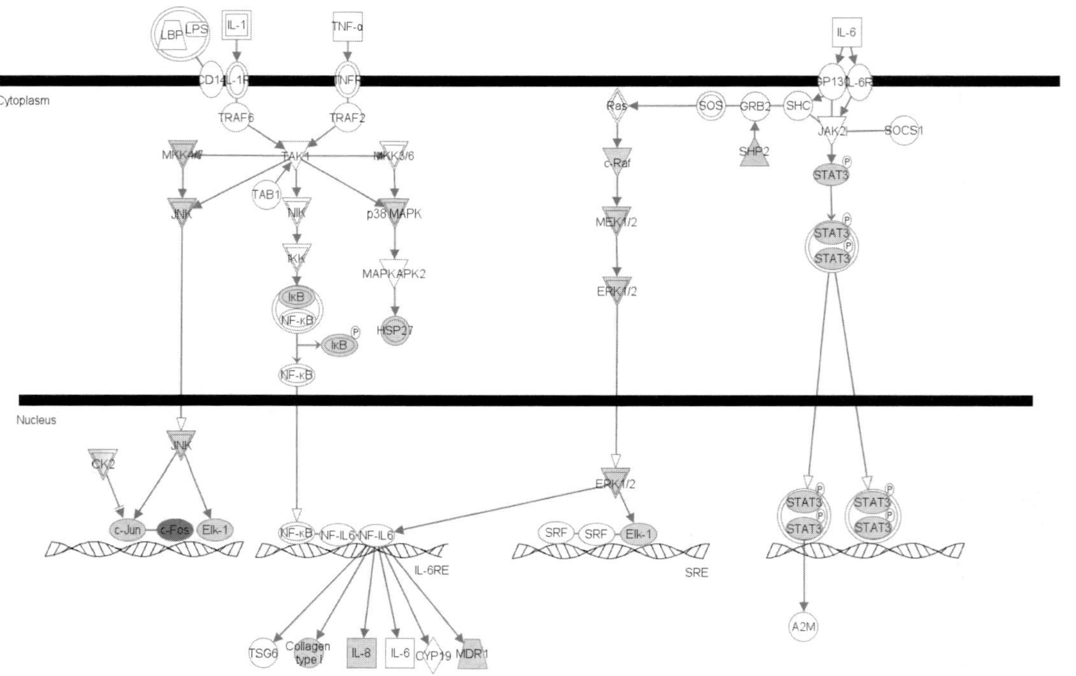

Fig. 7. Quantitative protein changes observed in the IL-6 signaling pathway in hypoxic GSC11 cells treated with WP1193 and IL-6, relative to hypoxic control cells. The image was created by overlaying the quantitative proteomic data on the IL-6 signaling pathway in Ingenuity Systems Analysis software. Dark indicates increased levels and gray means no quantitative change in proteins in cells treated with WP1193 and IL-6 relative to the hypoxic control cells.[92]

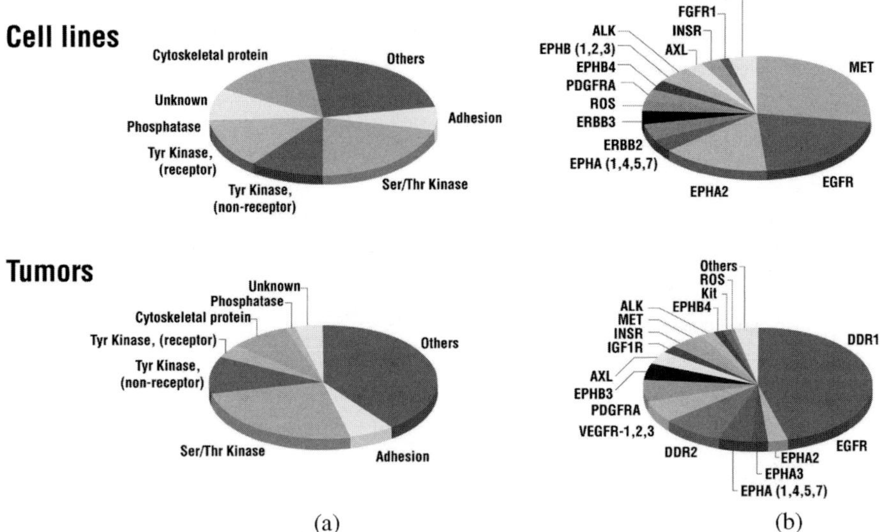

Fig. 8. Comparison of Tyrosine Kinase between Lung Cancer Cell Line and Tumors. (a) Distribution of phosphoproteins types. Each observed phosphoproteins was assigned a protein category from the PhosphoSite ontology; (b) Distribution of spectral counts among receptor tyrosine kinases (RTK). The total numbers of observed spectra assigned to each RTK over all of the cell lines (top) or the tumors (bottom) are represented as fractions of the total RTK spectra observed.[82]

tyrosine kinase phosphorylation showed higher levels in tumors (Fig. 8(b)). This work also identified known oncogenic kinases such as EGFR and c-Met as well as novel ALK and ROS fusion proteins by analyzing phosphotyrosine signaling profiles.

Quantitative phosphoproteomics experiments emerged in recent years, triggering the development of drugs for specific biological pathways. One successful example of cancer-specific drugs is the development of Herceptin, a monoclonal antibody against the HER2 receptor for breast cancer therapy. Wolf-Yadlin *et al.* have used phosphoproteomics strategy to investigate the role of phosphorylation in the effects of HER2 overexpression on EGF- and HRG-mediated signaling of erbB receptors. Specific combination of phosphorylation sites that correlate with cell proliferation and migration were identified, which may serve as potential targets for therapeutic intervention.[93,94]

Chen *et al.* utilized iTRAQ technology to reveal 57 tyrosine phosphorylated proteins altered during breast cancer development.[64] Among them,

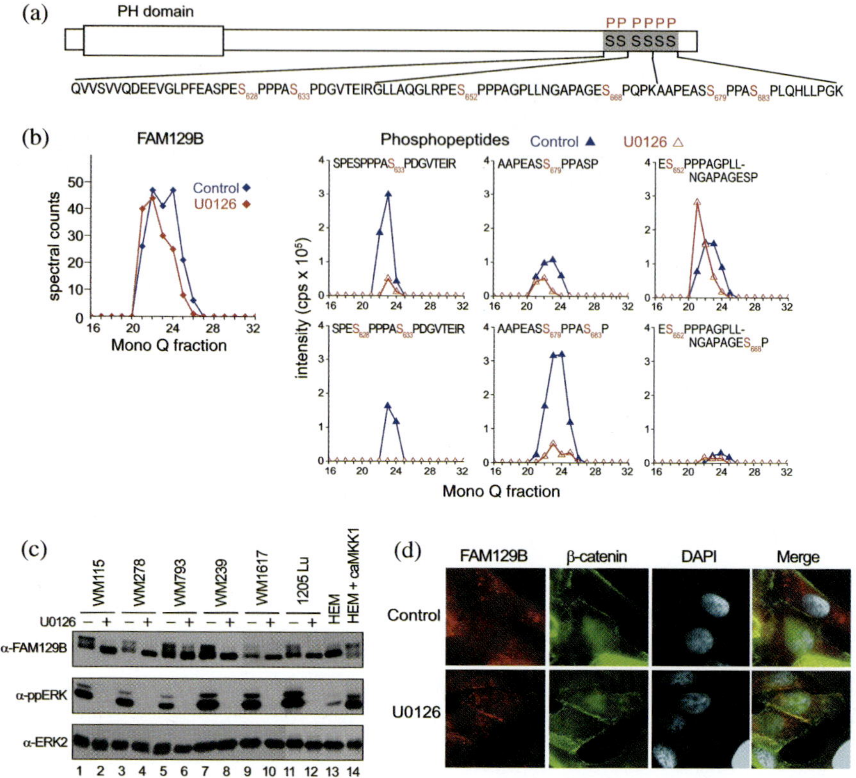

Fig. 9. FAM129B Controls Melanoma Cell Invasion.[96] (a) Domain structure of open reading frame FAM129B, showing observed phosphopeptide sequences (black) and phosphorylated residues (red); (b) (Left) Spectral counts show little change in protein abundance with U0126 treatment. (Right) Phosphorylation at S628, S633, S679, and S683 is downregulated by U0126, while phosphorylation at S652 and S668 is unaffected; (c) Dephosphorylation of FAM129B by U0126 treatment correlates with faster gel mobility in WM115 and five other melanoma cell lines (lanes 1–12). In contrast, FAM129B shows faster gel mobility in HEM-lp melanocytes (lane 13) and slower mobility when ERK signaling is activated by active mutant MKK1 (lane 14); (d) Dephosphorylation of FAM129B correlates with membrane association. WM115 melanoma cells were treated with or without U0126 for 24 hr and immunostained for FAM129B and b-catenin. In control cells, FAM129B reactivity was uniformly dispersed, while b-catenin localized to cell–cell junctions, consistent with cadherin interactions. In U0126-treated cells, FAM129B colocalized with b-catenin at cell–cell junctions; (e) Stable knockdown and rescue of FAM129B. (Lanes 1–3) Endogenous FAM129B expression in untreated WM239 cells (lane 1), cells stably expressing vector control (lane 2), or shRNA-FAM129B for knockdown (lane 3). (Lanes 4–7) shRNA knockdown cells stably expressing FAM129B-WT or mutant FAM129B-S/A, -S/E, or -S/D substituted at six phosphorylation sites. (Lanes 8 and 9) Naive WM239 cells stably expressing FAM129B-WT or FAM129B-S/A.

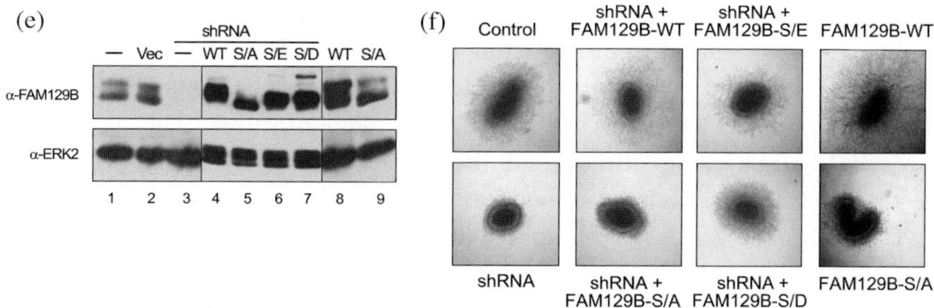

Fig. 9. (*Continued*)

SLC4A7 and TOLLIP have been discovered for the first time as novel tyrosine substrates that are associated with human cancer development. Breast cancer metastasis-associated phosphoproteins were revealed by conducting a comparative phosphoproteomics analysis of an isogenic pair of human breast tumor cell with opposite metastatic phenotype. Their finding suggests that the phosphorylation of isoform A of lamin A/C and GTPase activating protein binding protein 1 is associated with metastatic propensity.[95] Old *et al.* identified 90 phosphorylation events regulated by oncogenic B-Raf signaling, based on their responses to treating melanoma cells with MKK1/2 inhibitor. The importance of MINERVA/FAM129B protein and its MAP-kinase-dependent phosphorylation in controlling melanoma cell invasion was investigated (Fig. 9). The results suggested that phosphorylated FAM129B enhances invasion and represents an important mechanism for modulating cancer cell behavior.[96] Chen *et al.* investigated rapamycin-dependent phosphoproteomics and discovered that 250 phospho-sites in 161 proteins. A siRNA-dependent screen of these proteins showed that AKT induction by rapamycin was attenuated by depleting cellular CDC25B phosphatase. CDC25B depletion in Human cancer cell line, Hela, MCF-7, and Du145 enhanced the anticancer effect of rapamycin. This study shows that CDC25B may serve as a drug target for improving mTOR-targeted cancer therapies.[97]

5. Future Perspectives

The technology for globally analyzing proteins phosphorylation is still a critical component for better understanding of the signaling network alterations involved in cancer. To characterize aberrant signaling pathways in cancer, nowadays, quantitative cancer phosphoproteomics analysis was mostly carried out in cell models of different origins. The key signaling molecules were validated in the clinical samples. Although the diversity of the clinical samples may result in much more uncertainty in the quantitative measurement, to address the clinical implementations, the procedure for reliably analyzing clinical samples should be established. With the need of better sensitivity, further improvement in the sample preparation and enrichment methods for phosphorylated peptides and proteins are required. With the improvement of mass spectrometer, phosphopeptide enrichment procedures and computational algorithms, it can be envisaged that the protein phosphorylation involved in the cancer progress could be comprehensively deciphered, bringing impacts to the biomedical science. Once a maturation phase in method development has been achieved, the application of phosphoproteomics will be beyond signal transduction and cancer research and other fields such as system biology and drug development. Cancer markers, derived from tumor phosphoproteome, will help shift the scope to individualized medicine, which is one of the ultimate goals for application of phosphoproteomics in the clinical setting.

References

1. Hunter T. (1995) Protein kinases And phosphatases: The yin and yang of protein phosphorylation and signaling. *Cell* **80**: 225–236.
2. Blume-Jensen P, Hunter T. (2001) Oncogenic kinase signalling. *Nature* **411**: 355–365.
3. Bardwell AJ, Frankson E, Bardwell L. (2009) Selectivity of docking sites in MAPK kinases. *J Biol Chem* **284**: 13165–13173.
4. Futreal PA, Coin L, Marshall M, Down T, Hubbard T, Wooster R *et al.* (2004) A census of human cancer genes. *Nat Rev Cancer* **4**: 177–183.
5. Sathornsumetee S, Reardon DA. (2009) Targeting multiple kinases in glioblastoma multiforme. *Expert Opin Investig Drugs* **18**: 277–292.
6. Rane SG, Reddy EP. (2000) Janus kinases: Components of multiple signaling pathways. *Oncogene* **19**: 5662–5679.
7. Giamas G, Stebbing J, Vorgias CE, Knippschild U. (2007) Protein kinases as targets for cancer treatment. *Pharmacogenomics* **8**: 1005–1016.
8. Ralph SJ. (2007) An update on malignant melanoma vaccine research: Insights into mechanisms for improving the design and potency of melanoma therapeutic vaccines. *Am J Clin Dermatol* **8**: 123–141.
9. Malumbres M, Barbacid M. (2007) Cell cycle kinases in cancer. *Curr Opin Genet Dev* **17**: 60–65.
10. Traxler P, Bold G, Buchdunger E, Caravatti G, Furet P, Manley P *et al.* (2001) Tyrosine kinase inhibitors: From rational design to clinical trials. *Med Res Rev* **21**: 499–512.

11. Oppermann FS, Gnad F, Olsen JV, Hornberger R, Greff Z, Keri G *et al.* (2009) Large-scale proteomics analysis of the human kinome. *Mol Cell Proteomics* **8**: 1751–1764.

12. Sawyers CL. (2002) Rational therapeutic intervention in cancer: Kinases as drug targets. *Curr Opin Genet Dev* **12**: 111–115.

13. Yu LR, Issaq HJ, Veenstra TD. (2007) Phosphoproteomics for the discovery of kinases as cancer biomarkers and drug targets. *Proteomics Clin Appl* **1**: 1042–1057.

14. Ernst E. (2001) Mistletoe for cancer? *Eur J Cancer* **37**: 9–11.

15. Cohen MH, Williams GA, Sridhara R, Chen G, Pazdur R. (2003) Fda drug approval summary: Gefitinib (ZD1839) (Iressa) tablets. *Oncologist* **8**: 303–306.

16. Johnson JR, Cohen M, Sridhara R, Chen YF, Williams GM, Duan J *et al.* (2005) Approval summary for erlotinib for treatment of patients with locally advanced or metastatic non-small cell lung cancer after failure of at least one prior chemotherapy regimen. *Clin Cancer Res* **11**: 6414–6421.

17. Saglio G, Kim DW, Issaragrisil S, Le Coutre P, Etienne G, Lobo C *et al.* (2010) Nilotinib versus imatinib for newly diagnosed chronic myeloid leukemia. *N Engl J Med* **362**: 2251–2259.

18. Sawyers CL. (2010) Even better kinase inhibitors for chronic myeloid leukemia. *N Engl J Med* **362**: 2314–2315.

19. Vogelstein B, Kinzler KW. (2004) Cancer genes and the pathways they control. *Nat Med* **10**: 789–799.

20. Wang SE, Narasanna A, Perez-Torres M, Xiang B, Wu Fy, Yang S *et al.* (2006) HER2 kinase domain mutation results in constitutive phosphorylation and activation of HER2 And EGFR and resistance to EGFR tyrosine kinase inhibitors. *Cancer Cell* **10**: 25–38.

21. Lu Y, Zi X, Zhao Y, Mascarenhas D, Pollak M. (2001) Insulin-like growth factor-I receptor signaling and resistance to trastuzumab (Herceptin). *J Natl Cancer Inst* **93**: 1852–1857.

22. Hanahan D, Weinberg RA. (2000) The hallmarks of cancer. *Cell* **100**: 57–70.

23. Boersema PJ, Mohammed S, Heck AJ. (2009) Phosphopeptide fragmentation and analysis by mass spectrometry. *J Mass Spectrom* **44**: 861–878.

24. Syrstad EA, Turecek F. (2005) Toward a general mechanism of electron capture dissociation. *J Am Soc Mass Spectrom* **16**: 208–224.

25. Syka JE, Coon JJ, Schroeder MJ, Shabanowitz J, Hunt DF. (2004) Peptide and protein sequence analysis by electron transfer dissociation mass spectrometry. *Proc Natl Acad Sci USA* **101**: 9528–9533.

26. Gunawardena HP, He M, Chrisman PA, Pitteri SJ, Hogan JM, Hodges BD *et al.* (2005) Electron transfer versus proton transfer in gas-phase ion/ion reactions of polyprotonated peptides. *J Am Chem Soc* **127**: 12627–12639.

27. Molina H, Horn DM, Tang N, Mathivanan S, Pandey A. (2007) Global proteomic profiling of phosphopeptides using electron transfer dissociation tandem mass spectrometry. *Proc Natl Acad Sci USA* **104**: 2199–2204.

28. Zhang H, Zha X, Tan Y, Hornbeck PV, Mastrangelo AJ, Alessi DR *et al.* (2002) Phosphoprotein analysis using antibodies broadly reactive against phosphorylated motifs. *J Biol Chem* **277**: 39379–39387.

29. Rush J, Moritz A, Lee KA, Guo A, Goss VL, Spek EJ *et al.* (2005) Immunoaffinity profiling of tyrosine phosphorylation in cancer cells. *Nat Biotechnol* **23**: 94–101.

30. Thelemann A, Petti F, Griffin G, Iwata K, Hunt T, Settinari T *et al.* (2005) Phosphotyrosine signaling networks in epidermal growth factor receptor overexpressing squamous carcinoma cells. *Mol Cell Proteomics* **4**: 356–376.

31. Andersson L, Porath J. (1986) Isolation of phosphoproteins by immobilized metal (Fe3+) affinity chromatography. *Anal Biochem* **154**: 250–254.

32. Posewitz MC, Tempst P. (1999) Immobilized gallium(III) affinity chromatography of phosphopeptides. *Anal Chem* **71**: 2883–2892.

33. Nühse TS, Stensballe A, Jensen ON, Peck SC. (2003) Large-scale analysis of *in vivo* phosphorylated membrane proteins by immobilized metal ion affinity chromatography and mass spectrometry. *Mol Cell Proteomics* **2**: 1234–1243.

34. Neville DC, Rozanas CR, Price EM, Gruis DB, Verkman AS, Townsend RR. (1997) Evidence for phosphorylation of serine 753 in CFTR using a novel metal-ion affinity resin and matrix-assisted laser desorption mass spectrometry. *Protein Sci* **6**: 2436–2445.

35. Kokubu M, Ishihama Y, Sato T, Nagasu T, Oda Y. (2005) Specificity of immobilized metal affinity-based IMAC/C18 tip enrichment of phosphopeptides for protein phosphorylation analysis. *Anal Chem* **77**: 5144–5154.

36. Tsai CF, Wang YT, Chen YR, Lai CY, Lin PY, Pan KT *et al.* (2008) Immobilized metal affinity chromatography revisited: pH/acid control toward high selectivity in phosphoproteomics. *J Proteome Res* **7**: 4058–4069.

37. Gruhler A, Olsen JV, Mohammed S, Mortensen P, Faergeman NJ, Mann M *et al.* (2005) Quantitative phosphoproteomics applied to the yeast pheromone signaling pathway. *Mol Cell Proteomics* **4**: 310–327.

38. McNulty DE, Annan RS. (2008) Hydrophilic interaction chromatography reduces the complexity of the phosphoproteome and improves global phosphopeptide isolation and detection. *Mol Cell Proteomics* **7**: 971–980.

39. Gan CS, Guo T, Zhang H, Lim SK, Sze SK. (2008) A comparative study of electrostatic repulsion-hydrophilic interaction chromatography (ERLIC) versus SCX–IMAC-based methods for phosphopeptide isolation/enrichment. *J Proteome Res* **7**: 4869–4877.

40. Larsen MR, Thingholm TE, Jensen ON, Roepstorff P, Jorgensen TJ. (2005) Highly selective enrichment of phosphorylated peptides from peptide mixtures using titanium dioxide microcolumns. *Mol Cell Proteomics* **4**: 873–886.

41. Hoffmann MR, Martin ST, Choi W, Bahnemann DW. (1995) Environmental applications of semiconductor photocatalysis. *Chem Rev* **95**: 69–96.

42. Kweon HK, Hakansson K. (2006) Selective zirconium dioxide-based enrichment of phosphorylated peptides for mass spectrometric analysis. *Anal Chem* **78**: 1743–1749.

43. Wang Y, Chen W, Wu J, Guo Y, Xia X. (2007) Highly efficient and selective enrichment of phosphopeptides using porous anodic alumina membrane for MALDI-TOF MS analysis. *J Am Soc Mass Spectrom* **18**: 1387–1395.

44. Ficarro SB, Parikh JR, Blank NC, Marto JA. (2008) Niobium(V) oxide (Nb2O5): Application to phosphoproteomics. *Anal Chem* **80**: 4606–4613.

45. Ikeguchi YN, H. (1997) Determination of organic phosphates by column-switching high performance anion-exchange chromatography using on-line preconcentration on titania. *Anal Sci* **13**: 479–483.

46. Sano AN, H. (2004) Chemo-affinity of titania for the column-switching HPLC analysis of phosphopeptides. *Anal Sci* **20**: 565–566.

47. Sugiyama N, Masuda T, Shinoda K, Nakamura A, Tomita M, Ishihama Y. (2007) Phosphopeptide enrichment by aliphatic hydroxy acid-modified metal oxide chromatography for nano-LC-MS/MS in proteomics applications. *Mol Cell Proteomics* **6**: 1103–1109.

48. Kyono Y, Sugiyama N, Imami K, Tomita M, Ishihama Y. (2008) Successive and selective release of phosphorylated peptides captured by hydroxy acid-modified metal oxide chromatography. *J Proteome Res* **7**: 4585–4593.

49. Thingholm TE, Jensen ON, Robinson PJ, Larsen MR. (2008) SIMAC (sequential elution from IMAC), a phosphoproteomics strategy for the rapid separation of monophosphorylated from multiply phosphorylated peptides. *Mol Cell Proteomics* **7**: 661–671.

50. Lu AH, Salabas E, Schüth F. (2007) Magnetic nanoparticles: Synthesis, protection, functionalization, and application. *Angew Chem Int Ed* **46**: 1222–1244.

51. Chen CT, Chen YC. (2005) Fe3O4/TiO2 core/shell nanoparticles as affinity probes for the analysis of phosphopeptides using TiO2 surface-assisted laser desorption/ionization mass spectrometry. *Anal Chem* **77**: 5912–5919.

52. Lo CY, Chen WY, Chen CT, Chen YC. (2007) Rapid enrichment of phosphopeptides from tryptic digests of proteins using iron oxide nanocomposites of magnetic particles coated with zirconia as the concentrating probes. *J Proteome Res* **6**: 887–893.

53. Chen CT, Chen WY, Tsai PJ, Chien KY, Yu JS, Chen YC. (2007) Rapid enrichment of phosphopeptides and phosphoproteins from complex samples using magnetic particles coated with alumina as the concentrating probes for MALDI MS analysis. *J Proteome Res* **6**: 316–325.

54. Li Y, Xu X, QI D, Deng C, Yang P, Zhang X. (2008) Novel Fe3O4@TiO2 core-shell microspheres for selective enrichment of phosphopeptides in phosphoproteome analysis. *J Proteome Res* **7**: 2526–2538.

55. Li Y, Leng T, Lin H, Deng C, Xu X, Yao N *et al.* (2007) Preparation of Fe3O4@ZrO2 core-shell microspheres as affinity probes for selective enrichment and direct determination of phosphopeptides using matrix-assisted laser desorption ionization mass spectrometry. *J Proteome Res* **6**: 4498–4510.

56. Li Y, Liu Y, Tang J, Lin H, Yao N, Shen X *et al.* (2007) Fe3O4@Al2O3 magnetic core-shell microspheres for rapid and highly specific capture of phosphopeptides with mass spectrometry analysis. *J Chromatogr A* **1172**: 57–71.

57. Li Y, Lin H, Deng C, Yang P, Zhang X. (2008) Highly selective and rapid enrichment of phosphorylated peptides using gallium oxide-coated magnetic microspheres for MALDI-TOF-MS and nano-LC-ESI-MS/MS/MS analysis. *Proteomics* **8**: 238–249.

58. Li YC, Lin YS, Tsai PJ, Chen CT, Chen WY, Chen YC. (2007) Nitrilotri-acetic acid-coated magnetic nanoparticles as affinity probes for enrichment of histidine-tagged proteins and phosphorylated peptides. *Anal Chem* **79**: 7519–7525.

59. Zhao L, Wu R, Han G, Zhou H, Ren L, Tian R *et al.* (2008) The highly selective capture of phosphopeptides by zirconium phosphonate-modified magnetic nanoparticles for phosphoproteome analysis. *J Am Soc Mass Spectrom* **19**: 1176–1186.

60. Wei J, Zhang Y, Wang J, Tan F, Liu J, Cai Y *et al.* (2008) Highly efficient enrichment of phosphopeptides by magnetic nanoparticles coated with zirconium phosphonate for phosphoproteome analysis. *Rapid Commun Mass Spectrom* **22**: 1069–1080.

61. Wu HT, Hsu CC, Tsai CF, Lin PC, Lin CC, Chen YJ. (2011) Nanoprobe-based immobilized metal affinity chromatography for sensitive and comple-mentary enrichment of multiply phosphorylated peptides. *Proteomics* (DOI 10.1002/pmic.201000768)

62. Chien YY, Jan MD, Adak AK, Tzeng HC, Lin YP, Chen YJ *et al.* (2008) Globotriose-functionalized gold nanoparticles as multivalent probes for Shiga-like toxin. *Chembiochem* **9**: 1100–1109.

63. Ross PL, Huang YN, Marchese JN, Williamson B, Parker K, Hattan S *et al.* (2004) Multiplexed protein quantitation in *Saccharomyces cerevisiae* using amine-reactive isobaric tagging reagents. *Mol Cell Proteomics* **3**: 1154–1169.

64. Chen Y, Choong LY, Lin Q, Philp R, Wong CH, Ang BK *et al.* (2007) Differential expression of novel tyrosine kinase substrates during breast cancer development. *Mol Cell Proteomics* **6**: 2072–2087.

65. Wang Y, Ding SJ, Wang W, Jacobs JM, Qian WJ, Moore RJ *et al.* (2007) Profiling signaling polarity in chemotactic cells. *Proc Natl Acad Sci USA* **104**: 8328–8333.

66. Gerber SA, Rush J, Stemman O, Kirschner MW, Gygi SP. (2003) Absolute quantification of proteins and phosphoproteins from cell lysates by tandem MS. *Proc Natl Acad Sci USA* **100**: 6940–6945.

67. Larive RM, Urbach S, Poncet J, Jouin P, Mascre G, Sahuquet A *et al.* (2009) Phosphoproteomic analysis of Syk kinase signaling in human cancer cells reveals its role in cell–cell adhesion. *Oncogene* **28**: 2337–2347.

68. Pan C, Olsen JV, Daub H, Mann M. (2009) Global effects of kinase inhibitors on signaling networks revealed by quantitative phosphoproteomics. *Mol Cell Proteomics* **8**: 2796–2808.

69. Kruger M, Moser M, Ussar S, Thievessen I, Luber CA, Forner F *et al.* (2008) SILAC mouse for quantitative proteomics uncovers kindlin-3 as an essential factor for red blood cell function. *Cell* **134**: 353–364.

70. Geiger T, Cox J, Ostasiewicz P, Wisniewski JR, Mann M. (2010) Super-silac mix for quantitative proteomics of human tumor tissue. *Nat Methods* **7**: 383–385.

71. Chelius D, Bondarenko PV. (2002) Quantitative profiling of proteins in complex mixtures using liquid chromatography and mass spectrometry. *J Proteome Res* **1**: 317–323.

72. Bondarenko PV, Chelius D, Shaler TA. (2002) Identification and relative quantitation of protein mixtures by enzymatic digestion followed by capillary reversed-phase liquid chromatography-tandem mass spectrometry. *Anal Chem* **74**: 4741–4749.

73. Cutillas PR, Geering B, Waterfield MD, Vanhaesebroeck B. (2005) Quantification of gel-separated proteins and their phosphorylation sites by LC-MS using unlabeled internal standards: Analysis of phosphoprotein dynamics in a B cell lymphoma cell line. *Mol Cell Proteomics* **4**: 1038–1051.

74. Old WM, Meyer-Arendt K, Aveline-Wolf L, Pierce KG, Mendoza A, Sevinsky JR *et al.* (2005) Comparison of label-free methods for quantifying human proteins by shotgun proteomics. *Mol Cell Proteomics* **4**: 1487–1502.

75. Fang R, Elias DA, Monroe ME, Shen Y, McIntosh M, Wang P *et al.* (2006) Differential label-free quantitative proteomic analysis of Shewanella oneidensis cultured under aerobic and suboxic conditions by accurate mass and time tag approach. *Mol Cell Proteomics* **5**: 714–725.

76. Ono M, Shitashige M, Honda K, Isobe T, Kuwabara H, Matsuzuki H *et al.* (2006) Label-free quantitative proteomics using large peptide data sets generated by nanoflow liquid chromatography and mass spectrometry. *Mol Cell Proteomics* **5**: 1338–1347.

77. Jaffe JD, Mani DR, Leptos KC, Church GM, Gillette MA, Carr SA. (2006) PEPPeR, a platform for experimental proteomic pattern recognition. *Mol Cell Proteomics* **5**: 1927–1941.

78. Le Bihan T, Goh T, Stewart II, Salter AM, Bukhman YV, Dharsee M *et al.* (2006) Differential analysis of membrane proteins in mouse fore- and hindbrain using a label-free approach. *J Proteome Res* **5**: 2701–2710.

79. Cutillas PR, Vanhaesebroeck B. (2007) Quantitative profile of five murine core proteomes using label-free functional proteomics. *Mol Cell Proteomics* **6**: 1560–1573.

80. Lu YT, Han CL, Wu CL, Yu TM, Chien CW, Liu CL *et al.* (2008) Proteomic profiles of bronchoalveolar lavage fluid from patients with ventilator-associated pneumonia by gel-assisted digestion and 2-D-LC/MS/MS. *Proteomics Clin Appl* **2**: 1208–1222.

81. Quintana LF, Campistol JM, Alcolea MP, Banon-Maneus E, Sol-Gonzalez A, Cutillas PR. (2009) Application of label-free quantitative peptidomics for the identification of urinary biomarkers of kidney chronic allograft dysfunction. *Mol Cell Proteomics* **8**: 1658–1673.

82. Rikova K, Guo A, Zeng Q, Possemato A, Yu J, Haack H *et al.* (2007) Global survey of phosphotyrosine signaling identifies oncogenic kinases in lung cancer. *Cell* **131**: 1190–1203.

83. Wang YT, Hong TC, Tsou CC, Lin PY, Pan SH, Hong TM, Yang PC, Sung TY, Hsu WL, Chen YJ. (2010) An informatics-assisted label-free quantitation strategy that depicts phosphoproteomic profiles in lung cancer cell invasion. *J Proteome Res* **9**: 5582–5597.

84. Cohen P. (2002) The origins of protein phosphorylation. *Nat Cell Biol* **4**: 127–130.

85. Ashman K, Villar EL. (2009) Phosphoproteomics and cancer research. *Clin Transl Oncol* **11**: 356–362.

86. Diella F, Cameron S, Gemund C, Linding R, Via A, Kuster B *et al.* (2004) Phospho.ELM: A database of experimentally verified phosphorylation sites in eukaryotic proteins. *BMC Bioinformatics* **5**: 79.

87. Hornbeck PV, Chabra I, Kornhauser JM, Skrzypek E, Zhang B. (2004) PhosphoSite: A bioinformatics resource dedicated to physiological protein phosphorylation. *Proteomics* **4**: 1551–1561.

88. Gnad F, Ren S, Cox J, Olsen JV, Macek B, Oroshi M *et al.* (2007) PHOSIDA (phosphorylation site database): Management, structural and evolutionary investigation, and prediction of phosphosites. *Genome Biol* **8**: R250.

89. Obenauer JC, Cantley LC, Yaffe MB. (2003) Scansite 2.0: Proteome-wide prediction of cell signaling interactions using short sequence motifs. *Nucleic Acids Res* **31**: 3635–3641.

90. Schwartz D, Gygi SP. (2005) An iterative statistical approach to the identification of protein phosphorylation motifs from large-scale data sets. *Nat Biotechnol* **23**: 1391–1398.

91. Chen Y, Low TY, Choong LY, Ray RS, Tan YL, Toy W *et al.* (2007) Phosphoproteomics identified endofin, DCBLD2, and KIAA0582 as novel tyrosine phosphorylation targets of EGF signaling and Iressa in human cancer cells. *Proteomics* **7**: 2384–2397. Copyright Wiley-VCH Verlag GmbH & Co. KGaA. Reproduced with permission.

92. Nilsson CL, Dillon R, Devakumar A, Shi SD, Greig M, Rogers JC *et al.* (2010) Quantitative phosphoproteomic analysis of the STAT3/IL-6/HIF1alpha signaling network: An initial study in GSC11 glioblastoma stem cells. *J Proteome Res* **9**: 430–443.

93. Zhang Y, Wolf-Yadlin A, White FM. (2007) Quantitative proteomic analysis of phosphotyrosine-mediated cellular signaling networks. *Methods Mol Biol* **359**: 203–212.

94. Wolf-Yadlin A, Kumar N, Zhang Y, Hautaniemi S, Zaman M, Kim HD *et al.* (2006) Effects of HER2 overexpression on cell signaling networks governing proliferation and migration. *Mol Syst Biol* **2**: 54.

95. Xie X, Feng S, Vuong H, Liu Y, Goodison S, Lubman DM. (2010) A comparative phosphoproteomic analysis of a human tumor metastasis model using a label-free quantitative approach. *Electrophoresis* **31**: 1842–1852.

96. Old WM, Shabb JB, Houel S, Wang H, Couts KL, Yen CY *et al.* (2009) Functional proteomics identifies targets of phosphorylation by B-Raf signaling in melanoma. *Mol Cell* **34**: 115–131.

97. Chen RQ, Yang QK, Lu BW, Yi W, Cantin G, Chen YL *et al.* (2009) CDC25B mediates rapamycin-induced oncogenic responses in cancer cells. *Cancer Res* **69**: 2663–2668.

Chapter 10

Predicting MicroRNAs

Arthur Chun-Chieh Shih[*] and Tsunglin Liu[†]

1. Introduction

MicroRNAs (miRNAs) are 22-nucleotide endogenous small RNAs that regulate the expression of target genes by binding to the $3'$ un-translated regions ($3'$ UTRs) of mRNAs.[1-6] They suppress target gene expression by inhibiting the protein synthesis and/or degrading the mRNA transcripts. Unlike siRNAs, miRNAs do not normally bind to target sites with perfect sequence complementarity,[7-13] which poses a serious challenge in miRNA target prediction. A great deal of research has been devoted to predicting miRNA targets.[14-23] In this chapter, we focus on the prediction of miRNA genes, and direct the readers who are interested in miRNA target prediction to Refs. 16, 20–21, and 24–27.

It is known that many miRNAs are involved in the development, cell proliferation, apoptosis, tissue differentiation, and metabolism of a wide range of organisms.[1,28-32] Moreover, it has been reported that miRNAs play a role in several diseases, such as cardiac hypertrophy, Alzheimer's disease, cancer, autoimmune diseases, and diabetes.[33-41] For example, the famous tumor suppressor gene p53 has been found to directly upregulate miR-34, which in turns suppresses several downstream genes in the tumor suppressor network.[12,42]

Researchers estimate that miRNAs comprise 1%–5% of mammalian genes[16,29,43-44] and they may regulate a large proportion of protein-coding

[*]Institute of Information Science and Research Center for Information Technology Innovation, Academia Sinica, Taipei 115, Taiwan.
[†]Institute of Bioinformatics and Biosignal Transduction, National Cheng-Kung University, Tainan 701, Taiwan.

genes.[7,16,20,45] Currently, there are 940 human miRNAs (miRBase version 15), each of which may target a significant number of genes.[15,19,29] Initially, many miRNAs were identified by computational methods, most of which relied on evolutionary conservation as the main criterion for predicting new miRNAs.[29,46–49] However, recent studies have shown that many miRNA genes are not conserved even among closely related species.[50–52] Thus, the total number of miRNA genes in the human genome is still increasing.

Next-generation sequencing (NGS) has been used to identity novel miRNAs in recent years.[53–59] NGS technologies are capable of sequencing billions of DNA or RNA molecules per run within a week, thereby reducing the cost and sequencing time enormously. The technologies facilitate the study of genomics and transcriptomics, including sequencing small RNAs directly. We discuss the predictions of miRNAs with NGS and genomic data in the next section, and with genomic data only in Sec. 3. The last section contains some concluding remarks.

2. Predicting miRNAs with Next-Generation Sequencing and Genomic Data

NGS sequencing of small RNAs produces sequence segments (often called reads) of miRNAs, siRNAs, snoRNAs, tRNAs, and rRNAs. It has become a powerful experimental tool in miRNA studies, such as finding new miRNAs[53–59] or identifying expression changes in different conditions.[60,61] A survey of methods that use NGS data to find miRNA expression profiles can be found in Ref. 57. To predict miRNAs, it is necessary to distinguish miRNA reads from other reads. In this section, we explain how to process and analyze NGS data for predicting miRNAs.

2.1. *Small RNA sequencing by NGS*

NGS technologies are used in various platforms, such as Roche/454 FLX Pyrosequencer (http://www.454.com),[62] Illumina Genome Analyzer (formerly Solexa, http://www.illumina.com),[63] and Applied Biosystems SOLiDTM System (http://www3.appliedbiosystems.com/AB_Home/applicationstechnologies/SOLiD-System-Sequencing-B/index.htm). Since they were launched, the platforms have been utilized for massively parallel DNA/RNA read sequencing in genomics and transcriptomics (see Refs. 64–67, for more details). The data yields of the platforms are rather

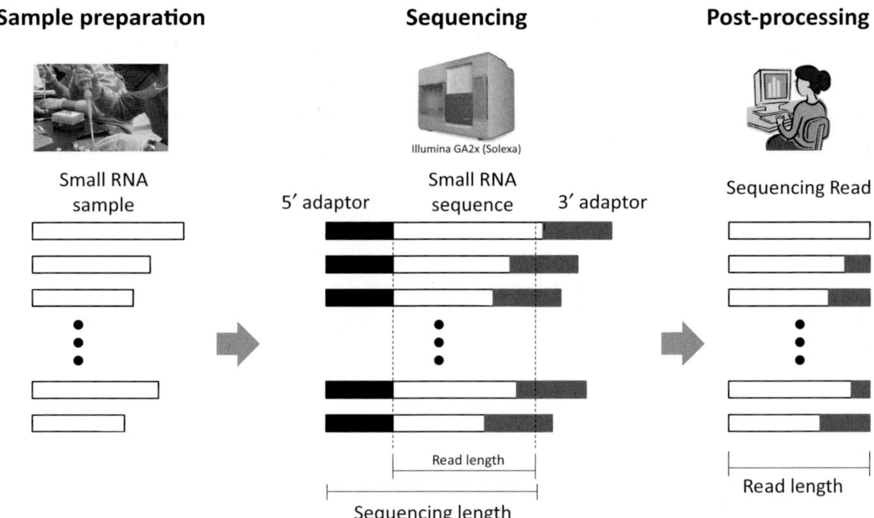

Fig. 1. Using NGS to obtain small RNA reads. In the first step, small RNA sequences (represented by the blank rectangles) are isolated and extracted from a sample by a series of experimental procedures and toolkits. Then, in the second step, the small RNAs are ligated with 5′ and 3′ RNA adaptors. The length of a ligated RNA sequence is the sum of the lengths of the 5′ and 3′ RNA adaptors plus the length of the flanked small RNA sequence. The black and gray rectangles in the middle column represent the 5′ and 3′ adaptors, respectively. After amplification by the Illumina Cluster Station and Genome Analyzer, the post-processing computing server trims the 5′ adaptor sequence of each sequencing read and outputs all of the reads with the same length. The output reads include not only small RNAs concatenated with partial 3′ adaptor sequences (the blank rectangles connected to the gray rectangles in the third column) but also other long RNA fragments (the full blank rectangles).

large. For example, the Illumina (Solexa) Genome Analyzer IIx (GA IIx) can produce about 120 million reads of short sequences (>100 bases), resulting in an output of 8–9 billion bases per run. After filtering out sequencing errors, there are still several billion high-quality bases for further analysis.

Obtaining small RNA raw reads involves three key steps: sample preparation, sequencing, and post-processing, as shown in Fig. 1. First, samples are processed by a series of experimental procedures, e.g. RNA isolation and size selection, before being sent to the sequencer. In the second step, sample RNAs of various lengths are connected to two adaptors at their 5′- and 3′-ends, and then amplified 15–20 times via PCR (Polymerase Chain Reaction) process to enhance the sequencing signals. An Illumina NGS sequencer produces fixed-length reads. If the length of the 5′ adaptor plus an RNA sequence is longer than the read, the RNA can only be sequenced partially; otherwise, the full RNA sequence can be sequenced completely. In the last step, a post-processing server removes the known $5'$ adaptor sequence and outputs the reads corresponding to either full or partial RNA molecules (the right-hand column in Fig. 1). Depending on the experimental protocol, the lengths of Illumina reads range from 35 bp to 150 bp. For miRNA studies, we are interested in reads shorter than 30 bases; thus, a protocol that produces short reads of 40 bp would be suitable.

In the post-processing step, it is also necessary to take care of the $3'$ adaptors. Since the length of all mature miRNAs is less than 25 bp, the reads derived from mature miRNAs should contain partial $3'$ adaptor sequences at the ends. The sequences must be trimmed before being mapped onto the target genome, as shown in Fig. 2. The exact sequences of $5'$ and $3'$ adaptors may not be universal over different platforms or protocols; therefore, the adaptor sequences should be recorded when receiving raw read data.

Adaptor trimming requires alignment between the reads and the adaptor sequence. Usually, the following parameters are needed to determine: (a) the allowed number of mismatches, (b) the minimum length aligned to the adaptor, and (c) the minimum length of the trimmed sequences. During alignment, one or two mismatches (but no gap) are usually regarded as tolerable sequencing errors. In addition, the alignment between a read and the $3'$ adaptor must be longer than a cutoff to avoid matching by chance. Finally, after trimming the adaptor sequences, the reads shorter and longer than the cutoff length are normally abandoned because they are less likely

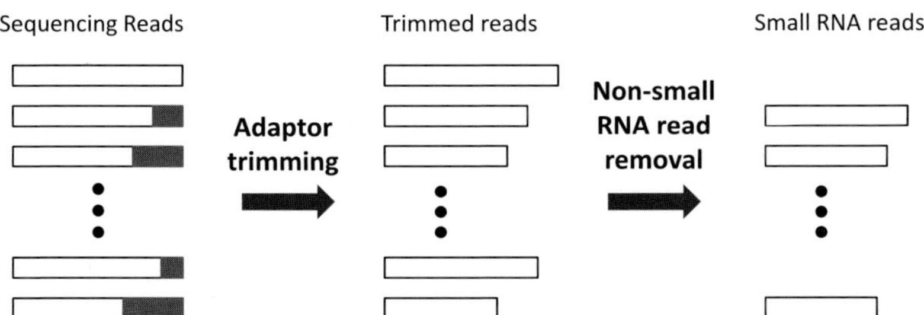

Fig. 2. The adaptor trimming procedure for original sequencing reads, which are the same length, but they may include partial $3'$ adaptor sequences (the black parts of the rectangles in the first column). After trimming the partial $3'$ adaptor sequences, the processed reads are of various lengths (the blank rectangles in the second column). Reads equal to or longer than 30 bp are removed, and the remaining reads are regarded as potential small RNA sequences.

to be from miRNAs. Since the length of most human miRNAs is 18–24 bases, trimmed reads of length 16–28 bases are normally retained. If other small RNAs, such like siRNAs and snoRNAs, are of interest, the cutoff length can be adjusted to include those reads.

2.2. *Mapping reads onto a target genome*

In addition to selection based on length, reads are mapped onto a target genome to obtain better miRNA candidates. Below, we explain how the mapping procedure and coverage profile calculation are performed. Then, we describe typical coverage profiles at the loci of miRNAs.

Traditional alignment tools like BLAST[68] and BLAT[69] are too slow to cope with the large amount of NGS data. Accordingly, many new alignment tools, e.g. SOAP,[70] MAQ,[71] Bowtie,[72] and BWA,[73,74] have been developed for mapping NGS reads. One way to further speed up the read alignment process is to cluster identical reads first. This avoids redundant mapping because only representative reads (often called unique reads, as shown in Fig. 3) are aligned. In addition to the issue of efficiency, NGS read mappings are confounded by two other factors: multiple hits and sequencing errors.

When a read aligns with more than one locus equally well, it is called a multiple hit. As it is often difficult to determine which locus of the original transcript a multiple hit belongs to, all possible locations are considered and the read is split evenly among them. In some cases, however, we may specify

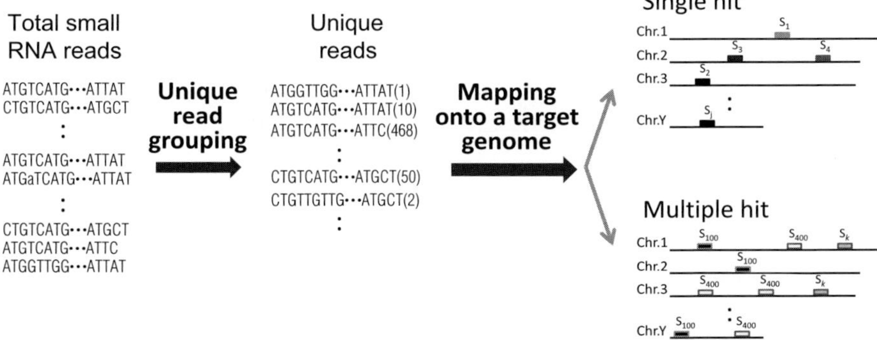

Fig. 3. Trimmed reads mapped onto a target genome. Identical reads are grouped together and represented by a unique read. The numbers in parentheses are the read counts of the unique reads. A read is called a single hit if it only maps to the target genome at one location and a multiple hit if it maps to multiple locations.

the true locus. For example, a read maps equally well to two regions, one of which also contains another single read. In this case, it can be assumed that the RNA was transcribed from that region. Expressed sequence tags (ESTs) may also be utilized to determine the true locus of a multiple hit, since a region with EST support is more likely to be transcribed than one without EST support. Although the above two processes help to narrow down the candidate loci, the actual locations should be determined and validated via experiments.

Sequencing errors also affect read mappings. Similar to trimming adaptors, it is necessary to allow a few mismatches during the mapping procedure. However, there is no benchmark for setting the number of allowed mismatches because experimental conditions vary. Fortunately, the advances in NGS technology have reduced the number of sequencing errors overall.

After mapping NGS reads onto a target genome, potential miRNA candidates are selected based on the mapping scenarios unique to miRNAs. Since small RNA samples also contain siRNAs, tRNAs, rRNAs, or snoRNAs, it is necessary to distinguish miRNAs from other RNAs. This can be done by classifying the origins of RNAs according to their read distributions (Fig. 4). As shown in Fig. 4(a), we can calculate the read distribution from x_1 to x_2 and x_3 to x_4 for each group. If the reads are derived from an miRNA gene, their distribution is usually similar to one of those shown in Figures 4(b)–4(d). However, these typical distributions are only for reference in the initial stage.

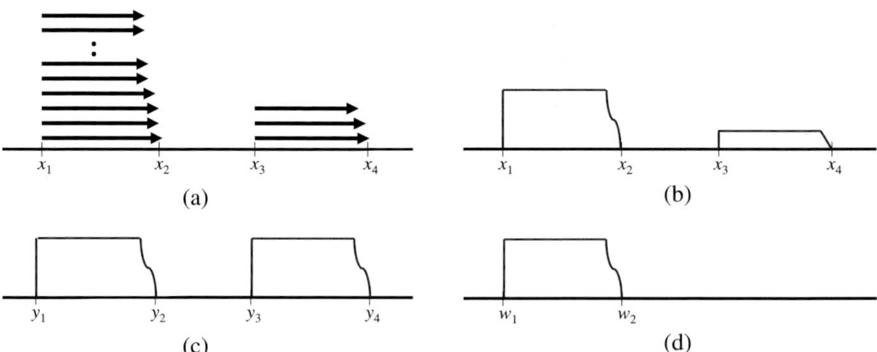

Fig. 4. Typical read distributions. (a) Each arrow represents a read. The reads are mapped onto a target genome in two neighboring regions: from x_1 to x_2 and from x_3 to x_4; (b) The coverage profile of the region between x_1 and x_4 in (a); (b)–(d) Different read coverage profiles of an miRNA stem-loop region.

Secondary structure is also a useful guide for selecting better miRNA candidates because miRNA precursors can fold into a stem-loop structure. However, folding a putative miRNA sequence is not straightforward because the range of the miRNA precursor is not clear. Thus, some rules are usually applied to determine the possible regions of miRNA genes. Based on the known animal miRNAs in the miRBase (http://www.mirbase.org/), the average length of a stem-loop region is less than 100 bp. Thus, it is possible to check whether two neighboring read distributions are close. For example, if the distance is less than 50 dp, both flanking regions can be extended by several dozens of bases so that the extended region can cover the two distributions, as shown in Fig. 5(a). Then, some computational tools, such as mfold[75] or RNAfold,[76] can be used to predict the potential secondary structure. If the predicted structure contains more than two branches or if there are many bases without pairings in the stem, the region may not be the precursor region of a miRNA gene. However, if most of the bases in the stem region have good pairings, even if there is a small bulge, the region should be checked further to determine whether most of the reads

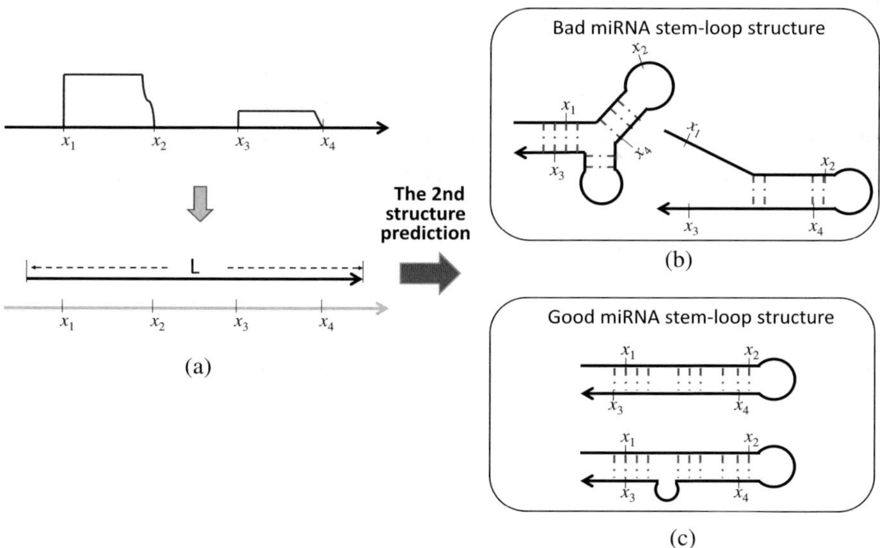

Fig. 5. A predicted miRNA stem-loop region and potential secondary structures; (a) the potential stem-loop sequence of a region in which two read groups are close; (b) the predicted secondary structures with a bad stem-loop structure or poor base-pairings; (c) the predicted secondary structures with a good stem-loop structure and base-pairings in the stem region.

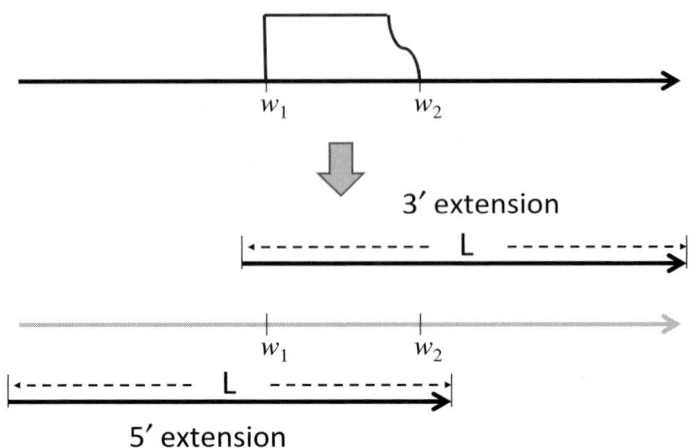

Fig. 6. The flanking region extensions of an isolated read group. In this example, an isolated read group is located between w_1 and w_2 and there are no other groups nearby. There are two possible extensions for this case: a $3'$ extension and a $5'$ extension. In the $3'$ extension, the stem-loop sequence is concatenated by a short flanking region on the left, the read group region, and a long region on the right. By contrast, in the $5'$ extension, the stem-loop sequence is concatenated by a long flanking region on the left, the read group region, and a short downstream region on the right.

locate in the stem region. If they do, the region is probably a miRNA precursor.

The prediction of miRNA secondary structures is more difficult when only one arm of the stem region appears as miRNAs in the sample. In this case, we do not know which part of the stem the read group located in before folding. Thus, we must consider the possibility that there may be two stem-loop region sequences: a $5'$ extension and a $3'$ extension, as shown in Fig. 6. Usually, the total extension length is 100–150 bp. After obtaining the extended sequences, the same approach can be used to predict the secondary structure, check if the predicted result is a good stem-loop structure, and then decide whether the reads were derived from a miRNA gene.

3. Predicting miRNAs with Genomic Sequences Alone

Predicting miRNAs with genomic sequences alone usually requires information about two key components: secondary structures and sequence conservation. Scanning a genome for all possible regions that fold into good stem-loop structures yields nearly all miRNAs, including novel ones. However, the false-positive rate is usually very high because there are many non-miRNA regions, e.g. tRNAs, rRNAs, and repeats, which also form good stem-loop structures. To reduce the number of false-positives, various filters must be applied.

The false-positive rate is much lower if miRNA candidates are identified as the homologs of known miRNAs. However, the homology-based approach cannot find novel miRNAs. Moreover, there is a risk of redundant searches of highly similar miRNA sequences, which usually belong to the same family.

The above structure-based and homology-based approaches complement each other; so they are often applied together to improve the accuracy of miRNA prediction. However, certain problems arise, depending on which approach is applied first. In the following subsections, we describe the procedure when the structure-based search or the homology-based search is applied first. We also discuss other methods, such as a reverse approach,[77] which have been used to predict novel miRNAs.

3.1. *Applying structure-based search first*

Figure 7 shows the flowchart of the approach. First, the target genomic sequences are downloaded from NCBI (http://www.ncbi.nlm.nih. gov/), UCSC Genome Browser (http://genome.ucsc.edu/), or Ensembl (http://www.ensembl.org/index.html). Then, RepeatMasker (http://www. repeatmasker.org/) is applied to screen DNA sequences for interspersed repeat and low-complexity regions. Although good stem-loop structures can be found in these regions, they are often filtered out because of the high false-positive rate. Repeats constitute over 40% of the human genome, but only a few of the known miRNAs are located in these regions. Using

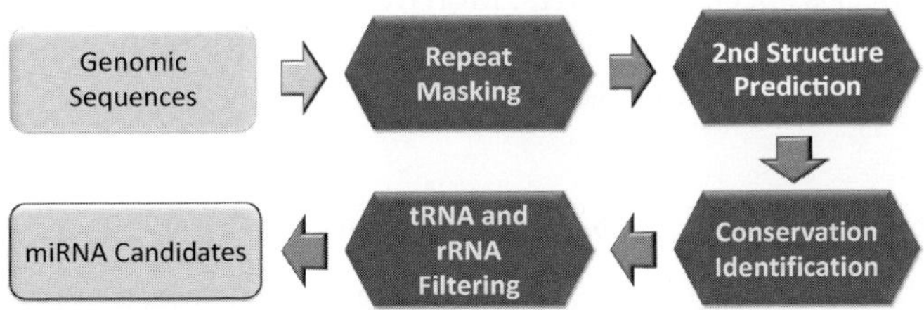

Fig. 7. Flowchart of the structure-based search approach, where the hexagons represent the four steps of the procedure: (1) masking repeat regions in the input genomic sequences; (2) predicting the secondary structures and checking whether they can fold into good stem-loop structures; (3) identifying conservation among species, and removing the regions containing tRNAs, rRNAs, and snoRNAs; and (4) reporting the predicted miRNA candidates.

a fixed-size sliding window, the masked genomic sequences are scanned for searching good stem-loop structures. The window size is usually set as 100–150 bp to match the length of miRNA precursors.

In a mammalian genome, a great many good stem-loop structures still exist in the nonrepeat regions. To reduce the number of candidates, it is necessary to examine conservation across species. The rationale behind this filtering step is that functional regions are more conserved than nonfunctional ones because of natural selection. In the UCSC Genome Browser (http://genome.ucsc.edu/), many sequence alignments between different vertebrate genomes are available. The alignment data can be used to check whether the candidate miRNAs are conserved among mammalians, vertebrates, or only in primate genomes.

Good stem-loop structures also occur in nonrepeat and conserved regions, e.g. regions containing tRNAs, rRNAs, and snoRNAs. To reduce the computation time, those regions can be filtered out based on the annotation tables of the target genome. The remaining stem-loop structures are considered good miRNA candidates.

To identify conserved functional RNAs in the human genome, Pedersen *et al.* (2006) used phylogenetic stochastic context-free grammars and an eight-way genome-wide alignment of the human, chimpanzee, mouse, rat, dog, chicken, zebra-fish, and puffer-fish genomes.[78] They found 48,479 candidates but estimated only 18,500 partially correct fRNAs (functional RNA structures, such as snoRNAs, miRNAs, splicing factors, and riboswitches). The fRNAs included 195 known miRNAs (in the miRNA Registry version 5.1), and 169 newly predicted miRNAs. The sensitivity of this approach seems to be quite high; however, many miRNAs are not cross-species conserved.[11,52] Therefore, their approach is not suitable for detecting species specific miRNAs. Moreover, the mature sequences of some miRNAs are not located in the stem regions of the precursors,[79] so the approach may miss such miRNAs.

3.2. *Applying homology-based search first*

Figure 8 shows the flowchart of the homology-based search approach which usually starts by collecting known miRNA mature sequences. In theory, the expression levels of the miRNAs that occur in both strands of the stem region in the stem-loop structure should be equal. However, in practice, the expression level of the miRNA from one strand is usually much higher than that from the other. In the miRBase, the former is called a mature sequence

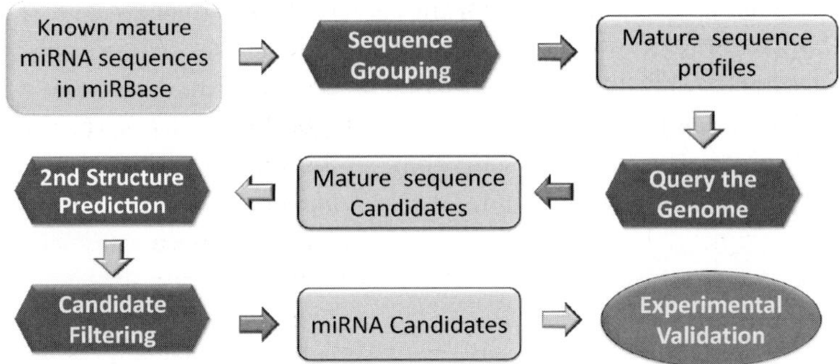

Fig. 8. Flowchart of the homology-based search approach, where the rectangles represent the input data and output results and the hexagons represent the processing steps. The input data are known miRNA mature sequences in miRBase. The steps of the process are as follows: (1) Mature sequences are classified and similar sequences are grouped together and collapsed into mature sequence profiles; (2) The profiles are used to query a target genome. The query results are called mature sequence candidates; (3) The candidates are subjected to secondary structure prediction; (4) The folded structures are checked to determine whether they are good stem-loop structures. Then, the predicted candidates are validated via experiments.

and the latter a star or minor sequence. Conservation analyses show that mature sequences are usually much more conserved among species than minor sequences. Thus, in most cases, only mature sequences are used as queries to search target genomes.

When collecting mature miRNAs, it is more efficient to select representative miRNAs because, in miRBase, many miRNA records classified as members of the same family have high sequence similarity. Thus, a miRNA representative can be one of the family members or the profile derived by merging the mature sequences in the same family.

During homology searches, it is advantageous to apply the sequence profiles of related miRNAs because the information of the whole family is available. There are various methods for generating profiles. For example, one method calculates the pairwise alignment scores and then uses a clustering algorithm, such as hierarchical clustering, to group similar sequences and generate the sequence motifs as a search basis. Currently, there is no gold benchmark method for generating miRNA sequence profiles. Thus, it is necessary to try some combinations of tools and parameter settings to optimize the performance.

Homology searches often output many miRNA candidates, from which better candidates can be selected by using structural information. Depending on the loci of miRNA candidates in the genome, we can extract their potential stem-loop regions from the genomic sequences and then use RNA folding tools to predict their secondary structures. After applying the filtering criteria (detailed in the previous section) we can determine whether the predicted candidates are potential miRNA candidates. If the stem-loop regions of some candidates overlap, we select the one whose predicted secondary structure has the minimal free energy as the final candidate.

3.3. *A reversed approach*

In contrast to the above two approaches, Chang *et al.* (2008) proposed a reverse approach to predict miRNAs by using tissue-selective frequent motifs. The rationale behind the method is that genes targeted by the same miRNA should share highly similar sequences in their 3′ UTRs (i.e. target sites). The method first identifies putative miRNA target sites in the 3′ UTRs of all genes. Novel miRNAs are then inferred for each resulting group. A six-step computational method is implemented to predict human tissue-selective binding motifs in 3′ UTRs and their potential regulatory miRNAs, as shown in Fig. 9.

This method starts by identifying a set of tissue-selective genes, which are the genes that express in only one or a few tissues, as mentioned in a number of works.[80−85] Then, microarray gene expression data is used to filter out highly expressed tissue-selective genes. This is necessary because a gene regulated by miRNAs usually has a low or moderate expression level rather than high one.[7−11] The remaining genes are called low-key tissue-selective genes. Among those genes, Chang *et al.* identified the motifs of seven nucleotides that appear frequently in the 3′ UTRs of the genes. They considered each motif as a potential miRNA target site and its complementary sequence as a potential miRNA seed. Then, they excluded the seeds that matched any known miRNA seeds and used the remaining motifs to check whether the motifs are located in the stem regions in the predicted secondary structures in the human genome that are good candidates for novel miRNAs.

Based on the proposed method, Chang *et al.* identified dozens of new human miRNAs and successfully validated some of them via experiments. Thus, it seems that the new method complements currently available methods.

3.4. *Other approaches*

A number of other methods have been proposed to improve the efficiency and effectiveness of the search process. For example, Xue *et al.* (2005)

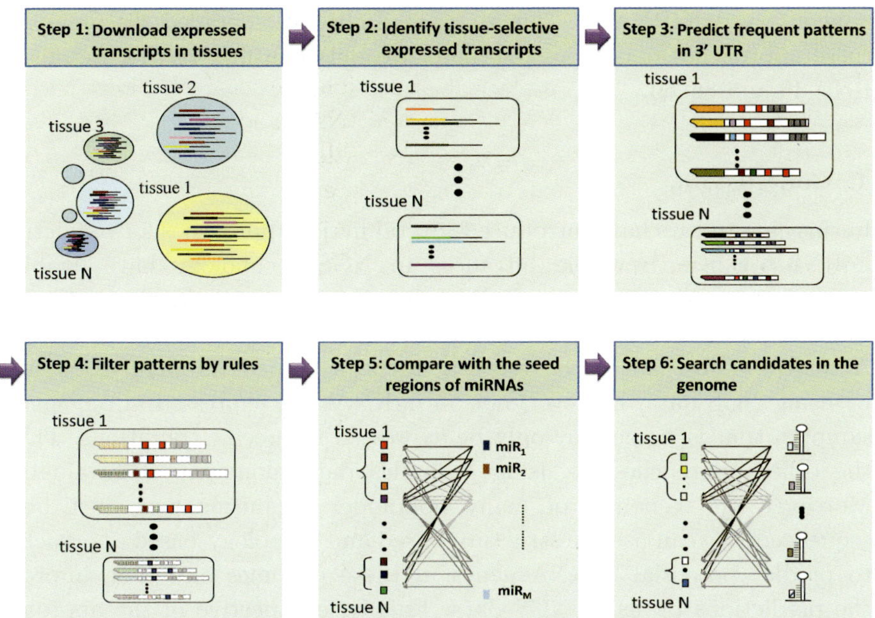

Fig. 9. The six steps for predicting miRNAs under Chang *et al.*'s reverse approach[77]: (1) The expression transcripts in different tissues (the bubbles in the upper left-hand figure) are downloaded from some public databases; (2) The tissue-selective transcripts (the lines in the rectangles) are identified as those expressed in only a few tissues; (3) A statistical approach is used to identify frequent motifs in the 3′ UTRs of the transcripts. The small blocks in the 3′ end of each transcript represent the predicted frequent motifs; (4) Some rules are applied to filter out motifs with low-confidence being miRNA binding sites. The gray blocks in the lower left-hand figure are the removed motifs; (5) The remaining motifs are compared with the seed regions of known miRNAs. The blocks in the right side of each brace and the blocks nearby the texts "miR"s represent the remaining motifs and the seed regions of known miRNAs, respectively. A line between these two data sets represents a motif that is mapped to a seed region. A motif may be connected with several miRNAs since the seed regions of these miRNAs are the same. Similarly, some miRNAs may connect with several motifs; (6) The remaining motifs without connecting any known miRNA are mapped to the stem regions of some known secondary structures in the human genome. Adapted from Ref. 77.

and Ng and Mishra (2007) used Support Vector Machine classifiers to identify possible miRNA candidates without relying on phylogenetic conservation.[86,87] Jiang *et al.* (2007) used a random forest prediction model to classify real and pseudo miRNA precursors.[88] Yousef *et al.* (2006) used a Naïve Bayes classifier and combined multispecies genomic data to identify novel and pseudo miRNAs.[49] Moreover, Nam *et al.* (2006) proposed a probabilistic method to predict clustered, nonclustered, conserved and nonconserved miRNAs.[89] Due to page limitation, detailed discussion of these approaches is not possible here. Readers can find further details in Refs. 49 and 86–89.

4. Conclusion

In this chapter, we have introduced several major approaches for predicting miRNAs. Since the rapid advances in NGS technology have enabled researchers to identify new miRNAs directly, the prediction by NGS and genomic data is the major approach to predicting miRNAs nowadays. Actually, almost all of the recently identified miRNAs were discovered by using NGS data. However, not all miRNAs are expressed in sequenced samples, that is, some may only be expressed in specific conditions. Thus, the other approaches by using genomics data alone are still valuable. Moreover, the genomes of many nonmodel organisms have just been sequenced. We can use the structure-based and homology-based approaches to predict potential miRNA genes in these genomes and then improve the predictions by using NGS data. Finally, irrespective of the approach adopted, the prediction results should be validated by experiments.

References

1. Ambros V. (2004) The functions of animal microRNAs. *Nature* **431**(7006): 350–355.
2. Bartel DP. (2004) MicroRNAs: Genomics, biogenesis, mechanism, and function. *Cell* **116**(2): 281–297.
3. Carrington JC, Ambros V. (2003) Role of microRNAs in plant and animal development. *Science* **301**(5631): 336–338.
4. Kim VN, Nam JW. (2006) Genomics of microRNA. *Trends Genet* **22**(3): 165–173.
5. Lai EC. (2003) MicroRNAs: Runts of the genome assert themselves. *Curr Biol* **13**(23): R925–R936.
6. Niwa R, Slack FJ. (2007) The evolution of animal microRNA function. *Curr Opin Genet Dev* **17**(2): 145–150.
7. Lim LP, Lau NC, Garrett-Engele P *et al.* (2005) Microarray analysis shows that some microRNAs downregulate large numbers of target mRNAs. *Nature* **433**(7027): 769–773.
8. Bagga S, Bracht J, Hunter S *et al.* (2005) Regulation by let-7 and lin-4 miRNAs results in target mRNA degradation. *Cell* **122**(4): 553–563.
9. Sood P, Krek A, Zavolan M *et al.* (2006) Cell-type-specific signatures of microRNAs on target mRNA expression. *Proc Natl Acad Sci USA* **103**(8): 2746–2751.

10. Stark A, Brennecke J, Bushati N *et al.* (2005) Animal microRNAs confer robustness to gene expression and have a significant impact on 3′UTR evolution. *Cell* **123**(6): 1133–1146.

11. Farh KK, Grimson A, Jan C *et al.* (2005) The widespread impact of mammalian microRNAs on mRNA repression and evolution. *Science* **310**(5755): 1817–1821.

12. He L, He X, Lim LP *et al.* (2007) A microRNA component of the p53 tumour suppressor network. *Nature* **447**(7148): 1130–1134.

13. Krol J, Loedige I, Filipowicz W. (2010) The widespread regulation of microRNA biogenesis, function and decay. *Nat Rev Genet* **11**(9): 597–610.

14. Rehmsmeier M, Steffen P, Hochsmann M *et al.* (2004) Fast and effective prediction of microRNA/target duplexes. *RNA* **10**(10): 1507–1517.

15. Lewis BP, Shih IH, Jones-Rhoades MW *et al.* (2003) Prediction of mammalian microRNA targets. *Cell* **115**(7): 787–798.

16. Lewis BP, Burge CB, Bartel DP. (2005) Conserved seed pairing, often flanked by adenosines, indicates that thousands of human genes are microRNA targets. *Cell* **120**(1): 15–20.

17. Kiriakidou M, Nelson PT, Kouranov A *et al.* (2004) A combined computational–experimental approach predicts human microRNA targets. *Genes Dev* **18**(10): 1165–1178.

18. Enright AJ, John B, Gaul U *et al.* (2003) MicroRNA targets in Drosophila. *Genome Biol* **5**(1): R1.

19. John B, Enright AJ, Aravin A *et al.* (2004) Human microRNA targets. *PLoS Biol* **2**(11): e363.

20. Krek A, Grun D, Poy MN *et al.* (2005) Combinatorial microRNA target predictions. *Nat Genet* **37**(5): 495–500.

21. Grun D, Wang YL, Langenberger D *et al.* (2005) MicroRNA target predictions across seven Drosophila species and comparison to mammalian targets. *PLoS Comput Biol* **1**(1): e13.

22. Cora D, Di Cunto F, Caselle M *et al.* (2007) Identification of candidate regulatory sequences in mammalian 3′ UTRs by statistical analysis of oligonucleotide distributions. *BMC Bioinformatics* **8**: 174.

23. Stark A, Brennecke J, Russell RB *et al.* (2003) Identification of Drosophila microRNA targets. *PLoS Biol* **1**(3): E60.

24. Barbato C, Arisi I, Frizzo ME *et al.* (2009) Computational challenges in miRNA target predictions: To be or not to be a true target? *J Biomed Biotechnol* **2009**: 803069.

25. Bentwich I. (2005) Prediction and validation of microRNAs and their targets. *FEBS Lett* **579**(26): 5904–5910.

26. Rajewsky N. (2006) MicroRNA target predictions in animals. *Nat Genet* **38**(Suppl): S8–S13.

27. Bartel DP. (2009) MicroRNAs: Target recognition and regulatory functions. *Cell* **136**(2): 215–233.

28. Johnston WK, Unrau PJ, Lawrence MS *et al.* (2001) RNA-catalyzed RNA polymerization: Accurate and general RNA-templated primer extension. *Science* **292**(5520): 1319–1325.

29. Lim LP, Glasner ME, Yekta S *et al.* (2003) Vertebrate microRNA genes. *Science* **299**(5612): 1540.
30. Berezikov E, Cuppen E, Plasterk RH. (2006) Approaches to microRNA discovery. *Nat Genet* **38**(Suppl): S2–S7.
31. Valencia-Sanchez MA, Liu J, Hannon GJ *et al.* (2006) Control of translation and mRNA degradation by miRNAs and siRNAs. *Genes Dev* **20**(5): 515–524.
32. Plasterk RH. (2006) Micro RNAs in animal development. *Cell* **124**(5): 877–881.
33. Hudder A, Novak RF. (2008) MiRNAs: Effectors of environmental influences on gene expression and disease. *Toxicol Sci* **103**(2): 228–240.
34. Zhang C. (2008) MicroRNomics: A newly emerging approach for disease biology. *Physiol Genomics* **33**(2): 139–147.
35. Chang TC, Mendell JT. (2007) MicroRNAs in vertebrate physiology and human disease. *Ann Rev Genomics Hum Genet* **8**: 215–239.
36. Li M, Marin-Muller C, Bharadwaj U *et al.* (2008) MicroRNAs: Control and loss of control in human physiology and disease. *World J Surg* **33**(4): 667–684.
37. Kloosterman WP, Plasterk RH. (2006) The diverse functions of microRNAs in animal development and disease. *Dev Cell* **11**(4): 441–450.
38. Bauersachs J, Thum T. (2007) MicroRNAs in the broken heart. *Eur J Clin Invest* **37**(11): 829–833.
39. Scalbert E, Bril A. (2008) Implication of microRNAs in the cardiovascular system. *Curr Opin Pharmacol* **8**(2): 181–188.
40. Latronico MV, Catalucci D, Condorelli G. (2007) Emerging role of microR-NAs in cardiovascular biology. *Circ Res* **101**(12): 1225–1236.
41. van Rooij E, Olson EN. (2007) MicroRNAs: Powerful new regulators of heart disease and provocative therapeutic targets. *J Clin Invest* **117**(9): 2369–2376.
42. He L, He X, Lowe SW *et al.* (2007) MicroRNAs join the p53 network — another piece in the tumour-suppression puzzle. *Nat Rev Cancer* **7**(11): 819–822.
43. Bentwich I, Avniel A, Karov Y *et al.* (2005) Identification of hundreds of conserved and nonconserved human microRNAs. *Nat Genet* **37**(7): 766–770.
44. Hammond SM. (2006) RNAi, microRNAs, and human disease. *Cancer Chemother Pharmacol* **58**(Suppl 1): s63–s68.
45. Brennecke J, Stark A, Russell RB *et al.* (2005) Principles of microRNA-target recognition. *PLoS Biol* **3**(3): e85.
46. Altuvia Y, Landgraf P, Lithwick G *et al.* (2005) Clustering and conservation patterns of human microRNAs. *Nucleic Acids Res* **33**(8): 2697–2706.
47. Berezikov E, Guryev V, van de Belt J *et al.* (2005) Phylogenetic shadowing and computational identification of human microRNA genes. *Cell* **120**(1): 21–24.
48. Wang X, Zhang J, Li F *et al.* (2005) MicroRNA identification based on sequence and structure alignment. *Bioinformatics* **21**(18): 3610–3614.
49. Yousef M, Nebozhyn M, Shatkay H *et al.* (2006) Combining multi-species genomic data for microRNA identification using a Naive Bayes classifier. *Bioinformatics* **22**(11): 1325–1334.

50. Berezikov E, Thuemmler F, van Laake LW *et al.* (2006) Diversity of microRNAs in human and chimpanzee brain. *Nat Genet* **38**(12): 1375–1377.

51. Berezikov E, van Tetering G, Verheul M *et al.* (2006) Many novel mammalian microRNA candidates identified by extensive cloning and RAKE analysis. *Genome Res* **16**(10): 1289–1298.

52. Lu J, Shen Y, Wu Q *et al.* (2008) The birth and death of microRNA genes in Drosophila. *Nat Genet* **40**(3): 351–355.

53. Stark MS, Tyagi S, Nancarrow DJ *et al.* (2010) Characterization of the Melanoma miRNAome by deep sequencing. *PLoS ONE* **5**(3): e9685.

54. Creighton CJ, Benham AL, Zhu H *et al.* (2010) Discovery of novel microRNAs in female reproductive tract using next generation sequencing. *PLoS ONE* **5**(3): e9637.

55. Wang WC, Lin FM, Chang WC *et al.* (2009) MiRExpress: Analyzing high-throughput sequencing data for profiling microRNA expression. *BMC Bioinformatics* **10**: 328.

56. Soares AR, Pereira PM, Santos B *et al.* (2009) Parallel DNA pyrosequencing unveils new zebrafish microRNAs. *BMC Genomics* **10**: 195.

57. Creighton CJ, Reid JG, Gunaratne PH. (2009) Expression profiling of microRNAs by deep sequencing. *Brief Bioinform* **10**(5): 490–497.

58. Lu YC, Smielewska M, Palakodeti D *et al.* (2009) Deep sequencing identifies new and regulated microRNAs in Schmidtea mediterranea. *RNA* **15**(8): 1483–1491.

59. Wheeler BM, Heimberg AM, Moy VN *et al.* (2009) The deep evolution of metazoan microRNAs. *Evol Dev* **11**(1): 50–68.

60. Landgraf P, Rusu M, Sheridan R *et al.* (2007) A mammalian microRNA expression atlas based on small RNA library sequencing. *Cell* **129**(7): 1401–1414.

61. Hsieh LC, Lin SI, Shih AC *et al.* (2009) Uncovering small RNA-mediated responses to phosphate deficiency in Arabidopsis by deep sequencing. *Plant Physiol* **151**(4): 2120–2132.

62. Margulies M, Egholm M, Altman WE *et al.* (2005) Genome sequencing in microfabricated high-density picolitre reactors. *Nature* **437**(7057): 376–380.

63. Bentley DR. (2006) Whole-genome re-sequencing. *Curr Opin Genet Dev* **16**(6): 545–552.

64. Shendure J, Ji H. (2008) Next-generation DNA sequencing. *Nat Biotechnol* **26**(10): 1135–1145.

65. Mardis ER. (2008) Next-generation DNA sequencing methods. *Ann Rev Genomics Hum Genet* **9**: 387–402.

66. Ansorge WJ. (2009) Next-generation DNA sequencing techniques. *N Biotechnol* **25**(4): 195–203.

67. Fullwood MJ, Wei CL, Liu ET *et al.* (2009) Next-generation DNA sequencing of paired-end tags (PET) for transcriptome and genome analyses. *Genome Res* **19**(4): 521–532.

68. Altschul SF, Madden TL, Schaffer AA *et al.* (1997) Gapped BLAST and PSI-BLAST: A new generation of protein database search programs. *Nucleic Acids Res* **25**(17): 3389–3402.

69. Kent WJ. (2002) BLAT–the BLAST-like alignment tool. *Genome Res* **12**(4): 656–664.

70. Li R, Li Y, Kristiansen K *et al.* (2008) SOAP: Short oligonucleotide alignment program. *Bioinformatics* **24**(5): 713–714.

71. Li H, Ruan J, Durbin R. (2008) Mapping short DNA sequencing reads and calling variants using mapping quality scores. *Genome Res* **18**(11): 1851–1858.

72. Langmead B, Trapnell C, Pop M *et al.* (2009) Ultrafast and memory-efficient alignment of short DNA sequences to the human genome. *Genome Biol* **10**(3): R25.

73. Li H, Durbin R. (2009) Fast and accurate short read alignment with Burrows–Wheeler transform. *Bioinformatics* **25**(14): 1754–1760.

74. Li H, Durbin R. (2010) Fast and accurate long-read alignment with Burrows–Wheeler transform. *Bioinformatics* **26**(5): 589–595.

75. Zuker M. (2003) Mfold web server for nucleic acid folding and hybridization prediction. *Nucleic Acids Res* **31**(13): 3406–3415.

76. Denman RB. (1993) Using RNAFOLD to predict the activity of small catalytic RNAs. *Biotechniques* **15**(6): 1090–1095.

77. Chang YM, Juan HF, Lee TY *et al.* (2008) Prediction of human miRNAs using tissue-selective motifs in $3'$ UTRs. *Proc Natl Acad Sci USA* **105**(44): 17061–17066.

78. Pedersen JS, Bejerano G, Siepel A *et al.* (2006) Identification and classification of conserved RNA secondary structures in the human genome. *PLoS Comput Biol* **2**(4): e33.

79. Cifuentes D, Xue H, Taylor DW *et al.* (2010) A novel miRNA processing pathway independent of Dicer requires Argonaute2 catalytic activity. *Science* **328**(5986): 1694–1698.

80. Hsiao LL, Dangond F, Yoshida T *et al.* (2001) A compendium of gene expression in normal human tissues. *Physiol Genomics* **7**(2): 97–104.

81. Megy K, Audic S, Claverie JM. (2002) Heart-specific genes revealed by expressed sequence tag (EST) sampling. *Genome Biol* **3**(12): RESEARCH0074.

82. Misra J, Schmitt W, Hwang D *et al.* (2002) Interactive exploration of microarray gene expression patterns in a reduced dimensional space. *Genome Res* **12**(7): 1112–1120.

83. Saito-Hisaminato A, Katagiri T, Kakiuchi S *et al.* (2002) Genome-wide profiling of gene expression in 29 normal human tissues with a cDNA microarray. *DNA Res* **9**(2): 35–45.

84. Huminiecki L, Lloyd AT, Wolfe KH. (2003) Congruence of tissue expression profiles from Gene Expression Atlas, SAGEmap and TissueInfo databases. *BMC Genomics* **4**(1): 31.

85. Liang S, Li Y, Be X *et al.* (2006) Detecting and profiling tissue-selective genes. *Physiol Genomics* **26**(2): 158–162.

86. Xue C, Li F, He T *et al.* (2005) Classification of real and pseudo microRNA precursors using local structure-sequence features and support vector machine. *BMC Bioinformatics* **6**: 310.

87. Ng KL, Mishra SK. (2007) De novo SVM classification of precursor microRNAs from genomic pseudo hairpins using global and intrinsic folding measures. *Bioinformatics* **23**(11): 1321–1330.
88. Jiang P, Wu H, Wang W *et al.* (2007) MiPred: Classification of real and pseudo microRNA precursors using random forest prediction model with combined features. *Nucleic Acids Res* **35**(Web Server issue): W339–W344.
89. Nam JW, Kim J, Kim SK *et al.* (2006) ProMiR II: A web server for the probabilistic prediction of clustered, nonclustered, conserved and nonconserved microRNAs. *Nucleic Acids Res* **34**(Web Server issue): W455–W458.

Chapter 11

MicroRNA Research in Cancer Biology: Databases and Tools

Hsien-Da Huang[*]

1. Overview

Among the many biological processes that microRNAs profoundly impact include cell cycle control, cellular growth and differentiation, apoptosis, as well as embryonic development. As small noncoding RNAs of approximately 22 nts, microRNAs (miRNAs) regulate gene expression post-transcriptionally through either suppressing mRNA translation or inducing mRNA degradation by hybridizing to the $3'$-untranslated regions ($3'$-UTR) of mRNAs. Thousands of miRNAs have been identified in mammalian cells over the past two decades. Previous studies suggest that miRNAs regulate 30% or more of all human protein-coding genes.[1,2]

Given the complexity of miRNA biogenesis and miRNA-target interaction, computational approaches are essential for calculating an enormous amount of data involved in miRNA registry, miRNA discovery, and identification of miRNA-target interaction. Nearly, all miRNA studies require bioinformatics resources, including multiple miRNA databases and various computational tools. Especially in cancer biology, investigations of miRNA functions and related mechanisms have attracted increasing interest in the recent decade. This chapter introduces the computational resources for studying miRNA regulations in cancer biology, including databases for accumulating miRNA genes and miRNA-target interactions, analysis methods for identifying miRNA-target interactions, and integrative analysis

[*]Department of Biological Science and Technology, and Institute of Bioinformatics and Systems Biology, National Chiao Tung University, Hsinchu 300, Taiwan.

                                    H.-D. Huang

of information from different data sources for maximizing prediction specificity of miRNA-target interactions. Results discussed herein provide a valuable insight into computational analysis in future miRNA studies. Furthermore, this chapter provides a valuable reference for biologists conducting computational analysis before experimental validations in their proposed studies of miRNAs.

Among the numerous miRNA-related database systems in recent years to elucidate miRNAs and their target genes (Fig. 1), miRBase[3] is the most complete repository for miRNA annotation and nomenclature. Also, miRTarBase,[41] TarBase,[4] miRecords,[5] and miR2Disease[6] contain experimentally validated miRNA-target interactions by reviewing pertinent literature. Additionally, miRGen,[7] miRGator,[8] miRDB,[9] microRNA.org[10] and miRNAMap[11,12] provide miRNA-target interactions based on combinations of extensively adopted target prediction programs. Moreover, several computational methods and web servers have been developed for computationally identifying target genes of miRNAs, i.e. miRNA-target interactions. For instance, miRTar, miRanda,[10] TargetScan,[2] RNAhybrid,[13] PicTar,[14] and PITA.[15] The above methods are extensively adopted, followed by experimental confirmation of the candidates for miRNA-target interactions.

The following sections introduce the databases for both miRNAs and miRNA-target interactions and the tools for identifying miRNA-target interaction.

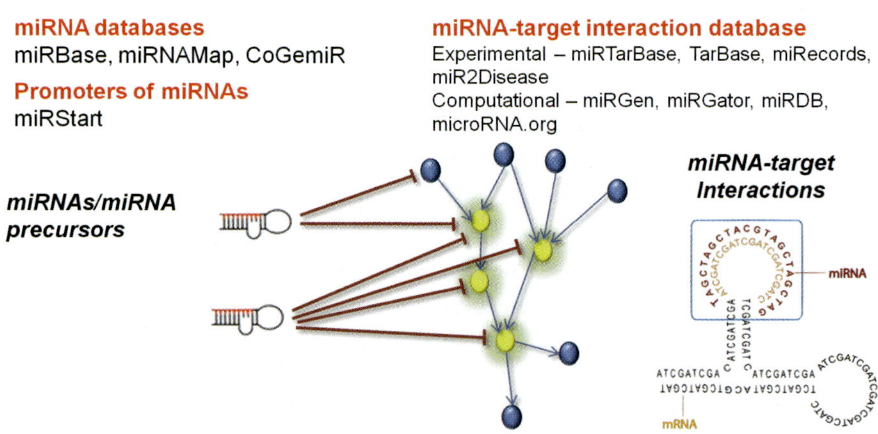

Fig. 1. Overview of miRNA resources.

2. Databases for MicroRNA Genes

Identification of lin-4 more than a decade ago ushered in the identification of thousands of miRNAs in many species, including worms, flies, plants, mice, rats, and humans. Among the many databases established for curating information of microRNA include known miRNA, putative miRNA, experimentally validated miRNA-target interactions, and putative miRNA-target interactions, as listed in Table 1.

The above miRNA databases generally provide the genomic locations of a miRNA precursor, accession number, miRNA family, their relation to protein coding genes, i.e. intergenic or intragenic miRNA, basic description, discovery based on specific methods, pertinent literature, and web links to putative miRNA-target interactions or known miRNA-target interactions. Moreover, the above resources provide user-friendly web interaction to facilitate the data accessibility of miRNA information by browsing or searching through use of keywords or miRNA accession, e.g. hsa-let-7a or hsa-mir-122.

miRBase: As the first and the most comprehensive microRNA database, miRBase contains known miRNAs along with their predicted source hairpin precursors and annotations related to their discovery, structure and

Table 1.　List of miRNA databases.

Resource names	Web links
miRNA genes	
miRBase [3]	http://microrna.sanger.ac.uk/
miRNAMap [11]	http://miRNAMap.mbc.nctu.edu.tw/
CoGemiR [16]	http://cogemir.tigem.it/
miRNA-target interactions (Experimentally validated)	
TarBase [4]	http://diana.cslab.ece.ntua.gr/tarbase/
miRecords [5]	http://mirecords.umn.edu/miRecords/
miRTarBase [41]	http://miRTarBase.mbc.nctu.edu.tw/
miR2Disease [6]	http://mirecords.biolead.org/
mirSel [17]	http://services.bio.ifi.lmu.de/mirsel/
miRNA-target interactions (Computationally predicted)	
miRGen [18]	http://www.diana.pcbi.upenn.edu/miRGen.html
miRGen 2.0 [7]	http://www.diana.pcbi.upenn.edu/miRGen.html
miRGator [8]	http://genome.ewha.ac.kr/miRGator/miRGator.html
miRDB [9]	http://mirdb.org/miRDB/
microRNA.org [10]	http://www.microrna.org/microrna/home.do
Promoters of miRNAs	
miRStart	http://mirStart.mbc.nctu.edu.tw/

functions. Also, the miRBase (version 15.0) contains 14,197 miRNA entries, with many more new sequences added continuously.

miRNAMap: As a genomic map of miRNA genes and their targets in metazoan genomes, miRNAMap accumulates known miRNAs, known miRNA targets, and putative miRNA/target relationships. Additionally, miRNA targets in 3'-UTR of genes, and known miRNA targets are identified using three computational methods, i.e. miRanda, RNAhybrid, and TargetScan. Based on three criteria, the putative miRNA target sites are filtered to retain a more likely miRNA target sites to reduce the rate of false-positive predictions of miRNA target sites. In particular, RNA accessibilities of the identified miRNA target site have been examined, providing further insight into the miRNA/target relationship.

To provide an overview of the conservation of microRNAs during evolution in various animal species, CoGemiR database gathers miRNA information on genomic location, conservation and expression data of both known and putative miRNAs.[16]

miRNA expression profiles shed light on relevant properties, including tissue specificity and differential expression in cancer(ous) and normal cells. Thus, the expression profiles of 224 human miRNAs in 18 major normal tissues in humans are monitored by conducting quantitative PCR experiments. The negative correlation between the miRNA expression profiles and the expression profiles of their target genes helps to elucidate the regulatory functions of the miRNA. Moreover, both textual and graphical web interfaces are redesigned to facilitate the retrieval of data from the miRNAMap.

Furthermore, elucidating transcriptional regulatory mechanisms of miRNAs, e.g. tumor suppressor miRNAs or oncogene miRNAs, warrants immediate attention, especially in the human genome. For this purpose, experimentally verified human microRNA promoters, i.e. transcriptional start sites (TSSs) are inadequate. Thus, the development group of miRStart (Table 1) identified human microRNA promoters by performing genome-wide analysis. That effort integrated several experimental datasets, including Cap analysis of gene expression (CAGE) tags, TSS Seq libraries,[19–22] and H3K4me3 chromatin signature derived from high-throughput analysis of gene transcription initiation[23,24] to provide direct evidence of miRNA TSSs. Moreover, a machine learning–based support vector machine (SVM) model was also developed to systematically identify high-confidence TSSs for each miRNA gene. miRStart, a web server, displays the identified transcriptional start sites of all human microRNAs. Above studies also successfully identified 940 human miRNA TSSs, including intergenic and intragenic miRNAs, by using abundant data generated from TSS-relevant experiments rather than using computational prediction methods.

3. Databases for microRNA-Target Interactions

miRNA research has received considerable emphasis recently. The accelerated rate of efforts to discover the miRNA gene necessitates clarification of the functions of these miRNAs. More than 20 databases and computational methods have been developed based on computational analysis to identify potential candidates for miRNA-target interactions (Fig. 2). Additionally, several databases that accumulate updated miRNA-target interactions (MTIs) with experimental support have been developed to investigate miRNA functions at various conditions and in different species.

Obtaining an experimentally validated miRNA-target interaction initially involves using computational methods to identify the target sites of miRNAs. Putative miRNA-target interactions are then validated based on molecular experiments, including reporter assay and western blotting analysis. Reporter assay and western blot are the conventional means of confirming the interaction between miRNA and its target mRNA. Additionally, the co-expression of predicted miRNA and mRNA target gene is often examined using Northern blot analysis, quantitative real-time PCR (qPCR), or *in situ* hybridization. In contrast with the conventional validation procedure, genome wide screenings approaches have been developed, e.g. microarray experiments with overexpression or knockdown of miRNAs, stable isotope labeling with amino acids in culture (SILAC), or pulsed

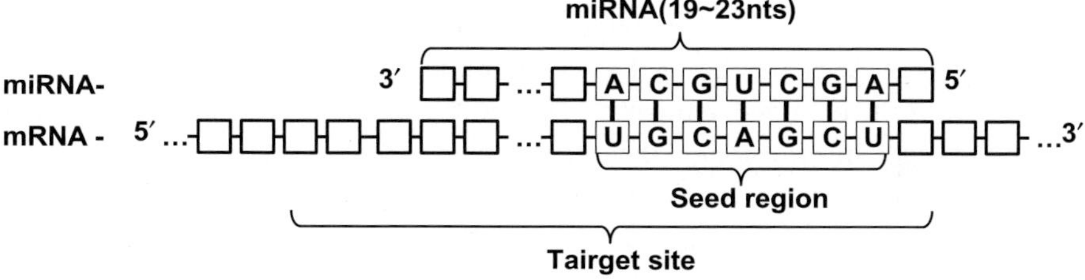

Fig. 2. miRNA-target interactions (MTIs).

SILAC (pSILAC). For instance, Selbach *et al.* determined the complement of all genes targeted by five miRNAs induced independently in HeLa cells by using microarrays and pSILAC[25]; more than 400 miRNA-target interactions were identified as well.

Over the past two years, a continuously growing number of identified miRNAs and their targets, combined with their major roles in biological systems, highlight the necessity of an accurate, up-to-date, easily accessible and centralized information repository. This subsection introduces several miRNA-target interaction databases, including miRTarBase, TarBase, miRecords, miR2Disease, and miRSel,[17] as listed in Table 1.

TarBase is the first resource that provides experimentally verified miRNA-target interactions by surveying pertinent literature. **miRecords** collects experimentally validated miRNA targets and computationally predicted miRNA targets. **miR2Disease** contains relationships among miRNAs, target genes, and human diseases. Additionally, **miRSel** systematically extracts miRNA-target relationships from PubMed abstracts by using a text mining method.

While containing the largest amount of validated MTIs, the **miRTar-Base** provides the most updated collection based on comparisons with other databases, e.g. TarBase, miRecords, and miR2Disease. Generally, the collected MTIs are experimentally validated based on reporter assay, Western blotting, or microarray experiments with the overexpression or knockdown of miRNAs.

In the release 2.4 of miRTarBase on Apr. 15, 2011, 3,969 curated miRNA-target interactions between 625 miRNAs and 2,433 target genes were accumulated from 884 articles. Especially, 2,819 human MTIs were collected between 269 miRNAs and 1,716 target genes with experimental support from 906 articles, and 513 and 1,418 interactions, which were experimentally confirmed by western blotting and reporter assay, respectively. Each human miRNA can target an average of five target genes.

miRTarBase recorded hsa-miR-122 as having 45 target genes, which were experimentally validated based on luciferase reporter assay or western blotting. As a liver-specific miRNA in humans, hsa-miR-122 is significantly downregulated in liver cancer.[26] Table 2 lists experimentally verified miRNA-target interactions collected in the miRTarBase.

Several databases have identified miRNA-target interactions based on computational methods such as miRGen, miRGator, miRDB, and microRNA.org. Notably, **miRGen** has unique functions of presenting miRNA clusters at any given inter-miRNA distance as well as providing

Table 2. Experimentally validated miRNA-target interactions of human miR-122 collected in miRTarBase.

Target genes	Descriptions	Validation methods
AACS	acetoacetyl-CoA synthetase	Reporter assay, RT-PCR
ADAM10	ADAM metallopeptidase domain 10	Reporter assay
ADAM17	ADAM metallopeptidase domain 17	Reporter assay, RT-PCR
AKT3	v-akt murine thymoma viral oncogene homolog 3 (protein kinase B, gamma)	Reporter assay, RT-PCR
ALDOA	aldolase A, fructose-bisphosphate	Northern blot, Reporter assay, RT-PCR
ANK2	ankyrin 2, neuronal	Reporter assay, RT-PCR
ANXA11	annexin A11	Reporter assay, RT-PCR
AP3M2	adaptor-related protein complex 3, mu 2 subunit	Reporter assay, RT-PCR
ATP1A2	ATPase, Na+/K+ transporting, alpha 2 polypeptide	Reporter assay, RT-PCR
BCL2L2	BCL2-like 2	Reporter assay, RT-PCR, Western blot
CCNG1	cyclin G1	Reporter assay
CYP7A1	cytochrome P450, family 7, subfamily A, polypeptide 1	qRT-PCR, Reporter assay
DSTYK	dual serine/threonine and tyrosine protein kinase	Reporter assay, RT-PCR
DUSP2	dual specificity phosphatase 2	Reporter assay, RT-PCR
ENTPD4	ectonucleoside triphosphate diphosphohydrolase 4	Reporter assay, RT-PCR
FAM117B	family with sequence similarity 117, member B	Reporter assay, RT-PCR
FOXJ3	forkhead box J3	Reporter assay, RT-PCR
FOXP1	forkhead box P1	Reporter assay, RT-PCR
FUNDC2	FUN14 domain containing 2	Reporter assay, RT-PCR
G6PC3	glucose 6 phosphatase, catalytic, 3	Reporter assay, RT-PCR
GALNT10	UDP-N-acetyl-alpha-D-galactosamine:polypeptide N-acetylgalactosaminyltransferase 10 (GalNAc-T10)	Reporter assay, RT-PCR
GTF2B	general transcription factor IIB	RT-PCR
GYS1	glycogen synthase 1 (muscle)	RT-PCR, Western blot, Northern blot, real time RT-PCR, Northern blot
IGF1R	insulin-like growth factor 1 receptor	Reporter assay
MAPK11	mitogen-activated protein kinase 11	Reporter assay, RT-PCR
MECP2	methyl CpG binding protein 2 (Rett syndrome)	Reporter assay, RT-PCR
NCAM1	neural cell adhesion molecule 1	Reporter assay, RT-PCR

(*Continued*)

mice, and rats) and fish.[32,33] By using this software program, 2,272 of 23,531 mammalian genes were identified as potential targets of 218 previously identified miRNAs, accounting for 10% of all protein-coding human genes.

As a combinatorial method, **PicTar** identifies miRNA target sites by searching for near-perfect seeds defined as a stretch of 7-nt starting at either position 1 or 2 from the 5′-end of the miRNA.[14] Importantly, **PITA** determines the accessibility of miRNA target site. Additionally, several luciferase reporter assays experiments have proven the correlation between the down-regulated effects and target site accessibility.[15] As a modification of RNA secondary structure prediction algorithm,[34] **RNAhybrid** calculates all possible MFE hybridizations between miRNA and a target gene by adopting the dynamic programming scheme. **DIANA-microT** identifies the putative miRNA target sites by using the dynamic programming scheme to calculate MFE with a sliding window of 38 nucleotides.[29]

In summary, most methods for identifying miRNA-target interactions have been developed by incorporating similar biological features. Many review articles briefly describe how various prediction methods differ from each other, as well as their prediction accuracies. Yue *et al.* thoroughly compared the features and the predictive performance of these methods.[35]

Given the lack of a method that can thoroughly elucidate all targets, several methods should be combined to identify the miRNA targets for experimental validation. A conventionally adopted strategy selects targets commonly identified by different methods, including TargetScan, miRanda, PicTar, and PITA. Theoretically, although algorithms must be compared based on metrics for sensitivity and selectivity, the limited number of known miRNA targets, i.e. true positives, and the even more limited number of miRNAs known to not interact with a target gene, i.e. true negatives, prevent a thorough evaluation. Ultimately, the predictions of various algorithms must be validated based on conventionally adopted experimental methods.

Table 4 provides an example of miRNA-target interactions of human hsa-mir-122 identified by miRanda, TargetScan, and RNAhybrid.

5. Integrative Analysis for Identifying More Effective Candidates of miRNA-Target Interactions

This section introduces an integrative analysis to identify miRNA-target interactions. According to Fig. 3, an integrated web server, **miRTar**, adopts six analysis scenarios for identifying miRNA target sites within the gene

Table 4. Example of computationally predicted miRNA-target interactions of human hsa-miR-122.

Target genes	Descriptions	Highest score
DLGAP4	Disks large-associated protein 4 (DAP-4) (SAP90/PSD-95-associated protein 4) (SAPAP4) (PSD-95/SAP90-binding protein 4).	184
CTNND2	Catenin delta-2 (Delta-catenin) (Neural plakophilin-related ARM-repeat protein) (NPRAP) (Neurojungin) (GT24).	182
NUMBL	Numb-like protein (Numb-R).	181
ITGA11	Integrin alpha-11 precursor.	180
CALCOCO1	coiled-coil transcriptional coactivator	178
BMP1	Bone morphogenetic protein 1 precursor (EC 3.4.24.19) (BMP-1) (Procollagen C-proteinase) (PCP) (Mammalian tolloid protein) (mTld).	178
BAI2	Brain-specific angiogenesis inhibitor 2 precursor.	176
DULLARD	dullard homolog	173
DDAH2	Chloride intracellular channel protein 1 (Nuclear chloride ion channel 27) (NCC27) (Chloride channel ABP) (Regulatory nuclear chloride ion channel protein) (hRNCC).	172
HMX2	homeo box (H6 family) 2	171
GPM6B	Neuronal membrane glycoprotein M6-b (M6b).	170
CRSP7	CRSP complex subunit 7 (Cofactor required for Sp1 transcriptional activation subunit 7) (Transcriptional coactivator CRSP70) (Activator-recruited cofactor 70 kDa component) (ARC70).	169
BICD1	Protein bicaudal D homolog 1 (Bic-D 1).	169
KIF11	Kinesin-like protein KIF11 (Kinesin-related motor protein Eg5) (Kinesin-like spindle protein HKSP) (Thyroid receptor-interacting protein 5) (TRIP-5) (Kinesin-like protein 1).	168
RAB3GAP1	Rab3 GTPase-activating protein catalytic subunit (RAB3 GTPase-activating protein 130 kDa subunit) (Rab3-GAP p130) (Rab3-GAP).	168
CACNA2D2	calcium channel	
VPS53	Vacuolar protein sorting-associated protein 53 homolog.	168
RIMS1	Regulating synaptic membrane exocytosis protein 1 (Rab3-interacting molecule 1) (RIM 1).	168
CUBN	Cubilin precursor (Intrinsic factor-cobalamin receptor) (Intrinsic factor-vitamin B12 receptor) (460 kDa receptor) (Intestinal intrinsic factor receptor).	167
STMN2	Stathmin-2 (SCG10 protein) (Superior cervical ganglion-10 protein).	166

computational models developed by considering a series of effective biological features of miRNA/mRNA duplex, capable of accurately predicting miRNA-target interaction with higher sensitivity and specificity. Researchers can adjust predictive parameters of prediction tools to maximize specificity in order to reduce false positive predictions; (2) selecting miRNAs of interest at the designed experimental conditions is essential to identifying miRNA-target interactions; and (3) smaller sets of coding genes can be selected based on expression analysis of coding gene expression profiles corresponding to the conditions of miRNA expression profiling. We recommend that researchers design experimental conditions for both miRNA expression profiling and coding gene expression profiling at the beginning of miRNA studies.

6. Summary

This chapter introduced various computational resources for studying miRNA regulations in cancer biology, including databases and tools for miRNAs and miRNA-target interactions. miRBase[3] is the most comprehensive repository for miRNA annotation and nomenclature. Several databases containing experimentally validated miRNA-target interactions are also introduced based on a literature survey, where miRTarBase is the most abundant one. According to this review, miRNAMap provides a user-friendly web server for miRNA-target interactions based on combinations of extensively adopted target prediction programs. Moreover, several computational methods are introduced to identify computationally miRNA-target interactions. We recommend TargetScan, miRanda, PicTar, and PITA to researchers for their first step for identifying miRNA-target interactions. Finally, an integrated resource, miRTar, is introduced to identify miRNA-target interactions by considering additional experimental information of miRNA expression profiling and coding gene expression profiling. Some strategies are also introduced to obtain better candidates of miRNA-target interactions in order to guide readers in the direction of designing computational and bench experiments for miRNA investigations in cancer biology.

References

1. Bartel DP. (2009) MicroRNAs: Target recognition and regulatory functions. *Cell* **136**(2): 215–233.

2. Lewis BP, Burge CB, Bartel DP. (2005) Conserved seed pairing, often flanked by adenosines, indicates that thousands of human genes are microRNA targets. *Cell* **120**(1): 15–20.

3. Griffiths-Jones S *et al.* (2008) miRBase: Tools for microRNA genomics. *Nucleic Acids Res* **36**(Database issue): D154–D158.

4. Papadopoulos GL *et al.* (2009) The database of experimentally supported targets: A functional update of TarBase. *Nucleic Acids Res* **37**(Database issue): D155–D158.

5. Xiao F *et al.* (2009) miRecords: An integrated resource for microRNA-target interactions. *Nucleic Acids Res* **37**(Database issue): D105–D110.

6. Jiang Q *et al.* (2009) miR2Disease: A manually curated database for microRNA deregulation in human disease. *Nucleic Acids Res* **37**(Database issue): D98–D104.

7. Alexiou P *et al.* (2010) miRGen 2.0: A database of microRNA genomic information and regulation. *Nucleic Acids Res* **38**(Database issue): D137–D141.

8. Nam S *et al.* (2008) miRGator: An integrated system for functional annotation of microRNAs. *Nucleic Acids Res* **36**(Database issue): D159–D164.

9. Wang X. (2008) miRDB: A microRNA target prediction and functional annotation database with a wiki interface. *RNA* **14**(6): 1012–1017.

10. Betel D *et al.* (2008) The microRNA.org resource: Targets and expression. *Nucleic Acids Res* **36**(Database issue): D149–D153.

11. Hsu SD *et al.* (2008) miRNAMap 2.0: Genomic maps of microRNAs in metazoan genomes. *Nucleic Acids Res* **36**(Database issue): D165–D169.

12. Hsu PW *et al.* (2006) miRNAMap: Genomic maps of microRNA genes and their target genes in mammalian genomes. *Nucleic Acids Res* **34**(Database issue): D135–D139.

13. Kruger J, Rehmsmeier M. (2006) RNAhybrid: MicroRNA target prediction easy, fast and flexible. *Nucleic Acids Res* **34**(Web Server issue): W451–W454.

14. Krek A *et al.* (2005) Combinatorial microRNA target predictions. *Nat Genet* **37**(5): 495–500.

15. Kertesz M *et al.* (2007) The role of site accessibility in microRNA target recognition. *Nat Genet* **39**(10): 1278–1284.

16. Maselli V, Di Bernardo D, Banfi S. (2008) CoGemiR: A comparative genomics microRNA database. *BMC Genomics* **9**: 457.

17. Naeem H *et al.* (2010) miRSel: Automated extraction of associations between microRNAs and genes from the biomedical literature. *BMC Bioinformatics* **11**(1): 135.

18. Megraw M *et al.* (2007) miRGen: A database for the study of animal microRNA genomic organization and function. *Nucleic Acids Res* **35** (Database issue): D149–D155.

19. Lee Y *et al.* (2004) MicroRNA genes are transcribed by RNA polymerase II. *EMBO J* **23**(20): 4051–4060.

20. Cai X, Hagedorn CH, Cullen BR. (2004) Human microRNAs are processed from capped, polyadenylated transcripts that can also function as mRNAs. *RNA* **10**(12): 1957–1966.

21. Borchert GM, Lanier W, Davidson BL. (2006) RNA polymerase III transcribes human microRNAs. *Nat Struct Mol Biol* **13**(12): 1097–1101.

22. Bortolin-Cavaille ML *et al.* (2009) C19MC microRNAs are processed from introns of large Pol-II, non-protein-coding transcripts. *Nucleic Acids Res* **37**(10): 3464–3473.

23. Marson A *et al.* (2008) Connecting microRNA genes to the core transcriptional regulatory circuitry of embryonic stem cells. *Cell* **134**(3): 521–533.

24. Corcoran DL *et al.* (2009) Features of mammalian microRNA promoters emerge from polymerase II chromatin immunoprecipitation data. *PLoS One* **4**(4): e5279.

25. Selbach M *et al.* (2008) Widespread changes in protein synthesis induced by microRNAs. *Nature* **455**(7209): 58–63.

26. Tsai WC *et al.* (2009) MicroRNA-122, a tumor suppressor microRNA that regulates intrahepatic metastasis of hepatocellular carcinoma. *Hepatology* **49**(5): 1571–1582.

27. Hammell M *et al.* (2008) mirWIP: MicroRNA target prediction based on microRNA-containing ribonucleoprotein-enriched transcripts. *Nat Methods* **5**(9): 813–819.

28. Wang X, El Naqa IM. (2008) Prediction of both conserved and nonconserved microRNA targets in animals. *Bioinformatics* **24**(3): 325–332.

29. Kiriakidou M *et al.* (2004) A combined computational-experimental approach predicts human microRNA targets. *Genes Dev* **18**(10): 1165–1178.

30. Lewis BP *et al.* (2003) Prediction of mammalian microRNA targets. *Cell* **115**(7): 787–798.

31. Grimson A *et al.* (2007) MicroRNA targeting specificity in mammals: Determinants beyond seed pairing. *Mol Cell* **27**(1): 91–105.

32. Enright AJ *et al.* (2003) MicroRNA targets in Drosophila. *Genome Biol* **5**(1): R1.

33. John B *et al.* (2004) Human MicroRNA targets. *PLoS Biol* **2**(11): e363.

34. Rehmsmeier M *et al.* (2004) Fast and effective prediction of microRNA/target duplexes. *RNA* **10**(10): 1507–1517.

35. Yue D, Liu H, Huang Y. (2009) Survey of computational algorithms for microRNA target prediction. *Curr Genomics* **10**(7): 478–492.

36. Benson DA *et al.* (2009) GenBank. *Nucleic Acids Res* **37**(Database issue): D26–D31.

37. Florea L *et al.* (1998) A computer program for aligning a cDNA sequence with a genomic DNA sequence. *Genome Res* **8**(9): 967–974.

38. Kapustin Y *et al.* (2008) Splign: Algorithms for computing spliced alignments with identification of paralogs. *Biol Direct* **3**: 20.

39. Wheelan SJ, Church DM, Ostell JM. (2001) Spidey: A tool for mRNA-to-genomic alignments. *Genome Res* **11**(11): 1952–1957.

40. Cartegni L, Chew SL, Krainer AR. (2002) Listening to silence and understanding nonsense: Exonic mutations that affect splicing. *Nat Rev Genet* **3**(4): 285–298.

41. Hsu SD, Lin FM, Wu WY, Liang C, Lee CJ, Huang CY, Tsou AP, Huang HD. (2011) miRTarBase: A database curates experimentally validated microRNA-target interactions, *Nucleic Acids Res* **39**: D163–D169.

Part III

Applications for Biomakers
and Drug Discovery

The Application of the Next-Generation Sequencing Technologies in Cancer Research

Lihua Cheng[†], Hong Xu[†] and Biaoyang Lin[*,†,‡,§]

1. Next-Generation Sequencing — A Brief Overview

In a historical perspective, the first implementation of the next-generation sequencing (NGS) is the Massively Parallel Signature Sequencing (MPSS) technology, which was invented by Sydney Brenner[1] and was developed into commercial applications by Lynx Therapeutics Inc. (acquired by Solexa Inc., and the latter by Illumina Inc.). In the last decade, a bout of technological innovations have made significant advancement in the next-generation landscape, and we have seen the implementations in many commercial products, with the most common platforms being the 454 sequencing (used in the 454 Genome Sequencers, Roche Applied Science, Basel), HiSeq 2000 (Genome Analyzer, Illumina, USA), the SOLiD platform (Applied Biosystems, Foster City, CA, USA), the HeliScope Single Molecule Sequencer technology (Helicos, Cambridge, MA, USA), and the integrated system [including DNA nanoballs, nanoarrays, and combinatorial probe-anchor ligation (cPAL[TM])] based DNA Sequencing by Complete genomics, Inc. (Mountain View, CA).

[*]Corresponding author.
[†]Systems Biology Division, Zhejiang-California International NanoSystems Institute, Zhejiang University, Hangzhou, Zhejiang 310029, China.
[‡]Swedish Neuroscience Institute, Swedish Medical Center, Seattle, WA 98122, USA.
[§]Department of Urology, University of Washington, Seattle, WA 98195, USA.

The NGS technologies will not be reviewed here, as comprehensive reviews of the NGS technologies were recently published,[2−4] and the technological descriptions are available from the companies website. New ideas and inventions are being constantly developed, and we will see continued technology advancement in the next-generation or the third generation sequencing platform.[5] For example, Pacific Biosciences Inc. has developed the single molecule real time (SMRT[TM]) DNA Sequencing technology.[6] In this technology, sequential enzymatic incorporation of fluorescently labeled deoxyribonucleoside triphosphates (dNTPs) into a growing DNA strand was observed optically in zero-mode waveguide (ZMW) nanostructure arrays, thus enabling parallel, simultaneous detection of thousands of single-molecule sequencing reactions.[6] Another promising technology is the nanopore sequencing,[7,8] in which fluorescent labeling and amplification are not needed. DNA can be sequenced by electrophoretically driving molecules in a solution through a nanoscale pore. Clarke *et al.* showed that a protein nanopore with a covalently attached adapter molecule could sequence DNA with accuracies averaging 99.8%.[8]

In the following sections, the focus would be on the applications of the NGS in cancer research. As this field is evolving rapidly, with new technology and platforms, and new applications evolving quickly, we may miss those new developments. This review covers publications up to July 2010 in this field. For readers interested in getting the most updated information on NGS, the SEQANSWER (http://seqanswers.com/) is a very good community-based information exchange forum.

2. Application of NGS in SNP Identification

Identification of single nucleotide polymorphisms (SNPs) can be achieved by complete genomic resequencing and targeted genomic sequencing. With the increased read length, pair-end sequencing, and higher density and multiplexing (with bar-coding), complete resequencing of the human genome including cancer genome has become practical at a reasonable cost. Pleasance *et al.* described a comprehensive catalogue of somatic mutations from a human cancer genome.[9] Using Illumina GAII genome analyzers, they obtained more than 40-fold average haploid genome coverage for the genomes of a malignant melanoma and a lymphoblastoid cell line from the same person, providing the first comprehensive catalogue of somatic mutations from an individual cancer. They identified 33,345 somatic base substitutions. A total of 32,325 were single-base and 510 were double-base substitutions (in which two adjacent bases show somatic mutations).

Another approach for sequencing the cancer genome is the targeted sequencing, including exome sequencing and sequencing of the complete genomic regions of interests. This approach could offer much higher coverage of areas of interest. Yeager *et al.* conducted a resequencing analysis of 136 kb (chr8: 128,473,000–128,609,802), a region identified as being associated with the risk of breast, colon, and prostate cancers, using the Roche/454 NGS technology in 39 prostate cancer cases and 40 controls of European origin. They identified common [major allele frequency (MAF) > 1%] SNPs within this region, including 442 novel SNPs that will be useful in analyzing association in a fine scale with prostate cancer and in assessing the functional consequences of select common variants.[10] Sugarbaker *et al.*[11] applied the 454 platform to characterize RNA mutations and expression level changes that are unique to malignant pleural mesotheliomas (MPMs) and not present in control tissues. They sequenced an average of 266 Mb of cDNA from each of four MPMs, a control pulmonary adenocarcinoma (ADCA), and a normal lung tissue. Previously observed differences in MPM RNA expression levels were confirmed. Point mutations were identified by using criteria that require the presence of the mutation in at least four reads and in both cDNA strands and the absence of the mutation from sequence databases, normal adjacent tissues, and other controls. In the four MPMs, 15 nonsynonymous mutations were discovered: 7 were point mutations, 3 were deletions, 4 were exclusively expressed as a consequence of imputed epigenetic silencing, and 1 was putatively expressed as a consequence of RNA editing. Notably, each MPM had a different mutation profile, and no mutated gene was previously implicated in MPM. Of the seven point mutations, three were observed in at least one tumor from 49 other MPM patients. The mutations were in genes that could be causally related to cancer and included *XRCC6*, *PDZK1IP1*, *ACTR1A*, and *AVEN*.

Recently, we described an approach of identifying SNPs in complete genomic regions of key genes including promoters, exons, introns, and downstream sequences by combining long-range PCR or NimbleGen sequence capture with NGS.[12] Using the *adenomatous polyposis coli* (*APC*) gene as an example, we identified 210 highly reliable SNPs by the NGS analysis program MAQ and Samtools, of which 69 were novel ones, in the 123 kb *APC* genomic region in 27 pairs of colorectal cancers and normal adjacent tissues. Although the *APC* gene is one of the mostly sequenced genes, we still identified many novel SNPs, thus illustrating the power of the NGS technology. We confirmed all of the eight randomly selected high-quality SNPs by allele specific PCR, suggesting that our false discovery rate is negligible. We identified 11 SNPs in the exonic region including one novel SNP that was not previously reported. Although 10 of them are synonymous, they were predicted to affect splicing by creating or removing exonic splicing enhancers or exonic splicing silencers.

The advantage of targeted sequencing of a big chunk of genomic regions versus exomes is the ability to identify SNPs affecting splicing in the introns and to identify SNPs in the promoter and enhancer regions. Human genes are regulated by diverse cis-acting elements to make the correctly spliced mRNAs at the right place and at the right time. SNPs affecting these regulatory elements of splicing will affect gene splicing, which was shown to contribute greatly to many human diseases including cancers.[13,14] These regulatory elements include the intronic and exonic splicing enhancer (ISE and ESE)[15] and suppressor (ESS and ISS) elements.[16] Exonic splicing enhancers (ESEs) are cis-regulatory elements that direct or facilitate accurate splicing of precursor RNAs[15] and exonic splicing silencers (ESSs) are cis-regulatory elements that inhibit the use of adjacent splice sites.[16] ESEs can bind SR proteins and recruit and stabilize proteins necessary to splicing in the spliceosome while ESSs can bind protein components of heterogeneous nuclear ribonucleoproteins (hnRNP) to repress exon usage.[14]

We also identified seven SNPs in the upstream region of the *APC* gene, three of which were only identified in the cancer tissues. Six of these upstream SNPs were predicted to affect transcription factor binding. For example, SNP at position chr5: 112070456 (G->A) (all positions refer to the human genome HG19 position) were predicted to have losses of two transcription factor bindings — c-Ets-1 [T00112] and R2 [T00712]. The SNP at position chr5: 112064475 (A->T) would result in the loss of PR B [T00696] and PR A [T01661] binding, but a gain of c-Myb [T00137] binding. The SNP at position chr5: 112063970 (G->C) would result in gains of binding to WT1 I-KTS [T00900], WT1-KTS [T01839], and ETF [T00270]. Bond *et al.* showed a single nucleotide polymorphism in the MDM2 promoter attenuates the p53 tumor suppressor pathway and accelerates tumor formation in humans.[17] Whether these SNPs in the *APC* promoter regions affect colorectal cancer initiation and progress remains to be studied.[12]

The advantage of targeted capture coupled with NGS is the ability to sequence at a much deeper coverage than whole genome sequencing, thus allowing the identification of rare alleles. However, the capture approach using commercial regents (such as Nimblegen or Agilent's capture arrays) is not perfect. They can miss some regions in the genomic sequence. We tested NimbleGens target capture approach for about 30 complete genomic regions of 30 genes, and the capture coverage ranged from about 60–95% (unpublished data). We also compared long-range PCR versus sequence capture for their ability to capture the genomic region of the *APC* gene. We plotted the sequence coverage across the *APC* gene using only the uniquely mapped sequence tags, and we found that the coverage was on average about the same. However, we did notice that for the GC-rich region, the PCR approach seemed to perform better than the NimbleGens target capture approach.

Efforts have been made to distinguish mutations between germline variants and somatic mutations by sequencing normal DNA from the same patient in parallel.[9,18−20] However, a critical issue in variation analysis for the cancer genome is to sort out the "driver mutations" from "passenger mutations." However, this is not an easy task. Recurrent mutations are more likely to be "driver mutations." Shah *et al.* showed that by analyzing the variants of granulosa-cell tumors (GCTs), which are the most common type of malignant ovarian sex cord-stromal tumor (SCST).[21] They identified a missense point mutation, 402C->G (C134W), in FOXL2. This mutation, identified by RNA-seq in the initial three samples, was present in 86 of 89 additional adult-type GCTs (97%), in 3 of 14 thecomas (21%), and in 1 of 10 juvenile-type GCTs (10%) but absent in 49 SCSTs of other types and in 329 unrelated ovarian or breast tumors. These data suggest this mutation is a driver mutation for GCTs.[21]

3. Application of NGS in Detecting Structural Variants

Tumor genomes often undergo somatic chromosomal alternations that result in copy-number variations (CNVs) [changes of gene dosage (gains and losses)], and gene fusion events caused by chromosomal translocation. These structural variants can be categorized into two classes: those that affect the copy numbers and those that do not. The former include insertions and deletions (indels) and are often referred to as copy-number variants, and the latter include inversions, gene fusion/chromosomal translocations. Traditionally, these types of analyses were performed using whole-genome array comparative genome hybridization or single-nucleotide polymorphism arrays.[22−24]

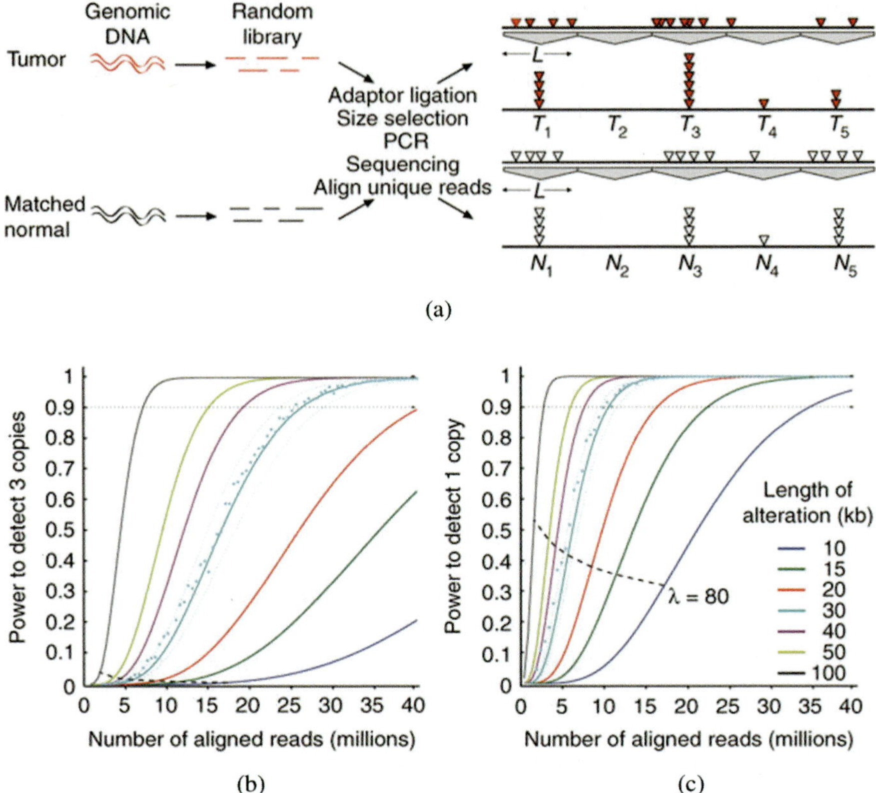

Fig. 1. Theoretical coverage required to detect single-copy gains and losses. (a) Schematic overview for detecting copy-number alterations by sequencing. (b) (c) Power calculations to detect copy-number alterations for a single copy gain and loss. We considered fixed windows L ranging from $L = 10\,\text{kb}$ to $L = 100\,\text{kb}$. Lines indicate approximated power based on the distribution of ratios of normally distributed random variables. For $L = 30\,\text{kb}$, we plotted simulation results for ratios of Poisson-distributed random variables (cyan dots). The approximation is accurate to within 10% (cyan dotted lines) for windows with average number of reads greater than 80 (dashed black line). The figures are adapted from Fig. 1 in Chiang *et al.*[26]

NGS can be used to study the CNV data by analyzing the distribution of aligned reads to a reference genome (Fig. 1(a)).[25,26] The idea is based on that the number of reads mapping to a region follows a Poisson distribution and is expected to be proportional to a copy number of chromosomal regions, assuming that the sequencing process is uniform. Chiang *et al.* estimated the power to detect copy-number alterations of a given size.[26] Assuming that sequence reads are randomly chosen from the genome, they argued that the number of reads aligning to a region will follow a Poisson distribution with a mean directly proportional to the size of the region and to the copy number. They estimated that with 10 million aligned reads, a region of 50-kilobase (kb) in the alignable portion of the human genome ($A = 2.2 \times 10^9$ for 36-bp reads) would be expected to have $50{,}000 \times 10^7/A \approx 230$ reads for two copies, $\sim$115 reads for one copy or $\sim$345 reads for three copies.[26] They then calculated the theoretical coverage required to detect single copy changes. Their simulation suggested that to detect a 50-kb region of a single-copy gain, at this stringency, one requires about 15 million aligned reads [Fig. 1(b)]; for a single-copy loss one needs about 6 million aligned reads [Fig. 1(c)]. They compared NGS and Affemetrix SNP arrays (Affymetrix SNP Array 6.0) in detecting CNVs of three pairs of matched normal and tumor cell lines and found that the copy-number segments detected by both approaches were highly concordant with respect to identifying the existence of a copy-number alteration, whereas massively parallel sequencing had somewhat better resolution for localizing the breakpoints. They showed that sequencing achieved a higher dynamic range for estimating copy-number alterations than arrays due to saturation effects of array hybridizations.[26]

Methods to use NGS for high-resolution multiplex analysis of CNV from nanogram quantities of DNA from formalin-fixed paraffin-embedded specimens were also developed by Wood *et al.* and by Schweiger *et al.*[27,28] A high Pearson correlation of 0.9362277 between the NGS and the aCGH data sets was observed,[27] suggesting that NGS can be used to replace aCGH in CNV analysis (Fig. 2).

With the development of mate-pair or paired-end read sequencing approaches, a more detailed analysis of structural changes in the cancer genome can be analyzed. In pair-end sequencing, two paired reads are generated and can be mapped to a reference genome. If the pairs map at a distance that is substantially different from the expected length, or with an anomalous orientation, they would suggest structural variants. This kind of paired-end mapping (PEM) technique has been successfully used to discover structural variants at a much higher resolution than array-based methods.[29] Medvedev *et al.* illustrated the approach of PEM (Fig. 3).[29]

Clark *et al.* generated greater than $\times 30$ genomic sequence coverage using a novel 50-base mate paired strategy with a 1.4 kb mean insert library for U87MG, a commonly studied grade-IV glioma cell.[30] They generated a total of 1,014,984,286 mate-ends T and identified 35 interchromosomal

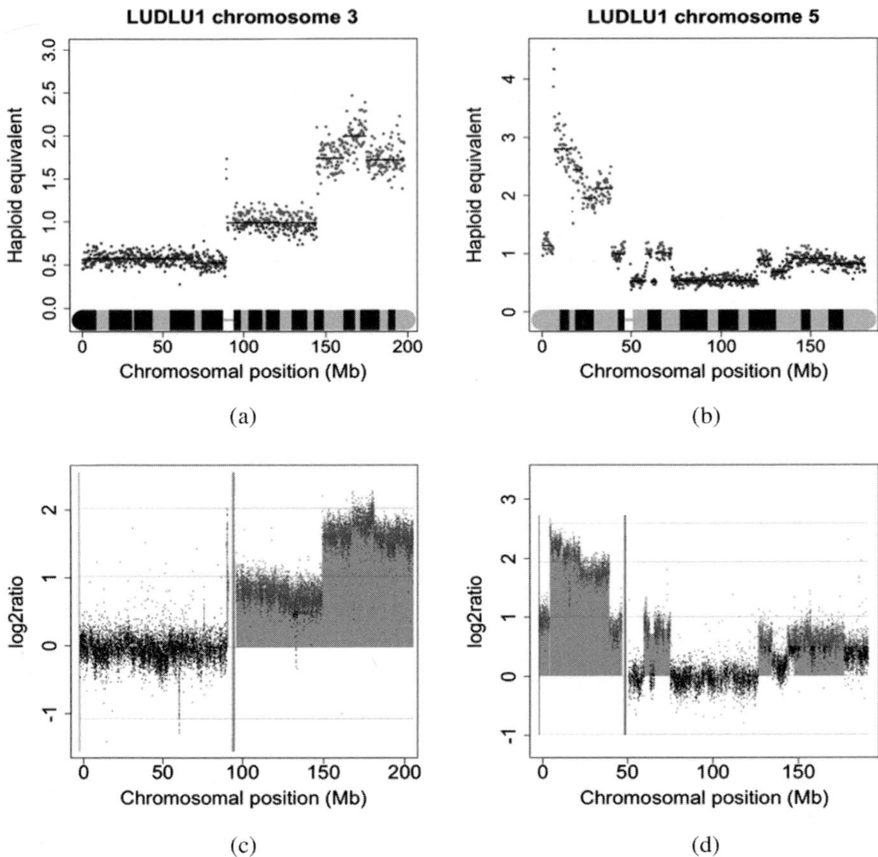

Fig. 2. Examples of a comparison of copy-number analysis by sequencing and aCGH. (a) Chromosome 3 analyzed by NGS; (b) chromosome 5 analyzed by NGS; (c) chromosome 3 analyzed by aGCH; (d) chromosome 5 analyzed by aCGH. A black and white copy of Fig. 1 from wood *et al.*[28]

translocation events, 1,315 structural variations ($>$100 bp), 191,743 small ($<$21 bp) insertions and deletions (indels), and 2,384,470 single nucleotide variations (SNVs). Furthermore, they showed that 512 genes were homozygously mutated, including 154 by SNVs, 178 by small indels, 145 by large microdeletions, and 35 by interchromosomal translocations.[30] Campbell *et al.* used massively parallel sequencing to generate sequence reads from both ends of short DNA fragments derived from the genomes of two individuals with lung cancer.[25] By investigating read pairs that did not align correctly with respect to each other on the reference human

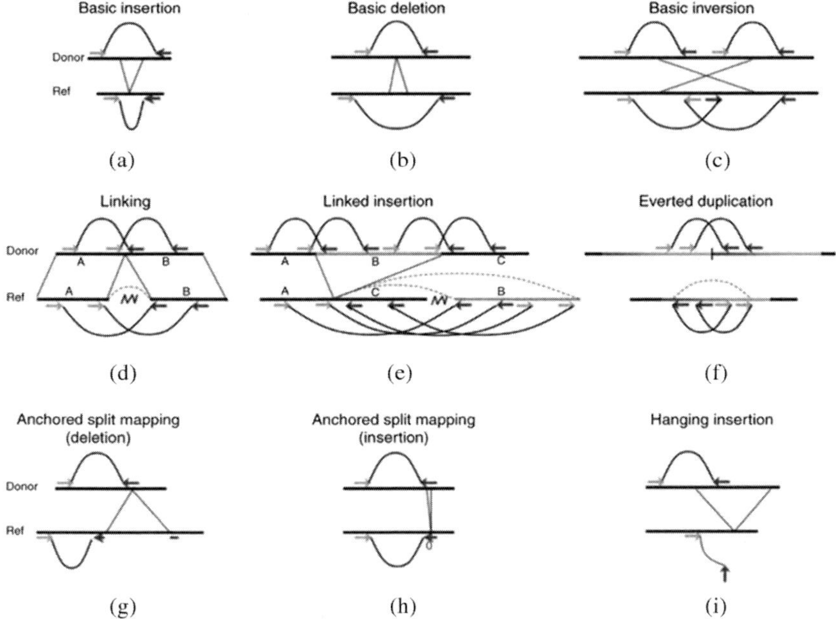

Fig. 3. A black and white copy of the Fig. 1 from Medvedev *et al.*[29] Illustrations of PEM signatures. Mate pairs are sampled from the donor, where they are ordered with opposite orientation (the black mate follows the gray), and are mapped to the reference (ref). Basic signatures include (a) insertions and (b) deletions, where the mapped distance is different from the insert size, as well as (c) inversions, where the order of the two mates is preserved but one of them changes orientation. (d) The linking signature has several discordant mate pairs with similar mapped distances identifying adjacency in the donor (dashed gray arrows) of two distal segments of the reference. The orientation and order of the mapped mate pairs depends on the orientation and order of the two segments in the reference; here, these are unchanged. (e) A linked insertion signature is composed of two linking signatures and arises when the inserted sequence (gray) is copied from another location in the genome. (f) A tandem duplication will create an everted duplication linking signature, with mates mapping out of order but with proper orientations. These mate pairs link the end of the duplicated region to its beginning. (g), (h) In the anchored split-mapping signature, one mate has a good mapping, whereas the other has a split-mapping. For a deletion (g), the prefix and suffix surround the deletion, whereas for an insertion (h), the split read has the prefix and suffix mapped to adjacent locations, while a middle part does not map. (i) When a novel genomic segment is inserted, a hanging insertion signature is created, in which only one of the mates has a good mapping.

genome, they characterized 306 germline structural variants and 103 somatic rearrangements to the base-pair level of resolution.[25] These studies illustrated the power of pair-end sequencing in detecting structural variants including small indels, microdeletions, chromosomal translocations, etc.

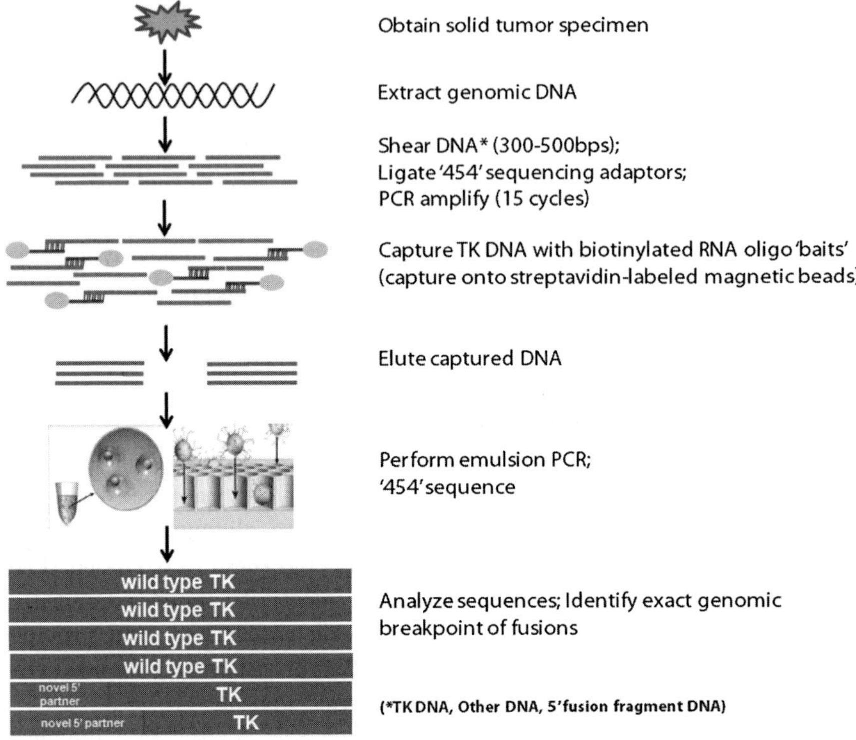

Fig. 5. A black and white copy of Fig. 2 from Chmielecki *et al.*[36] Schematic of experimental workflow of Chmielecki's approach to screen systematically for fusions involving the 90 human TKs.

time the genomic fusion sequences of CCDC6-RET in TPC-1 cells and FGFR1OP2-FGFR1 in KG-1 cells. They demonstrated the feasibility of their approach to identify TK fusions across multiple human cancers in a high-throughput, unbiased manner.

5. RNA-seq and End-Tag Sequencing for Cancer Expression Profiling

Both RNA-seq and end-tag sequencing (Taq-seq) have been applied to cancer expression profiling. Taq-seq is technically the sequencing of sequence tag at the $3'$d end of RNAs. The sequence tags are usually the sequence tags marked by a 4-bp cutter (e.g. NlaIII) and is closest to the $3'$ poly A tail as the technology usually employed oligo-dT to capture polyadenylated RNA and to start second strand cDNA synthesis.

For example, in one implementation using NlaIII and MmeI to generate end-tags, the tag is a 17-bp sequence tag adjacent to the 3′ most NlaIII site of an individual transcript.

The end-tag sequencing is similar to SAGE or LongSAGE except there is no ditag ligation step after the generation of taqs and the sequencing is done using high throughput NGS technologies. We were among the first to apply Tag-seq approaches for cancer expression profiling using the early implementation of the NGS technology, the MPSS technology.[37–39] With the newer NGS technology, Morrissy *et al.* also applied the Tag-seq approach for expression profiling of normal human tissues.[40] They showed that Tag-seq has a better ability to detecting lowly expressed transcripts than LongSAGE due to much deeper sequence depth in the Tag-seq approach. For example, using the same RNA (hESC), they showed that the LongSAGE identified 7,055 genes while the Tag-seq identified 11,165 genes, which included 93.5% of the genes found by LongSAGE.[40] Another example is the ability to detect transcription factors (TFs), which are usually lowly expressed. A total of 429 TFs were detected by LongSAGE, whereas 799 TFs were detected by Tag-seq.[40] They further compared Tag-seq to Affymetrix arrays and showed that the genes detected in common had a 13-fold greater dynamic range in Tag-seq (at 10 million sequencing read depth) versus Affymetrix.[40] Finally, they showed that Tag-seq approach were able to detect antisense tags, which are transcripts that transcribed from the opposite strand and intronic tags that may represent unannotated exons and UTRs within known genes. All the above demonstrate the superiority of Tag-seq over LongSAGE or DNA arrays.

Another of the advantages of Tag-seq is its ability to discriminate antisense transcripts.[40] Sense and antisense genes are encoded on the opposite strands of the same genomic locus and yield transcripts that have sequence complementarity. For gene expression profiling experiments, where accurate profiling of transcripts from both strands of the genome is required, Tag-seq data are superior to RNA-seq,[40] which will be described below. Tag-seq is able to identify transcript isoforms due to the alternative use of polyadenylation sites that generate different 3′ ends.

In RNA-seq, all RNAs of a sample or polyA$^+$ RNAs if one preferred, are randomly fragmented, reverse transcribed, ligated to adapters and then these fragments are sequenced. Figure 6 illustrated the RNA-seq procedures.

Compared to microarrays, RNA-seq has very low background signal because DNA sequences can be unambiguously mapped to reference

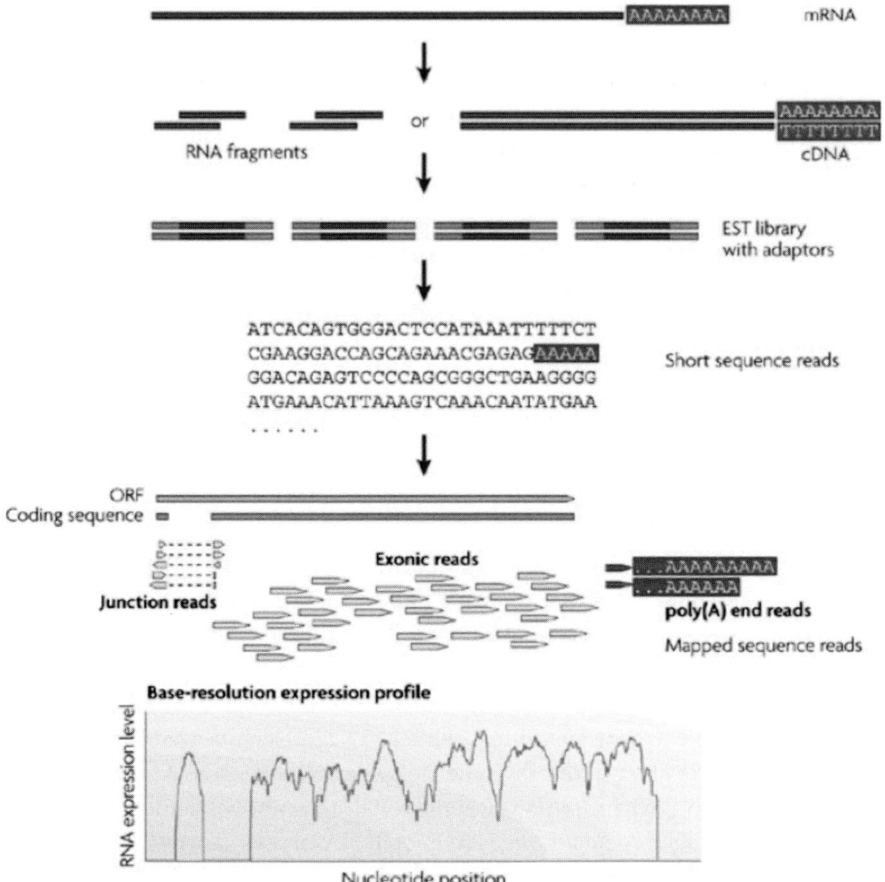

Fig. 6. A black and white copy of Fig. 1 from Wang *et al.*[41] Briefly, long RNAs are first converted into a library of cDNA fragments through either RNA fragmentation or DNA fragmentation. Sequencing adaptors (gray) are subsequently added to each cDNA fragment and a short sequence is obtained from each cDNA using high-throughput sequencing technology. The resulting sequence reads are aligned with the reference genome or transcriptome, and classified as three types: exonic reads, junction reads, and poly(A) end-reads. These three types are used to generate a base-resolution expression profile for each gene, as illustrated at the bottom of the figure; a yeast ORF with one intron is shown.

genes or genomes. It has a large dynamic range of over five orders of magnitude versus that of a few hundreds for microarrays.[41,42] RNA-seq is more sensitive demonstrating a superior ability in detecting lowly expressed and differentially expressed genes,[41,43] and in detecting RNA isoforms.[44] More recently, Fu *et al.* also showed that expression values from RNA-seq seemed to correlate better with protein levels than those from microarray.[45] Bradford *et al.* compared RNA-seq data generated on the Applied Biosystems SOLiD platform and data from Affymetrix Exon 1.0 ST arrays, and they found that RNA-seq more sensitive to detecting differentially expressed exons than the Exon Array, reflecting the wider dynamic range achievable on the SOLiD platform.[46]

RNA-seq can also be used to discover novel transcribed regions (previously unannoated) of the genome due to its sequence depth. Mortazavi *et al.* sequenced 140 million mapped sequence reads for three mouse tissues and identified ~17,000 candidate new parts of existing genes (novel exons, novel 5′ and 3′ extension, etc.), plus 596 new candidate transcripts.[42]

RNA-seq is an ideal method for measuring changes in transcriptomes including changes in gene-splicing isoforms. We will see RNA-seq to become the standard platform in cancer transcriptome analysis. The great heterogeneity encounter by cancer tissues and cells requires the ability to study expression profiling eventually at single-cell level. Recently, Tang *et al.* showed that it is possible to perform RNA-seq at single-cell level.[47,48] We will see the application of this single cell analysis in cancer research in the future.

SNPs and mutations can also be analyzed from RNA-seq data. Shah *et al.* applied RNA-seq to identify a recurrent missense point mutation 402C->G (C134W) in FOXL2 for Granulosa-cell tumors (GCTs).[21] GCTs are the most common type of malignant ovarian sex cord-stromal tumor (SCST). FOXL2 is a gene encoding a transcription factor known to be critical for granulosa-cell development. Confirmation analysis showed that the FOXL2 mutation was present in 86 of 89 additional adult-type GCTs (97%), in 3 of 14 thecomas (21%), and in 1 of 10 juvenile-type GCTs (10%). The mutation was absent in 49 SCSTs of other types and in 329 unrelated ovarian or breast tumors.[21] They illustrated the power of NGS in identifying a single, recurrent somatic mutation (402C->G) in FOXL2 that was present in almost all morphologically identified adult-type GCTs. Mutant FOXL2 is a potential driver in the pathogenesis of adult-type GCTs.[21]

Recently, Morin *et al.* applied NGS to identify recurrent somatic mutations in the polycomb-group oncogene EZH2 in 7.2% of Follicular lymphoma (FL) and in 21.7% of the GCB subtype of diffuse large B-cell lymphoma (DLBCL).[49] The polycomb-group oncogene EZH2 encodes a histone methyltransferase that is responsible for trimethylating Lys27 of histone H3 (H3K27).[49] They showed that the mutation resulted in the replacement of a single tyrosine in the SET domain of the EZH2 protein (Tyr641).[49]

6. Application of NGS in Genome-Wide Analysis of TF-Binding Site (ChIP-Seq)

ChIP-seq is the technology that is replacing ChIP-chip in global analysis of DNA-binding sites. Both require the ChIP (chromatin immunoprecipitation). In ChIP, DNA-bound proteins (e.g. a transcription factor) are crosslinked to the DNA using formaldehyde. Then, cells are lysed and the genomic DNAs are sonicated into short (<1 kbp) pieces. An antibody for the protein of interest is, then, used to immunoprecipiate the DNA–protein complexes. The crosslinks are reversed. The DNAs are, then, subjected to NGS in the ChIP-seq protocol.

ChIP-seq technology is rapidly emerging as a cost effective, global technology for high-resolution, genome-wide unbiased mapping of protein–DNA interactions in chromatin-mapped complexes. Since its early application in the global identification of *in vivo* binding of the insulator binding protein, CTCF,[50] the neuron-restrictive silencer factor (NRSF), and STAT1,[51,52] ChIP-seq is becoming the standard technology in studying TF–DNA binding.

Ji *et al.* recently showed that ChIP-seq has better ability than ChIP-chip in identifying TF-binding regions for NRSF.[53] A summary of the advantages of ChIP-seq versus ChIP-chip is shown in Table 1 (taken from the Illumina's website, www.illumina.com) including higher resolution in ChIP-seq and not depending on pre-printed chips, for which the sequences need to be defined first.

We applied ChIP-seq to study the growth inhibition response program of the androgen receptor (AR).[54] The AR plays important roles in the development of the male phenotype and in different human diseases, including prostate cancers. The AR can act either as a promoter or a tumor suppressor depending on cell types.[55] The AR proliferative response program has been well studied, whereas its prohibitive response program has not yet been thoroughly studied. The PC-3 was initiated from a bone metastasis of a grade IV prostatic adenocarcinoma from a 62-year-old male Caucasian.[56] Several studies have demonstrated that prostate cancer PC3 cells expressing the transfected androgen receptor (PC3-AR) show growth inhibition and reduced invasion capabilities.[55,57–60] ChIP-seq

Table 1.　ChIP-seq compared to Chip-chip analysis.

	ChIP-seq	ChIP-chip	ChIP-seq advantage
Starting material	Low: Down to 10 ng	4 μg	Hundreds-fold lower DNA input requirements means fewer IP reactions
Flexibility	Yes: Genome-wide assay of any sequenced organism	Limited: Dependent on available products	Not limited to contents available on arrays
Positional resolution	± 50 bp	$\pm 500-1000$ bp	Site mapping can be an order of magnitude more precise
Sensitivity	Widely tunable: Increase counts to increase sensitivity	Poor: Based on hybridization and ratios	Simply increase the number of counts to obtain desired sensitivity
Cross-hybridization	None: Each DNA is individually sequenced	Significant	Higher quality data even in complex genomes

analysis revealed 6,629 AR-binding regions in the cancer genome of PC3 cells with an FDR (false discovery rate) cut off of 0.05.

We found that about 22.4% (638 of 2,849) can be mapped to within 2 kb of the transcription start site (TSS).[54] Using Gibbs Motif Sampling algorithm of the CisGenome program,[53] we identified three novel AR-binding motifs in the AR-binding regions of PC3-AR cells (Fig. 7), and two of them share a core consensus sequence CGAGCTCTTC, which together mapped to 27.3% of AR-binding regions (1,808/6,629). In contrast, only about 2.9% (190/6,629) of AR-binding sites contains the canonical AR matrix M00481, M00447 and M00962 (from the Transfac database), which is derived mostly from AR proliferative responsive genes in androgen dependent cells.

7. Application of NGS in Epigenomic Analysis of Cancers

Epigenetic changes in the human genome including: post-translational modifications of histone subunits, methylation of cytosines in DNA, and various RNA-mediated mechanisms, play important roles in carcinogenesis. NGS has been applied to the study of global epigenetic changes in the

                    L. Cheng, H. Xu and B. Lin

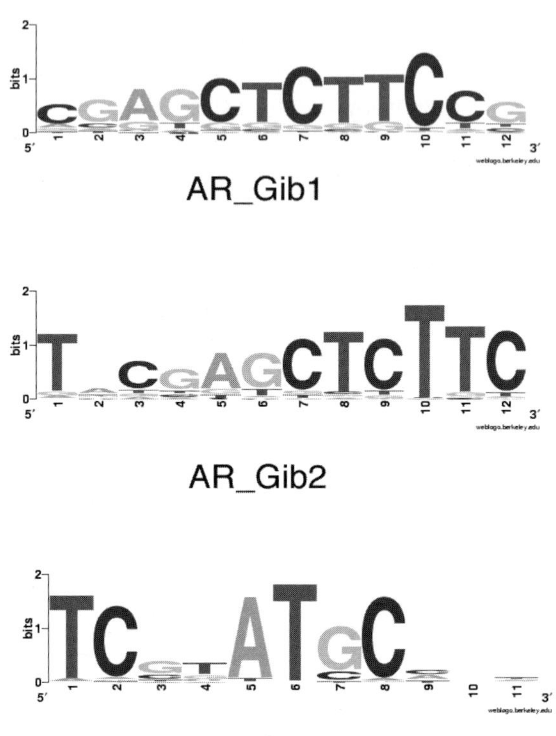

Fig. 7. The consensus sequences of the three novel AR matrice identified, which can be mapped to 1,448, 1,012, 317 AR-binding regions, respectively. The consensus logos were generated with the Web logo program (http://weblogo.berkeley.edu/). From Fig. 4 of Lin *et al.*[54]

human genome: such as the post-translational histone modifications, DNA methylation, mapping of open chromatin regions. We will review the studies in these analyses.

7.1. *Application of NGS in studying histone modifications*

Histones bind to DNA in eukaryotic cells. Specific residues (typically lysines) in the histones are subject to post-translational modifications such as methylation, acetylation, ubiquitination, and phosphorylation, and form the so-called histone marks that mark the genomics regions into different functional regions (e.g. transcribed versus untranscribed) in cell-type or developmental stage specific manner. Methylation of specific histone lysine residues, including H3K4, H3K36, H3K79, H3K9, H3K27, and H4K20, regulate chromatin structures and gene activities. The methylation at these lysine side chains can be either mono-, di-, or trimethylated forms, and each exert distinct effects on transcription. For example, H3K4me3 (trimethylation of lysine residue 4 in the histone 3) marks the active promoters and H3K27me3 (trimethylation of lysine 27 in histone 3) mark repressed promoters and bivalent modification of both H3K4me3 and H3K27me3 marks poised promoter (not active but ready to be reactive to stimuli).[57] H3K9me3 (H3 lysine 9 trimethylation) is associated to heterochromatin and gene silencing.[58]

Barski *et al.* generated high-resolution maps for the genome-wide distribution of 20 histone lysine and arginine methylations as well as histone variant H2A.Z, RNA polymerase II, and the insulator-binding protein CTCF across the human genome using NGS.[50] They identified typical patterns of histone methylations exhibited at promoters, insulators, enhancers, and transcribed regions. They showed that H3K4me1, H3K4me2, H3K4me3, and H3K36me3 signals correlated with gene activation. They found that H3K27me2 and H3K27me3 were correlated with gene silencing, and also observed a modest correlation between H3K9me3 and H3K9me2 levels and gene silence with some exceptions noted. Figure 8 shows an example of active gene regions (STAT1 and STAT4) and an inactive gene MOB1. The activated genes are marked by H3K4 methylations while the inactivated genes are marked by H3K27 methylations (Fig. 8). RNA polymerase II also marks the site of transcription initiation and CTCF marks the boundaries of histone methylation domains.[50]

Zhang *et al.* developed the Human Histone Modification Database (HHMD, http://bioinfo.hrbmu.edu.cn/hhmd) that stores and integrates histone modification datasets.[59] The latest collections include 43 location-specific histone modifications in human covering methylation, acetylation, and ubiquitylation probed by ChIP-seq, ChIP-chip, and qChIP. The database can be searched by: histone modification, gene ID, functional categories, chromosome location, and cancer name. It also includes a user-friendly visualization tool named HisModView, by which genome-wide histone modification maps can be shown.

One thing worth pointing out is that, as histone marks can be diffusely enriched over thousands of base pairs, algorithms developed for TF ChIP-seq analysis might not be the best as TF binding are often more localized and show strong peaks. Zhang *et al.* developed an algorithm called spatial clustering approach for the identification of ChIP-enriched regions (SICER) that identifies spatial clusters of signals

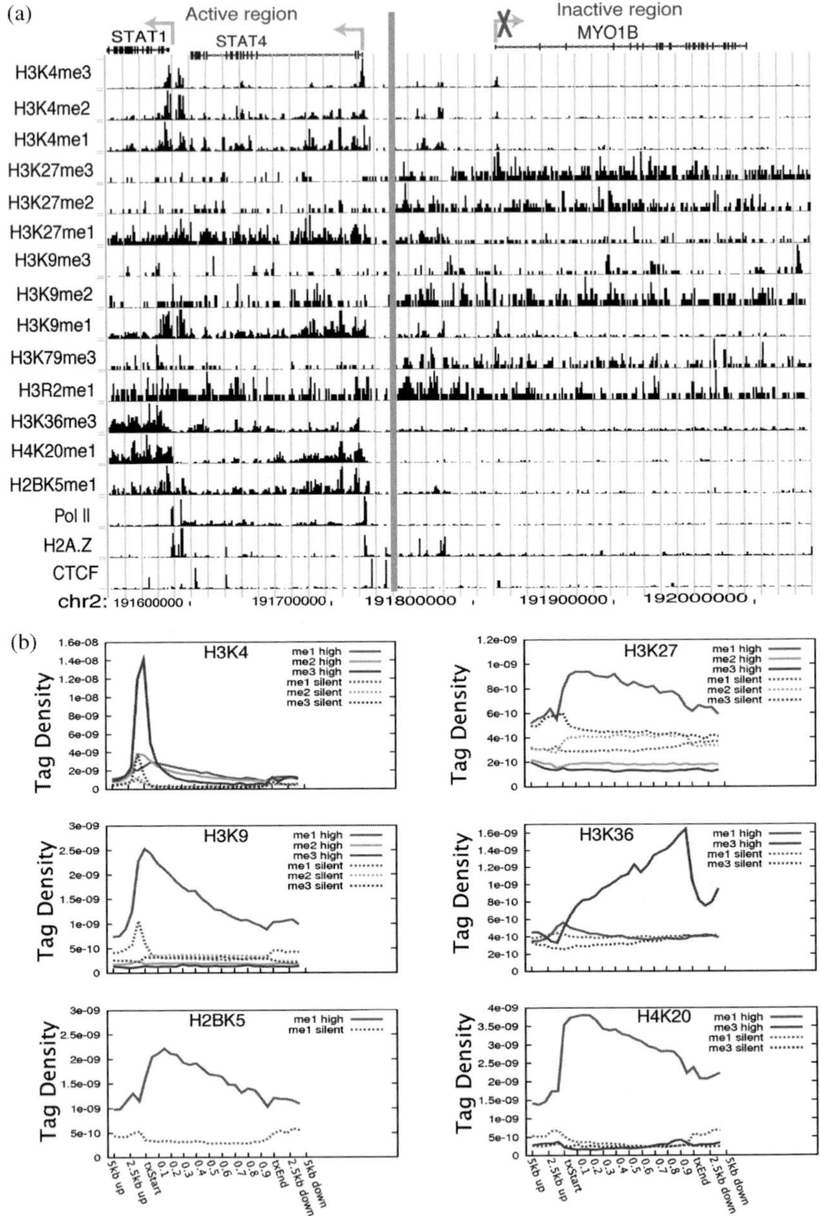

Fig. 8. A black and white copy of Fig. 4. from Barski *et al.*[50] Histone methylation patterns of active and inactive Genes. (a) A typical example of histone methylation patterns at active and inactive genes shown as custom tracks on the UCSC genome browser. (b) Profiles of histone methylation patterns of 1,000 highly active or silent genes across the gene bodies. Density plots extend 5 kb 5′ and 3′ of the gene bodies.

unlikely to appear by chance.[60] Other programs tailored to histone modification ChIP analysis include ChIPDiff,[61] ChromaSig,[62] and NPS (http://liulab.dfci.harvard.edu/NPS/.).[63,64]

7.2. *Application of NGS in studying DNA methylation*

5-methylcytosine (5mC) accounts for 1–6% of the nucleotides within mammalian and plant genomes[65] and plays important roles in cancers.[66] There are about 29 million CpGs in the human genome. The methylation status of these CpGs (methylated, hydroxymethylated, or unmethylated state) is collectively referred to as the DNA methylome. With the NGS, it is now possible to study the DNA methylome at a single nucleotide resolution level. There are three approaches for DNA methylome analysis. The first one is the endonuclease restriction-based methods named Methyl-seq, where fragmentation patterns generated through DNA digestion by methylation-sensitive enzymes are used for NGS.[67] An alternative technology is the MeDIP (methylated DNA immunoprecipitation)-seq. In MeDIP-seq, methylated DNAs are immunoprecipited with an antibody against 5mC and then subject to NGS.[68] Recently, Butcher and Beck developed the AutoMeDIP-seq: a high-throughput, whole genome, DNA methylation assay.[69]

The third technology combines NGS with bisulfite conversion (BS-seq) to study the DNA methylome globally.[70,71] In BS-seq, DNA is treated with bisulfite. Unmethylated cytosine (C) nucleotides are converted into uracil (U), which is then converted to thymine (T). After sequencing, methylated cytosine residues will appear in the obtained sequence and will read as Cs, while unmethylated cytosine residues will appear as Ts.[70,71]

As DNA methylation contributing to the remodeling of chromatin via recruitment of methyl–CpG-binding domain (MBD) proteins, methods were developed to isolate or enrich methyl-CpG containing DNAs using MBD, i.e. MDB-seq. Serre *et al.* developed a novel technique, MBD-isolated Genome Sequencing (MiGS), which combines precipitation of methylated DNA by recombinant methyl-CpG binding domain of MBD2 protein and sequencing of the isolated DNA by NGS.[72] They applied MiGS to analyze the methylation patterns of three isogenic cancer cell lines with varying degrees of DNA methylation, and were able to successfully detect previously known methylated regions as well as identify hundreds of novel methylated regions.[72]

Various technologies for methylome analyses have both advantages and disadvantages. For example, MeDIP-seq and MDB-seq are enrichment-based; therefore, the sequencing cost is a fraction of that required for BS-seq. MeDIP is not limited to restriction sites or CpG islands and is, therefore, relatively unbiased at capturing methylated genomic DNA. BS-seq is able to achieve a single nucleotide resolution of the methylated DNA. However, general problems associated with bisulfite conversion (e.g. incomplete conversion etc.) still exist for BS-seq. BS-seq is also very costly as it requires whole genome sequencing (no-enrichment). To save cost, Smith *et al.* developed a reduced representation bisulfite sequencing (RRBS) strategy that generates nucleotide resolution DNA methylation bisulfite sequencing libraries that enrich for CpG-dense regions by methylation-insensitive restriction digestion.[73] RRBS enriches its libraries by digesting genomic DNA with restriction endonucleases that are specific for CpG containing motifs (Fig. 9). At the same sequence depth, RRBS enhances the coverage for the CpG dinucleotide.

Ruike *et al.* applied methylated DNA immunoprecipitation combined with high-throughput sequencing MeDIP-seq to obtain whole-genome DNA methylation profiles for eight human breast cancer cell (BCC) lines and for normal human mammary epithelial cells (HMEC).[74] They found that BCC lines were a massively reduced methylation level particularly in CpG-poor regions when compared to HMEC.

8. Other Applications of NGS

The applications of NGS in biological and medical studies keep expanding and novel applications are being constantly proposed. These include the application of NGS in studying RNAs, such as in identifying targets of RNA-binding proteins and in identifying microRNA targets.

CLIP (Crosslinking immunoprecipitation)-seq is a technology where RNAs are crosslinked with RNA-binding proteins, and then isolated by immunoprecipitation with antibodies against the RNA-binding proteins. Stanford *et al.* applied CLIP-seq to identify 23,632 binding sites for essential splicing factor SFRS1 in the transcriptome of cultured human embryonic kidney cells.[75] They identified a purine-rich octamer containing a GAAGAA core that is present in a majority of exonic, intronic, ncRNA, and intergenic transcripts coprecipitated by SFRS1. They also found enrichment of binding sites between 21 and 40 nt from the 5′ or 3′ splice sites in exonic RNA fragments bound by SFRS1. Searching the Human Gene Mutation

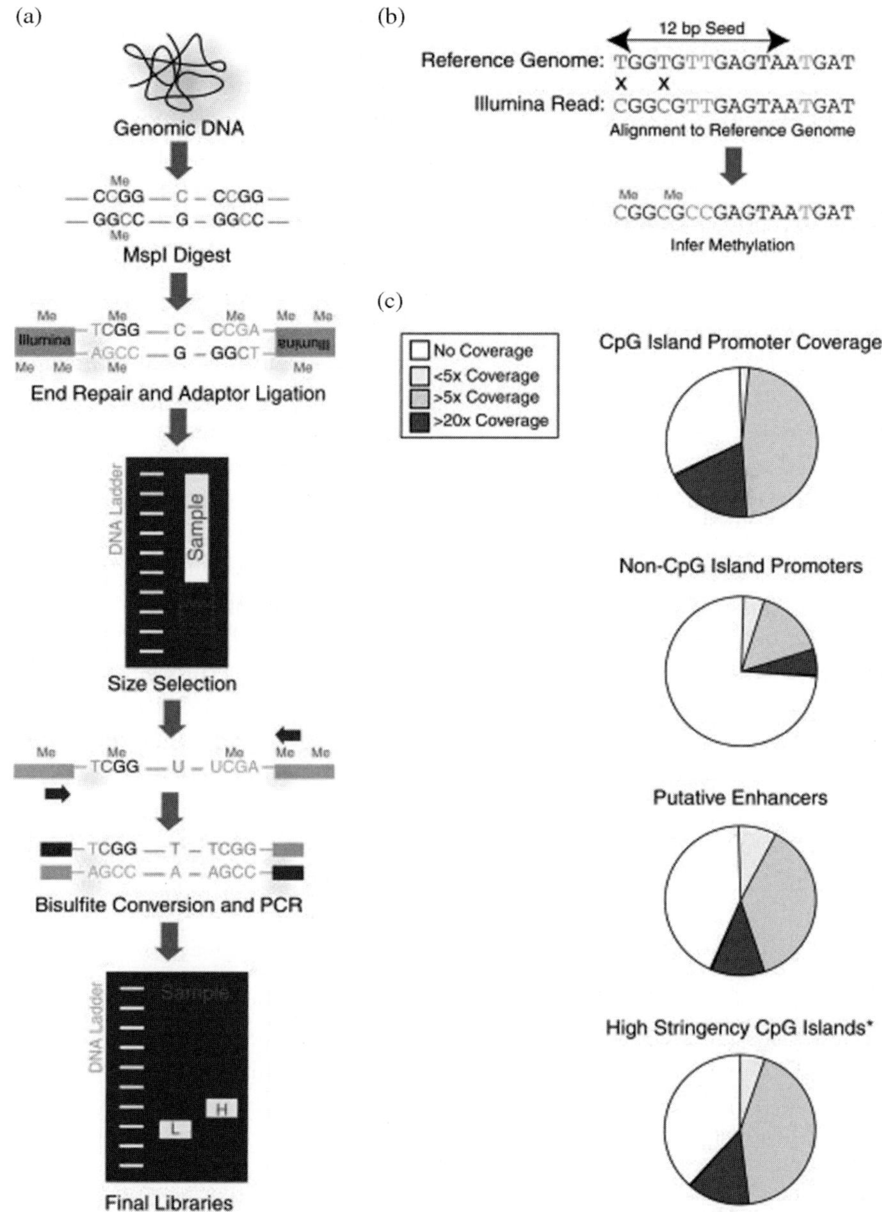

Fig. 9. A black and White copy of Fig. 1 from Smith *et al.*[73] Reduced representation bisulfite sequencing. (a) General outline of RRBS library construction protocol including all key steps. (b) Overview of RRBS read alignment. (c) Sequencing coverage at selected regions of interest within the mouse genome as generated from a representative V6.5 mouse embryonic stem (ES) cell library.

Database, they found that mutations in the SFRS1 binding motifs for many genes that have been implicated in human diseases.[75]

Hafner *et al.* developed photoactivatable ribonucleoside-enhanced crosslinking and immunoprecipitation (PAR-CLIP) protocol to overcome the shortcomings of CLIP,[76,77] which include low efficiency of UV 254 nm RNA-protein crosslinking, difficulty in identifying the location of the crosslink, and to separate UV-crosslinked target RNA segments from background noncrosslinked RNA fragments. In PAR-CLIP, photoreactive ribonucleoside analogs, such as 4-thiouridine (4-SU) and 6-thioguanosine (6-SG) were added to live cells and were incorporated into nascent RNA transcripts by living cells. UV-crosslinking at 365 nm is highly efficient. Crosslinked and coimmunoprecipitated RNAs with an RNA-binding protein were isolated and subjected to NGS.[76,77] The advantage of the PAR-CLIP-seq is the ability to precisely identify the position of crosslinking by mutations residing in the sequenced cDNAs. When using 4-SU, crosslinked sequences have thymidine to cytidine transitions, whereas using 6-SG, cross-linked sequences have guanosine to adenosine mutations. The presence of the mutations in crosslinked sequences makes it possible to separate them from the background RNAs.[76,77] They successfully applied PAR-CLIP-seq in the identification of mRNA targets of RNA-binding proteins such as UM2, QKI, and IGF2BP1-3.[76,77]

miRNAs play important roles in cancers by regulating targeted mRNAs through base-pairing interactions involving mostly 6–8 nucleotide seed sequences at the 5′ end of the miRNA.[78,79] NGS has been applied in expression profiling of microRNAs.[80,81] It has been used to identify differentially expressed miRNA in cancers.[82–84]

9. Future Challenges

There are many challenges in the future application of NGS to cancer research including: sample preparations, comparisons of different NGS plat-forms, data analysis, and bioinformatics challenges, data storage challenges, etc. We will not discuss them all and will only focus on two challenges unique to cancer researches.

9.1. *To distinguish driver mutations from passenger mutations*

The power of NGS has allowed the discovery of tens of thousands of genetic alterations in some of the sequenced cancer genomes; a formidable challenge is to sort out active "driver" mutations from inactive "passenger" mutations in order to understand the biology of cancer and to develop novel targeted therapeutic approaches. The catalogue of Somatic Mutations in Cancer (COSMIC) database records published somatic mutations in cancers, allowing the comparison and cataloguing of the mutations identified, and the discovery of recurrently mutated genes.[85–87] The database (v48) consists of 141,212 mutations (unique mutations 23,907) from 2,760,220 experiments and 10,383 curated papers. It also includes data from 28 whole Cancer Genomes sequencing data (http://www.sanger.ac.uk/cosmic/).

Carter *et al.* developed a computational method, called Cancer-specific High-throughput Annotation of Somatic Mutations (CHASM), to identify and prioritize those missense mutations most likely to generate functional changes that enhance tumor cell proliferation.[88] The method has high sensitivity and specificity. When applied to 607 missense mutations found in a recent glioblastoma multiforme (GBM) study, they were able to identify 49 of the 607 missense mutations (8%) to be putative driver mutations,[88] including the mutation in IDH1 that was recently shown to be a driver mutation n GBM.[89] They further applied the method to 963 of the missense somatic missense mutations discovered in 24 pancreatic cancers, and identified putative driver mutations (false discovery rate $</=0.3$) including three known pancreatic cancer driver genes (P53, SMAD4, and CDKN2A), and an additional 15 genes with putative driver mutations include genes coding for kinases (PIK3CG, DGKA, STK33, TTK, and PRKCG), for cell cycle related proteins (NEK8), and for proteins involved in cell adhesion (CMAS, PCDHB2).[90]

9.2. *To deal with cancer heterogeneity and archived formalin-fixed paraffin-embedded samples*

Cancer is a complex disease. Not only are clinical cancer samples generally mixed with normal cells such as fibroblasts (tumor stroma) and infiltrating lymphocytes, but also clinical cancer samples probably represent cancer cells at different stage of developments with different underlying molecular profiles. It may be necessary to partition the cells in a clinical sample into different populations. However, determining which properties to base on for the partition is a challenge. It may be ultimately necessary to sequence single cells to comprehend the heterogeneity of cancer cells. Fortunately, some NGS applications based on single cells are now possible such as the single-cell RNA-seq analysis.[47,91]

Archive formalin-fixed paraffin-embedded (FFPE) samples are abundantly available and often with long clinical follow-ups. Studies using such samples often have great clinical significances. Weng *et al.* performed miRNA deep-sequencing analysis on paired frozen and FFPE specimens and found that they showed very similar results.[84] Comparing microarray, deep-sequencing and RT-PCR methodologies, they found that deep-sequencing and RT-PCR methodologies showed a higher correlation among the three platforms.[84] Wood *et al.* also developed a method for applying NGS in high-resolution multiplex analysis of copy number variation from nanogram quantities of DNA from FFPE specimens.[28] Beck *et al.* developed the 3′-End Sequencing for Expression Quantification (3SEQ) from archival tumor samples using NGS.[92] They performed gene expression profiling by 3SEQ and microarray on both frozen tissue and FFPE tumors from 23 soft tissue tumors (desmoid type fibromatosis (DTF) and solitary fibrous tumor (SFT)).[92] They found that 3SEQ were able to identify many genes differentially expressed between the tumor types (FDR < 0.01) on both the frozen tissue ($\sim$9.6 K genes) and FFPE tumors ($\sim$8.1 K genes). However, microarrays were only able to identify differentially expressed genes from frozen tissue at fewer numbers ($\sim$4.64 K), but failed to identify any significant number (only 69 were identified) of differentially expressed genes on FFPE samples.[92] With such advances, we will see NGS being used in much larger cohort studies in cancer research.

In summary, we will see NGS as a major driver in understanding cancer genome and cancer biology at global scale. Recently, NCI (National Cancer Institute, USA) announced that it will expand the Cancer Genome Atlas (TCGA) (http://cancergenome.nih.gov/) project to up to 25 tumor types and it will apply NGS to at least 100 tumor samples for each of several types of childhood cancers. The International Cancer Genome Consortium (ICGC) (http://www.icgc.org/) set the goal to coordinate large-scale cancer genome studies in tumors from 50 major cancer types and/or subtypes. They proposed systematic studies of more than 25,000 cancer genomes at the genomic, epigenomic, and transcriptomic levels to reveal the repertoire of oncogenic mutations and their consequences, and to define clinically relevant subtypes for prognosis and therapeutic management, and eventually to develop novel cancer therapies. The normal human genetic variation is also important for identifying and understanding cancer mutations. The 1000 Genomes Project (http://www.1000genomes.org/) was launched in January 2008 as an international effort aiming to establish a comprehensive catalogue of human genetic variation.

References

1. Brenner S, Johnson M, Bridgham J *et al.* (2000) Gene expression analysis by massively parallel signature sequencing (MPSS) on microbead arrays. *Nat Biotechnol* **18**(6): 630–634.
2. Ansorge WJ. (2009) Next-generation DNA sequencing techniques. *Nat Biotechnol* **25**(4): 195–203.
3. Mardis ER. (2008) Next-generation DNA sequencing methods. *Annu Rev Genomics Hum Genet* **9**: 387–402.
4. Shendure J, Ji H. (2008) Next-generation DNA sequencing. *Nat Biotechnol* **26**(10): 1135–1145.
5. Lin B, Wang J, Cheng Y. (2008) Recent patents and advances in the next-generation sequencing technologies. *Recent Patents on Biomedical Engineering* **1**: 60–67.
6. Eid J, Fehr A, Gray J *et al.* (2009) Real-time DNA sequencing from single polymerase molecules. *Science* **323**(5910): 133–138.
7. Branton D, Deamer DW, Marziali A *et al.* (2008) The potential and challenges of nanopore sequencing. *Nat Biotechnol* **26**(10): 1146–1153.
8. Clarke J, Wu HC, Jayasinghe L *et al.* (2009) Continuous base identification for single-molecule nanopore DNA sequencing. *Nat Nanotechnol* **4**(4): 265–270.
9. Pleasance ED, Cheetham RK, Stephens PJ *et al.* (2010) A comprehensive catalogue of somatic mutations from a human cancer genome. *Nature* **463**(7278): 191–196.
10. Yeager M, Xiao N, Hayes RB *et al.* (2008) Comprehensive resequence analysis of a 136 kb region of human chromosome 8q24 associated with prostate and colon cancers. *Hum Genet* **124**(2): 161–170.
11. Sugarbaker DJ, Richards WG, Gordon GJ *et al.* (2008) Transcriptome sequencing of malignant pleural mesothelioma tumors. *Proc Natl Acad Sci USA* **105**(9): 3521–3526.
12. Cheng Y, Wang J, Shao J *et al.* (2010) Identification of novel SNPs by next-generation sequencing of the genomic region containing the APC gene in colorectal cancer patients in China. *OMICS* **14**(3): 315–325.
13. Cartegni L, Chew SL, Krainer AR. (2002) Listening to silence and understanding nonsense: Exonic mutations that affect splicing. *Nat Rev Genet* **3**(4): 285–298.
14. Wang GS, Cooper TA. (2007) Splicing in disease: Disruption of the splicing code and the decoding machinery. *Nat Rev Genet* **8**(10): 749–761.
15. Cartegni L, Wang J, Zhu Z *et al.* (2003) ESEfinder: A web resource to identify exonic splicing enhancers. *Nucleic Acids Res* **31**(13): 3568–3571.
16. Wang Z, Rolish ME, Yeo G *et al.* (2004) Systematic identification and analysis of exonic splicing silencers. *Cell* **119**(6): 831–845.
17. Bond GL, Hu W, Bond EE *et al.* (2004) A single nucleotide polymorphism in the MDM2 promoter attenuates the p53 tumor suppressor pathway and accelerates tumor formation in humans. *Cell* **119**(5): 591–602.
18. Pleasance ED, Stephens PJ, O'Meara S *et al.* (2010) A small-cell lung cancer genome with complex signatures of tobacco exposure. *Nature* **463**(7278): 184–190.

19. Ley TJ, Mardis ER, Ding L *et al.* (2008) DNA sequencing of a cytogenetically normal acute myeloid leukaemia genome. *Nature* **456**(7218): 66–72.

20. Mardis ER, Ding L, Dooling DJ *et al.* (2009) Recurring mutations found by sequencing an acute myeloid leukemia genome. *N Engl J Med* **361**(11): 1058–1066.

21. Shah SP, Kobel M, Senz J *et al.* (2009) Mutation of FOXL2 in granulosa-cell tumors of the ovary. *N Engl J Med* **360**(26): 2719–2729.

22. McCarroll SA, Kuruvilla FG, Korn JM *et al.* (2008) Integrated detection and population-genetic analysis of SNPs and copy number variation. *Nat Genet* **40**(10): 1166–1174.

23. Cooper GM, Zerr T, Kidd JM *et al.* (2008) Systematic assessment of copy number variant detection via genome-wide SNP genotyping. *Nat Genet* **40**(10): 1199–1203.

24. Carter NP. (2007) Methods and strategies for analyzing copy number variation using DNA microarrays. *Nat Genet* **39**(7 Suppl): S16–S21.

25. Campbell PJ, Stephens PJ, Pleasance ED *et al.* (2008) Identification of somatically acquired rearrangements in cancer using genome-wide massively parallel paired-end sequencing. *Nat Genet* **40**(6): 722–729.

26. Chiang DY, Getz G, Jaffe DB *et al.* (2009) High-resolution mapping of copy-number alterations with massively parallel sequencing. *Nat Methods* **6**(1): 99–103.

27. Schweiger MR, Kerick M, Timmermann B *et al.* (2009) Genome-wide massively parallel sequencing of formaldehyde fixed-paraffin embedded (FFPE) tumor tissues for copy-number- and mutation-analysis. *PLoS One* **4**(5): e5548.

28. Wood HM, Belvedere O, Conway C *et al.* (2010) Using next-generation sequencing for high resolution multiplex analysis of copy number variation from nanogram quantities of DNA from formalin-fixed paraffin-embedded specimens. *Nucleic Acids Res* **38**(14): e151.

29. Medvedev P, Stanciu M, Brudno M. (2009) Computational methods for discovering structural variation with next-generation sequencing. *Nat Methods* **6**(11 Suppl): S13–S20.

30. Clark MJ, Homer N, O'Connor BD *et al.* (2010) U87MG decoded: The genomic sequence of a cytogenetically aberrant human cancer cell line. *PLoS Genet* **6**(1): e1000832.

31. Zhao Q, Caballero OL, Levy S *et al.* (2009) Transcriptome-guided characterization of genomic rearrangements in a breast cancer cell line. *Proc Natl Acad Sci USA* **106**(6): 1886–1891.

32. Maher CA, Palanisamy N, Brenner JC *et al.* (2009) Chimeric transcript discovery by paired-end transcriptome sequencing. *Proc Natl Acad Sci USA* **106**(30): 12353–12358.

33. Berger MF, Levin JZ, Vijayendran K *et al.* (2010) Integrative analysis of the melanoma transcriptome. *Genome Res* **20**(4): 413–427.

34. Wang XS, Prensner JR, Chen G *et al.* (2009) An integrative approach to reveal driver gene fusions from paired-end sequencing data in cancer. *Nat Biotechnol* **27**(11): 1005–1011.

35. Levin JZ, Berger MF, Adiconis X *et al.* (2009) Targeted next-generation sequencing of a cancer transcriptome enhances detection of sequence variants and novel fusion transcripts. *Genome Biol* **10**(10): R115.
36. Chmielecki J, Peifer M, Jia P *et al.* (2010) Targeted next-generation sequencing of DNA regions proximal to a conserved GXGXXG signaling motif enables systematic discovery of tyrosine kinase fusions in cancer. *Nucleic Acids Res* **38**(20): 6985–6996.
37. Lin B, White JT, Lu W *et al.* (2005) Evidence for the presence of disease-perturbed networks in prostate cancer cells by genomic and proteomic analyses: A systems approach to disease. *Cancer Res* **65**(8): 3081–3091.
38. Cheng L, Lu W, Kulkarni B *et al.* (2010) Analysis of chemotherapy response programs in ovarian cancers by the next-generation sequencing technologies. *Gynecol Oncol* **117**(2): 159–169.
39. Grigoriadis A, Mackay A, Reis-Filho JS *et al.* (2006) Establishment of the epithelial-specific transcriptome of normal and malignant human breast cells based on MPSS and array expression data. *Breast Cancer Res* **8**(5): R56.
40. Morrissy AS, Morin RD, Delaney A *et al.* (2009) Next-generation tag sequencing for cancer gene expression profiling. *Genome Res* **19**(10): 1825–1835.
41. Wang Z, Gerstein M, Snyder M. (2009) RNA-seq: A revolutionary tool for transcriptomics. *Nat Rev Genet* **10**(1): 57–63.
42. Mortazavi A, Williams BA, McCue K *et al.* (2008) Mapping and quantifying mammalian transcriptomes by RNA-seq. *Nat Methods* **5**(7): 621–628.
43. Marioni JC, Mason CE, Mane SM *et al.* (2008) RNA-seq: An assessment of technical reproducibility and comparison with gene expression arrays. *Genome Res* **18**(9): 1509–1517.
44. Sultan M, Schulz MH, Richard H *et al.* (2008) A global view of gene activity and alternative splicing by deep sequencing of the human transcriptome. *Science* **321**(5891): 956–960.
45. Fu X, Fu N, Guo S *et al.* (2009) Estimating accuracy of RNA-Seq and microarrays with proteomics. *BMC Genomics* **10**: 161.
46. Bradford JR, Hey Y, Yates T *et al.* (2010) A comparison of massively parallel nucleotide sequencing with oligonucleotide microarrays for global transcription profiling. *BMC Genomics* **11**: 282.
47. Tang F, Barbacioru C, Nordman E *et al.* (2010) RNA-seq analysis to capture the transcriptome landscape of a single cell. *Nat Protoc* **5**(3): 516–535.
48. Tang F, Barbacioru C, Bao S *et al.* (2010) Tracing the derivation of embryonic stem cells from the inner cell mass by single-cell RNA-Seq analysis. *Cell Stem Cell* **6**(5): 468–478.
49. Morin RD, Johnson NA, Severson TM *et al.* (2010) Somatic mutations altering EZH2 (Tyr641) in follicular and diffuse large B-cell lymphomas of germinal-center origin. *Nat Genet* **42**(2): 181–185.
50. Barski A, Cuddapah S, Cui K *et al.* (2007) High-resolution profiling of histone methylations in the human genome. *Cell* **129**(4): 823–837.
51. Robertson G, Hirst M, Bainbridge M *et al.* (2007) Genome-wide profiles of STAT1 DNA association using chromatin immunoprecipitation and massively parallel sequencing. *Nat Methods* **4**: 651–657.

52. Johnson DS, Mortazavi A, Myers RM *et al.* (2007) Genome-wide mapping of *in vivo* protein–DNA interactions. *Science* **316**(5830): 1497–1502.
53. Ji H, Jiang H, Ma W *et al.* (2008) An integrated software system for analyzing ChIP-chip and ChIP-seq data. *Nat Biotechnol* **26**(11): 1293–1300.
54. Lin B, Wang J, Hong X *et al.* (2009) Integrated expression profiling and ChIP-seq analyses of the growth inhibition response program of the androgen receptor. *PLoS One* **4**(8): e6589.
55. Niu Y, Altuwaijri S, Lai KP *et al.* (2008) Androgen receptor is a tumor suppressor and proliferator in prostate cancer. *Proc Natl Acad Sci USA* **105**(34): 12182–12187.
56. Kaighn ME, Narayan KS, Ohnuki Y *et al.* (1979) Establishment and characterization of a human prostatic carcinoma cell line (PC-3). *Invest Urol* **17**(1): 16–23.
57. Bernstein BE, Mikkelsen TS, Xie X *et al.* (2006) A bivalent chromatin structure marks key developmental genes in embryonic stem cells. *Cell* **125**(2): 315–326.
58. Franz H, Mosch K, Soeroes S *et al.* (2009) Multimerization and H3K9me3 binding are required for CDYL1b heterochromatin association. *J Biol Chem* **284**(50): 35049–35059.
59. Zhang Y, Lv J, Liu H *et al.* (2010) HHMD: The human histone modification database. *Nucleic Acids Res* **38**(Database issue): D149–D154.
60. Zang C, Schones DE, Zeng C *et al.* (2009) A clustering approach for identification of enriched domains from histone modification ChIP-Seq data. *Bioinformatics* **25**(15): 1952–1958.
61. Xu H, Wei CL, Lin F *et al.* (2008) An HMM approach to genome-wide identification of differential histone modification sites from ChIP-seq data. *Bioinformatics* **24**(20): 2344–2349.
62. Hon G, Ren B, Wang W. (2008) ChromaSig: A probabilistic approach to finding common chromatin signatures in the human genome. *PLoS Comput Biol* **4**(10): e1000201.
63. Zhang Y, Shin H, Song JS *et al.* (2008) Identifying positioned nucleosomes with epigenetic marks in human from ChIP-Seq. *BMC Genomics* **9**: 537.
64. Grosso AR, Martins S, Carmo-Fonseca M. (2008) The emerging role of splicing factors in cancer. *EMBO Rep* **9**(11): 1087–1093.
65. Montero LM, Filipski J, Gil P *et al.* (1992) The distribution of 5-methylcytosine in the nuclear genome of plants. *Nucleic Acids Res* **20**(12): 3207–3210.
66. Ballestar E, Esteller M. (2008) SnapShot: The human DNA methylome in health and disease. *Cell* **135**(6): 1144–e1.
67. Brunner AL, Johnson DS, Kim SW *et al.* (2009) Distinct DNA methylation patterns characterize differentiated human embryonic stem cells and developing human fetal liver. *Genome Res* **19**(6): 1044–1056.
68. Down TA, Rakyan VK, Turner DJ *et al.* (2008) A Bayesian deconvolution strategy for immunoprecipitation-based DNA methylome analysis. *Nat Biotechnol* **26**(7): 779–785.

69. Butcher LM, Beck S. (2010) AutoMeDIP-seq: A high-throughput, whole genome, DNA methylation assay. *Methods* **52**(3): 223–231.

70. Tost J, Gut IG. (2007) DNA methylation analysis by pyrosequencing. *Nat Protoc* **2**(9): 2265–2275.

71. Cokus SJ, Feng S, Zhang X *et al.* (2008) Shotgun bisulphite sequencing of the Arabidopsis genome reveals DNA methylation patterning. *Nature* **452**(7184): 215–219.

72. Serre D, Lee BH, Ting AH. (2010) MBD-isolated Genome Sequencing provides a high-throughput and comprehensive survey of DNA methylation in the human genome. *Nucleic Acids Res* **38**(2): 391–399.

73. Smith ZD, Gu H, Bock C *et al.* (2009) High-throughput bisulfite sequencing in mammalian genomes. *Methods* **48**(3): 226–232.

74. Ruike Y, Imanaka Y, Sato F *et al.* (2010) Genome-wide analysis of aberrant methylation in human breast cancer cells using methyl-DNA immunoprecipitation combined with high-throughput sequencing. *BMC Genomics* **11**: 137.

75. Sanford JR, Wang X, Mort M *et al.* (2009) Splicing factor SFRS1 recognizes a functionally diverse landscape of RNA transcripts. *Genome Res* **19**(3): 381–394.

76. Hafner M, Landthaler M, Burger L *et al.* (2010) PAR-CliP — a method to identify transcriptome-wide the binding sites of RNA binding proteins. *J Vis Exp* (41) DoI: 10.3791/2034.

77. Hafner M, Landthaler M, Burger L *et al.* (2010) Transcriptome-wide identification of RNA-binding protein and microRNA target sites by PAR-CLIP. *Cell* **141**(1): 129–141.

78. Tong AW, Nemunaitis J. (2008) Modulation of miRNA activity in human cancer: A new paradigm for cancer gene therapy? *Cancer Gene Ther* **15**(6): 341–355.

79. Ruvkun G. (2006) Clarifications on miRNA and cancer. *Science* **311**(5757): 36–37.

80. Willenbrock H, Salomon J, Sokilde R *et al.* (2009) Quantitative miRNA expression analysis: Comparing microarrays with next-generation sequencing. *RNA* **15**(11): 2028–2034.

81. Hackenberg M, Sturm M, Langenberger D *et al.* (2009) miRanalyzer: A microRNA detection and analysis tool for next-generation sequencing experiments. *Nucleic Acids Res* **37**(Web Server issue): W68–W76.

82. Wyman SK, Parkin RK, Mitchell PS *et al.* (2009) Repertoire of microRNAs in epithelial ovarian cancer as determined by next generation sequencing of small RNA cDNA libraries. *PLoS One* **4**(4): e5311.

83. Weng L, Wu X, Gao H *et al.* (2010) MicroRNA profiling of clear cell renal cell carcinoma by whole-genome small RNA deep sequencing of paired frozen and formalin-fixed, paraffin-embedded tissue specimens. *J Pathol* **222**(1): 41–51.

84. Szczyrba J, Loprich E, Wach S *et al.* (2010) The microRNA profile of prostate carcinoma obtained by deep sequencing. *Mol Cancer Res* **8**(4): 529–538.

85. Bamford S, Dawson E, Forbes S *et al.* (2004) The COSMIC (Catalogue of Somatic Mutations in Cancer) database and website. *Br J Cancer* **91**(2): 355–358.

86. Forbes SA, Bhamra G, Bamford S *et al.* (2008) The catalogue of somatic mutations in cancer (COSMIC). *Curr Protoc Hum Genet* **Chapter 10**: Unit 10.1.

87. Forbes SA, Tang G, Bindal N *et al.* (2010) COSMIC (the Catalogue of Somatic Mutations in Cancer): A resource to investigate acquired mutations in human cancer. *Nucleic Acids Res* **38**(Database issue): D652–D657.

88. Carter H, Chen S, Isik L *et al.* (2009) Cancer-specific high-throughput annotation of somatic mutations: Computational prediction of driver missense mutations. *Cancer Res* **69**(16): 6660–6667.

89. Yan H, Parsons DW, Jin G *et al.* (2009) IDH1 and IDH2 mutations in gliomas. *N Engl J Med* **360**(8): 765–773.

90. Carter H, Samayoa J, Hruban RH *et al.* (2010) Prioritization of driver mutations in pancreatic cancer using cancer-specific high-throughput annotation of somatic mutations (CHASM). *Cancer Biol Ther* **10**(6): 582–587.

91. Tang F, Barbacioru C, Bao S *et al.* (2010) Tracing the derivation of embryonic stem cells from the inner cell mass by single-cell RNA-seq analysis. *Cell Stem Cell* **6**(5): 468–478.

92. Beck AH, Weng Z, Witten DM *et al.* (2010) 3′-end sequencing for expression quantification (3SEQ) from archival tumor samples. *PLoS One* **5**(1): e8768.

Chapter 13

Membrane Proteomics for the Opportunity of Cancer Biomarker and Drug Target Discovery

Chia-Li Han[†], Chih-Wei Chien[‡], Pei-Mien Chen[§]
I-Hsuan Chen[¶] and Yu-Ju Chen[*,†]

1. Significance of Membrane Proteins as Diagnostic and Therapeutic Biomarkers in Cancer

In eukaryotes, the biological membrane performs as a natural barrier which compartmentalizes not only the cells from external environment but also intracellular organelles. The cell surface membrane, or plasma membrane, maintains the essential boundaries between the cytoplasm and the extracellular environment. Membrane proteins are molecules that are attached to, or associated with the membrane of a cellular organelle. Membrane-associated proteins include proteins physically embedded in the lipid bilayers, such as ion channels and transmembrane proteins, and proteins anchored to the membrane that sense external signals, transport specific molecules, and connect the membrane to the cytoskeleton, the extracellular matrix (ECM), and adjacent cells.[1] Among various cellular components, dysregulation of membrane proteins has been linked to variety of human cancers. The composition and characteristics of membrane proteins of tumor cells are modified during malignant transformation and make

[*]Corresponding author.

[†]Institute of Chemistry, Academia Sinica, Taipei 115, Taiwan.

[‡]Department of Chemistry, National Tsing Hua University, Hsinchu 300, Taiwan.

[§]Agricultural Biotechnology Research Center, Academia Sinica, Taipei 115, Taiwan.

[¶]Department of Chemistry, National Taiwan Normal University, Taipei 106, Taiwan.

them likely candidates for cancer biomarkers.[2] Thus, membrane proteins, especially plasma membrane proteins, present the most potential roles as disease biomarker candidates. In addition, because of their localization on the surface of cells and organelles, membrane proteins are also most promising drug targets.[3]

Most of the FDA-approved cancer biomarkers and $\sim$70% of all known pharmaceutical drug targets are mainly membrane proteins.[4] Almost two-thirds of the currently used pharmaceutical drug targets are among these plasma membrane proteins.[5] Biomarkers are biological molecules that are indicators of physiologic state and also of change during a disease process.[6] The use of biomarkers in cancer aims at measuring the biologically relevant exposure more validly and precisely. Most tumor markers in current clinical use (Table 1),[7] such as human chorionic gonadotropin-β, α-fetoprotein, carcinoembryonic antigen, prostate-specific antigen, HER-2, and mucins (e.g. CA 19.9, CA 125, and CA 15.3) are membrane proteins and would secret to the serum for diagnostic use.[8–10] Some mucins, e.g. the secreted and transmembrane mucins, are intimately involved in inflammation and cancer. Moreover, diverse human malignancies overexpress transmembrane mucins to exploit their role in signaling cell growth and survival. Mucins have thus been identified as markers of adverse prognosis and attractive therapeutic targets. Notably, the findings that certain transmembrane mucins induce transformation and promote tumor progression have provided the experimental basis for demonstrating that inhibitors of their function are effective as anti-tumor agents in preclinical models.[11] Among the US FDA-approved cancer biomarkers, MUC1 (CA15-3) serum assay is to monitor clinical course in breast Cancer[12] and MUC16 (CA125) serum assay is used to detect early-stage disease and monitor clinical course in ovarian cancer.[13] Elevated levels of CA 19-9, an intracellular adhesion molecule, occur primarily in patients with pancreatic and biliary tract cancers.[14]

According to the list of drugs and corresponding targets obtained from the DrugBank database,[15] Yildirim *et al.* built a bipartite network composed of US Food and Drug Administration (FDA)-approved drugs or experimental drugs (drugs in the pipeline or not yet approved by the FDA) and proteins linked by drug–target binary associations.[5] During the analysis, about 60% of approved drug targets are membrane proteins and the percentage is down to $\sim$40% when experimental targets are added. Despite the increase in diversity of experimental drug targets, targets of recently FDA-approved drugs in the past 10 years (1996–2006) are still

Table 1. List of US FDA-approved membrane proteins biomarkers (from Ref. 7).

Biomarker	Type	Source	Cancer type	Clinical use
α-Fetoprotein	Glycoprotein	Serum	Nonseminomatous testicular	Staging
Human chorionic gonadotropin-β	Glycoprotein	Serum	Testicular	Staging
CA19-9	Carbohydrate	Serum	Pancreatic	Monitoring
CA125	Glycoprotein	Serum	Ovarian	Monitoring
CEA	Protein	Serum	Colon	Monitoring
Epidermal growth factor receptor	Protein	Serum	Colon	Selection of therapy
Thyroglobulin	Protein	Serum	Thyroid	Monitoring
PSA (total)	Protein	Serum	Prostate	Screening and monitoring
PSA (complex)	Protein	Serum	Prostate	Screening and monitoring
PSA (free PSA %)	Protein	Serum	Prostate	Benign prostatic hyperplasia versus cancer diagnosis
CA15-3	Glycoprotein	Serum	Breast	Monitoring
CA27-29	Glycoprotein	Serum	Breast	Monitoring
Oestrogen receptor and progesterone	Protein (IHC)	Breast tumour	Breast	Selection for hormonal therapy
HER2/NEU	Protein (IHC)	Breast tumour	Breast	Prognosis and selection of therapy
HER2/NEU	Protein	Serum	Breast	Monitoring
BTA	Protein	Urine	Bladder	Monitoring
High molecular weight CEA and mucin	Protein (Immunofluorescence)	Urine	Bladder	Monitoring

BTA, bladder tumour-associated antigen; CA, cancer antigen; CEA, carcinoembryonic antigen; IHC, immunohistochemistry; PSA, prostate-specific antigen.

mostly membrane proteins (Fig. 1).[5] Notably, membrane proteins also present major classification in disease-related proteins.

For example, the epidermal growth factor receptor (EGFR), one of four members of the EGFR/HER family of receptor tyrosine kinases (RTKs), plays an important role in normal organogenesis and in neoplastic steps involved in membrane proteomics studies. The foundation of any biomarker discovery effort is based on identification of proteins that show differentially expression between disease and control samples. For membrane proteomics, there are two different proteomics strategies, gel-based and solution-based methods, to perform in-depth identification of membrane proteome (Fig. 2). In *gel-based* approaches, the membrane proteins are purified, separated by 2D-PAGE (two-dimensional polyacrylamide gel electrophoresis, Fig. 2(a)), IEF (isoelectric focusing) or SDS (sodium dodecyl sulfate)-PAGE (Fig. 2(b)), performed in-gel digestion, analyzed by MS, and obtained the protein identification through database searching. In *solution-based* approaches, also called *shotgun proteomics*, membrane protein sample is directly digested and then subjected to off-line or on-line multi-dimensional LC-MS/MS analysis (Fig. 2(c)). To perform efficient analysis of membrane proteome, various methods including improved purification, solubilization, digestion, separation, and quantitation methods have been reported and the brief introduction are given in the following section.

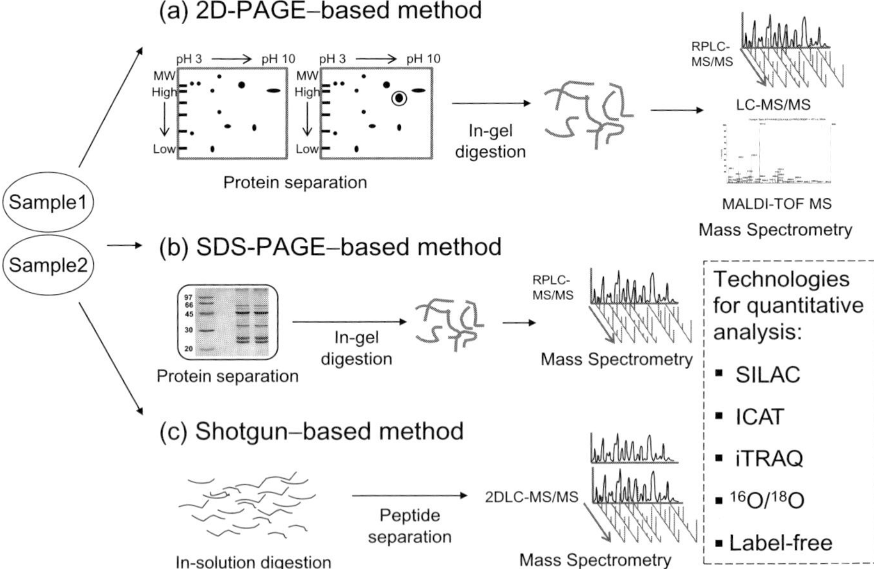

Fig. 2. Overview of quantitative membrane proteomics including (a) 2D-PAGE-based method; (b) SDS-PAGE-based method; and (c) shotgun-based method.

3.1. *Purification of membrane proteins*

There are several methods for purification of plasma membrane proteins (also see review in Ref. 27 for more details).[1,26,27] The most frequently used membrane isolation methods employ a combination of *differential centrifugation* and *sucrose density gradient centrifugation* methods. Because of the differences in lipid and protein compositions, membranes have different densities and can be separated from other organelles. Differential centrifugation can remove soluble proteins, the majority of mitochondria, and nuclei from the cell homogenates. Sucrose density gradients can further separate membranes with different densities. In this method, cells or tissues are resuspended in high-salt buffer for disruption of ionic bond, blend the cells or tissues by homogenizer, ultracentrifuge the suspension at 100,000 to 900,000 × g for the membrane pellet. Additional washing step using carbonate buffer will be applied to remove the membrane-associated proteins and the insoluble pellet will be collected after ultracentrifugation. This technique has also been used to enrich various cellular organelles. The most serious problem in this method is that plasma membrane fractions are often contaminated by other proteins from organelle membranes, including mitochondria, ER, and Golgi apparatus and by non-membrane proteins that associate with membrane proteins, such as the high abundance cytoskeletal proteins.[28] This method is also more complicated and time-consuming compared to other methods.

The *detergent/polymer two-phase partitioning system* is an attractive enrichment method for plasma membrane proteins.[29,30] Aqueous two-phase partition is a technique that separates proteins, on the basis of different solubility in surface properties, by partitioning the cell lysate between two immiscible, aqueous solutions of polymers. The two aqueous polymer solutions most commonly used are dextran and poly(ethyleneglycol), which, when mixed at certain concentrations, form the two immiscible phases. The upper phase is rich in poly(ethyleneglycol) and is relatively more hydrophobic than the lower phase, which is rich in dextran. While membrane proteins are enriched in the detergent phase, the soluble proteins are isolated in the polymer phase. To improve separation, a secondary partitioning step employs a plasma membrane–specific affinity ligand such as wheat germ agglutinin conjugated to dextran. Ideally, any contaminating membranes left over from the first step will partition into the poly(ethyleneglycol) phase, while the affinity ligand causes the target membrane to partition into the dextran phase. Due to its sensitivity to salt, temperature, affinity ligands, and exact polymer concentration, application

Table 2.   Summary of purification and digestion methods for membrane proteomics.

Name	Targeted proteins	Sample	Reference
Purification			
Aqeoues-polymer two phase partitioning	Plasma membrane proteins	Cells and tissues	29, 30
Affinity purification (biotinyla-tion)	Plasma membrane proteins	Cells	31
Affinity purification (glycoslya-tion)	Plasma membrane proteins	Cells and tissues	34

Name	Buffer	Digest region	Reference
Digestion			
Proteinase K	High-pH buffer	Extramembrane loop	41
Lys-C	4 M urea, 100 mM Tris HCl, pH 8.0	Extramembrane loop	42
Trypsin	2 M Urea	Extramembrane loop	43
60% MeOH/ trypsin	60% methanol	Intergral membrane proteins	44
hppK-CNBr/FA	90% FA	Transmembrane α-helical domain	45
Tube-gel digestion	Ammonium bicarbonate and 2% SDS	Intergral membrane proteins and transmembrane α-helical domain	46
Gel-assisted digestion	6 M urea, 5 mM EDTA, 2% SDS, 0.1 M TEABC	Intergral membrane proteins and transmembrane α-helical domain	47

3.3. *Protein/peptide separation methods for membrane proteomic study*

To obtain an efficient proteome-scale analysis, pre-fractionation of proteins/ peptides prior to MS detection is crucial for comprehensive identification and quantification of membrane proteins. The commonly used separation methods for membrane proteins can be classified in two different categories: the *electrophoresis method* and the *chromatography method*. In electrophoresis method, 2D-PAGE and SDS-PAGE separation are operated on protein level, and IEF has been used on both protein level and peptide level.

In chromatography method, strong cation exchange chromatography (SCX) and reversed phase (RP) chromatography were mostly common used on peptide level separation.

3.3.1. *Electrophoresis method (I): Two-dimensional polyacrylamide gel electrophoresis (2D-PAGE)*

In 2D-PAGE, membrane proteins are separated according to charge (pI) by isoelectric focusing (IEF) in the first dimension and size by SDS-PAGE in the second dimension. 2D-PAGE has unique advantage for the resolving complex mixtures of proteins, permitting the simultaneous visualization and analysis of hundreds or even thousands of gene products. Resolved protein spots are excised from the gel, subjected to in-gel tryptic digestion, and analyzed by MS to identify proteins of interest. 2D-PAGE had demonstrated success in profiling the sub-proteomes of the bacterial outer membrane,[48] organelle membranes,[49] and plasma membrane.[50] However, despite the unique features of 2D-PAGE, the serious under-representation of membrane proteins reported in the above studies reveal its drawback of low sensitivity, low tolerance with detergents, and limited detection of highly hydrophobic proteins.

3.3.2. *Electrophoresis method (II): Sodium dodecyl sulfate polyacrylamide gel electrophoresis (SDS-PAGE)*

SDS-PAGE is a technique used to separate membrane proteins according to their molecular weight (MW). Protein samples are first denatured by sample buffer (including 0.05 M Tris-HCl, 1% SDS, 10% glycerol, 0.01% bromophenol Blue, pH 6.8) and then loaded to the SDS-PAGE for separation. The protein bands are cut out and performed in-gel digestion with trypsin. The peptides are separated and analyzed by LC-MS/MS (also called "GeLC-MS").[51] Because of the use of highly denaturing sample buffer, the combination of SDS-PAGE and RPLC-MS/MS has been widely applied to membrane proteomics analysis.[52] This drawbacks of the method include sample lose during wash and elution steps and low throughput.

3.3.3. *Electrophoresis method (III): Isoelectric Focusing (IEF)*

Isoelectric focusing (IEF) was first established using carrier ampholytes for the use as the first dimension of 2D-PAGE.[53] IEF can separate proteins/peptides based on their pIs, which can be used as a filter criterion for protein identification. High resolution, improved reproducibility as well as higher loading capacity are the key advantages of IPG strip used in IEF separation.[54] Similar to 2D-PAGE, it has disadvantage of low detection sensitivity of hydrophobic protein. Thus, IEF was mostly applied in the peptide-level separation, followed by LC-MS/MS for protein identification. McQuade *et al.*[55] analyzed membrane proteomics of human embryonic stem cells by peptide IPG-IEF and LC-MS/MS. A total of 2,292 nonredundant proteins were identified with two or more high accuracy peptide matches from 260 μg human embryonic stem cell membrane enriched fraction. A total of 1,279 (44.9%) proteins were annotated to be membrane proteins, of which 395 proteins were predicted to be derived from the plasma membrane compartment. The TMHMM predicted 904 integral membrane proteins with up to 16 TM domains from the list.

3.3.4. *Chromatography methods (I): Strong cation exchange chromatography (SCX)*

For membrane proteomics analysis, the most commonly used shotgun proteomics method is Multidimensional Protein Identification Technology which comprises online or off-line SCX followed by RPLC-MS/MS.[39,56] SCX is established by a silica-based material with a bonded coating of a hydrophilic and anionic polymer, poly(2-sulfoethyl aspartamide)(SO_3 functional group), which can interact with peptides with positive charges and elute the peptides with a salt gradient, such as KCl. There are some successful studies that used SCX as first dimensional fractionation method followed by RPLC-MS/MS. Durr *et al.*[57] identified 450 proteins in luminal endothelial cell plasma membranes isolated from rat lungs and from cultured rat lung microvascular endothelial cells. Blonder *et al.*[58] demonstrated the identification of 2563 proteins from ∼200 μg nature killer cell microsomal proteins, of which 876 (34%) are classified as membrane proteins.

3.3.5. *Chromatography methods (II): Reversed phase liquid chromatography*

Reversed phase liquid chromatography (RPLC) is always a separation method prior to MS analysis. RPLC separates peptides based on the hydrophobicity of peptides and elutes peptides by high polarity buffer (e.g. acetonitrile). The C18 material is used for peptide separation and C4 or C8 are used for protein separation. Manadas *et al.*[59] analyzed 200 μg membrane-enriched fraction form cortical brain tissue using RP at high pH as first-dimensional separation followed by RPLC-MS/MS. A total of 308 proteins were identified, of which 222 were annotated as membrane proteins.

3.3.6. *Chromatography methods (III): Solution IEF*

Solution-IEF method using "OFFGEL" was introduced by Agilent Technologies.[60] Samples (peptides or proteins) in the liquid phase will be charged and separated according to their pI values. Compared with the IPG-IEF method, the samples from OFFEGL can be directly introduced to LC-MS/MS analysis without further sample processing or clean-up. Solution-IEF has proven to be a competitive fractionation tool in shotgun proteomics for different samples.[59] Elschenbroich *et al.*[61] analyzed membrane-enriched fraction isolated from murine C2C12 cells by OFFGEL (12 fractions, pH 3–10) followed by RPLC-MS/MS. A total of 1,398 proteins were reported including 465 proteins predicted to contain TM domains.

3.4. *Methodologies for quantitative membrane proteomics*

Recent advancements in quantitative proteomics technologies for membrane proteins show great promise on the identification of differentially expressed proteins between cancerous and normal samples and provide a panel of candidate molecules for cancer diagnosis and therapy. In gel-based proteomics methods, comparison 2D-PAGE and its more sophisticated technique of differential gel electrophoresis (DIGE) have some success on the quantitation of the membrane proteins in cancers.[62] Typically, a 2D-DIGE experiment is carried out by labeling the proteins from the diseased and the control samples with Cy3, and Cy5 *N*-hydroxysuccinimidyl ester fluorescent dyes. Cy2 dye is used to label the internal standard, which consists of a pooled sample comprising equal amounts of the two samples to be compared, enabling both inter- and intra-gel matching (Fig. 3). The pooled-sample internal standard represents every protein present across all samples. Scanning the gel at Cy2, Cy3, and Cy5 excitation wavelengths using a fluorescence imager allows visualization of the differentially labeled proteins. Resolved protein spots are excised from the gel, subjected to in-gel digestion and analyzed by MS to identify proteins of interest. However, because of several disadvantages mentioned in the Sec. 3.3 of this

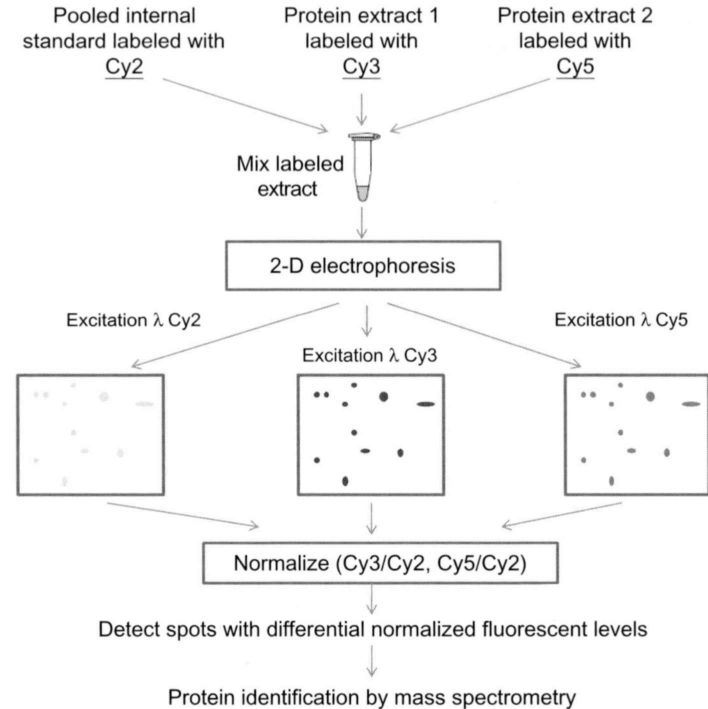

Fig. 3. Workflow of 2D DIGE (modified from Ref. 62).

chapter, 2D-PAGE-based quantification is of limited utility for membrane proteomics.

In solution-based method, either stable isotopic labeling or label-free quantitation is applied to the membrane proteomics studies in cancers. The integration of the multi-dimensional solution-based protein/peptide separation techniques and isotopic labeling methods has helped to minimize the insolubilization encountered in membrane proteomics studies using gel-based methods and provided the more comprehensive characterization of membrane proteins. Stable isotopic labeling can be introduced into amino acids in three ways: (i) metabolic labeling, (ii) chemical labeling, or (iii) enzymatic labeling.

3.4.1. *Metabolic labeling*

The most widely used metabolic labeling is stable isotope labeling by amino acid in cell culture (SILAC). The SILAC strategy was introduced by Ong

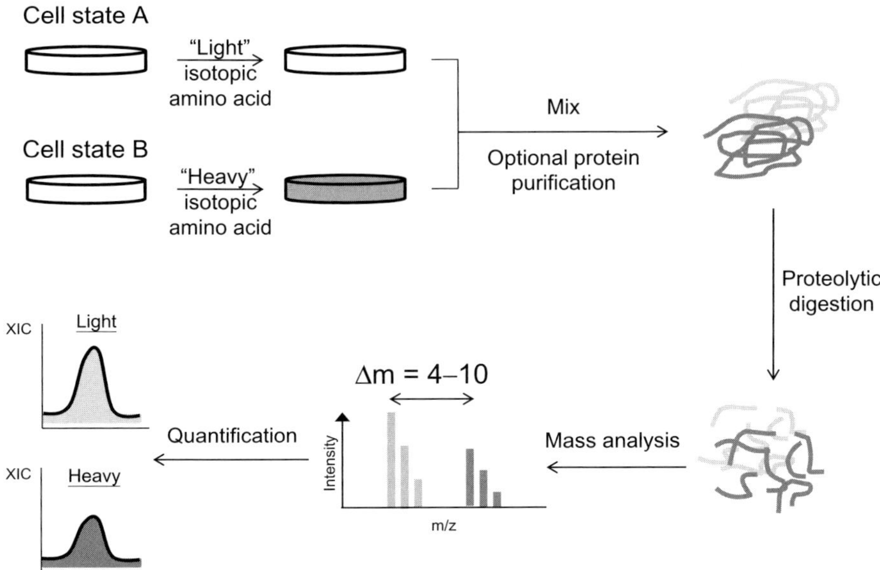

Fig. 4. Experimental workflow for SILAC.

et al.[63] and Du *et al.*[64] as a simple and accurate approach to expression proteomics (Fig. 4). In brief, isotopically synthesized light and heavy amino acids are separately added to amino acid deficient cell culture media in samples to be compared and are therefore incorporated into all proteins after cell culture. The cells labeled with light and heavy amino acids are combined and preceded with shotgun approach. The mass difference due to the light and heavy isotope can be measured in a MS scan. The relative quantification is thus determined by the intensity ratio of the peptides pairs. For design of multiplexed quantitation, SILAC offers options for selection of isotopically labeled amino acids such as arginine, leucine, lysine, serine, methionine, and tyrosine.[65] This method is compatible with virtually all cell culture conditions and show superior accuracy for protein quantitation. The SILAC is mostly limited in the analysis of cell lines, the incorporation of isotopes into clinical tissues still remain further investigation.

3.4.2. *Chemical labeling*

Quantitative proteomics using chemical and enzymatic labeling are sensitive to variations in protein purification between the compared samples

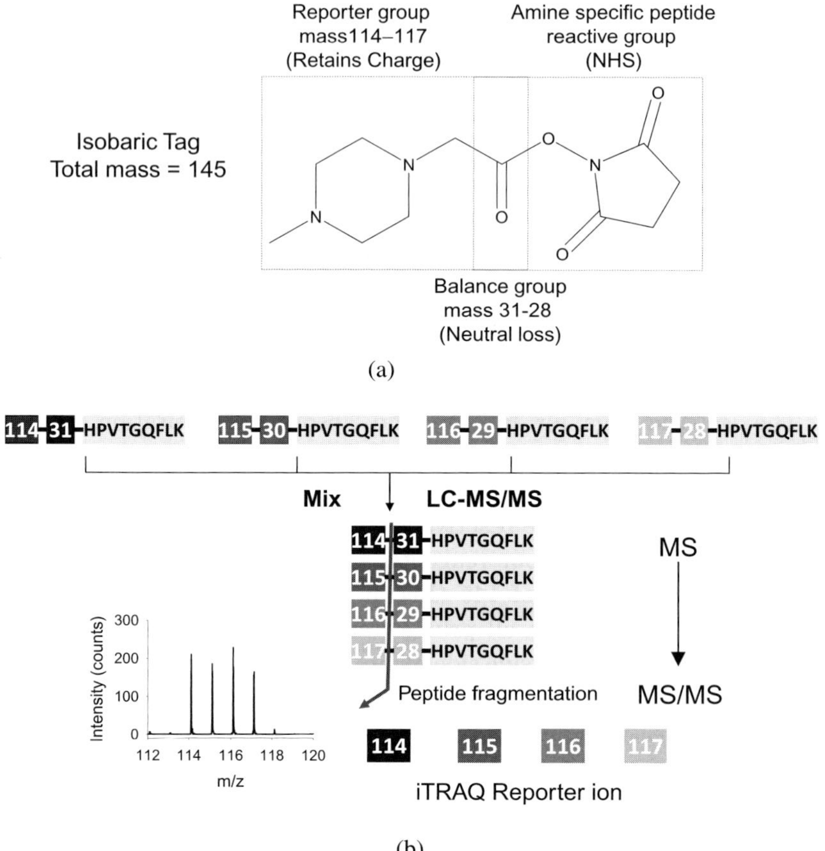

Fig. 6. (a) The components of the multiplexed isobaric tagging chemistry. The complete molecule consists of a reporter group, a mass balance group, and a peptide-reactive group. (b) The illustration of the isotopic tagging of four isobaric combinations and the peptide/protein quantitation was obtained from MS/MS stage.

oxygen (^{18}O) can be introduced into at the C-terminus of peptides through enzymatic labeling by digesting the proteins in the presence of H$_2^{18}$O using trypsin, LysC or GluC. Generally, one or two ^{18}O molecules will be introduced into the peptides. After enzymatic labeling, the labeled and unlabeled peptide pairs are measured by high resolution LC-MS/MS. The peak intensity of the peptide pair can be used for quantification. This technique is a simple and straightforward method for various kinds of sample types. The major drawback is that the labeling is not homogeneous, i.e. sometimes one ^{18}O is incorporated, while at other times two ^{18}O are incorporated.

3.4.4. *Label-free quantitation*

An alternative to the stable isotope-based strategies is the label-free approach, which is based on ion intensity quantification either by counting the fragment spectra of the peptides identifying the different proteins or by measuring the chromatographic peaks of the peptides. The experiment workflow is relatively simple compared to the labeling methods; proteins are first performed protease digestion and then directly analyzed by LC-MS/MS. Currently, there are three categories of label-free quantitation strategies: (1) spectral counting method: counting and comparing the number of MS/MS spectra of the identified peptides of a given protein; (2) peak intensity method: measuring and comparing the MS intensity of peptide precursor ions; and (3) extracted ion chromatogram method: integrating and comparing MS peak areas that are integrated over the chromatographic time scale of each peptides. This technique is cheaper to perform but requires the experiment to be repeated more times in a very stable LC instrument, more robust computing power, and computation algorithms.

4. The Applications of Quantitative Membrane Proteomics Strategies for Discovery of Cancer Biomarker and Therapeutic Target Candidates

Proteomics analysis holds promise to be valuable for identification of tissue and serum biomarkers associated with human cancers. Targeting with the membrane proteins, various quantitative membrane proteomics strategies had been applied to analyze protein expression levels in tissues, tumor cells, or body fluids from cancer patients to discover membrane protein biomarkers that will potentialy improve early diagnosis, prognosis, and prediction of treatment response.

4.1. *Tissue membrane proteomics*

For tissue membrane proteomics, Alfonso *et al.* used the 2D DIGE analysis for the membrane proteomes of paired tumoral and normal mucosal tissues from 6 colorectal cancer patients.[70] By further validation with immunohistochemistry using tissue microarray, annexin A2, annexin A4, and VDAC showed the potential as markers for colorectal cancer diagnosis and therapy. Pawlik and co-workers used ICAT to compare nipple aspirate fluid samples from 18 women patients with early breast cancer versus 4 healthy controls.[71] Among 36 differentially expressed proteins, they found and validated higher content of vitamin-D binding protein from breast cancer patients. Focusing on membrane-associated proteins, Rajcevic *et al.* applied iTRAQ-based approach to analyze serially transplanted glioma xenografts in rats during the progression from an invasive to an angiogenic phenotype.[72] Of the 1460 quantified proteins, known and novel candidate proteins were identified that characterize the switch from a non-angiogenic to a highly angiogenic phenotype and pointed to enhanced intercellular cross-talk and metabolic activity adopted by tumor cells in the angiogenic compared with the non-angiogenic phenotype. The expression levels of Calnexin, AnxA5, and AnxA2 were evaluated in patient specimens of different glioma types (grade I to IV) using a tissue microarray containing about 180 clinical cores. All three proteins showed increased expression in high grade compared with low grade gliomas (Fig. 7). These data suggest

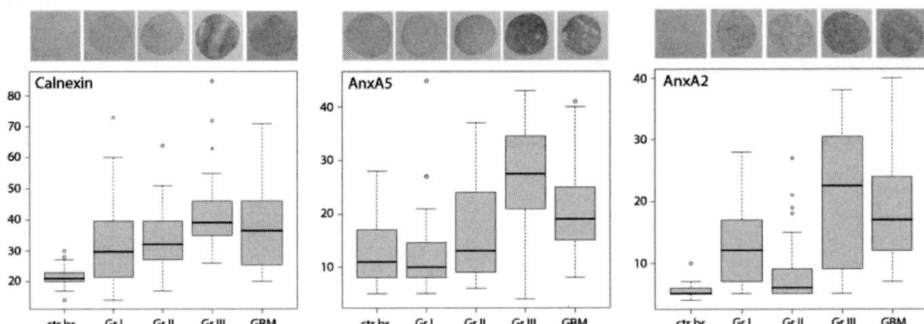

Fig. 7. Immunohistochemistry and quantification of Calnexin, AnxA5, and AnxA2 on a tissue microarray (adapt from Ref. 72). Immunohistochemistry and quantification thereof of candidate proteins Calnexin, AnxA5, and AnxA2 on a TMA containing more than 200 clinical gliomas (grades I–IV) are shown. The *y axis* of the box plots corresponds to mean pixel intensity (relative units between 0 and 255). A significant difference in expression was found between low grade (I and II) gliomas and high grade (III and IV) gliomas for Calnexin, AnxA5, and AnxA2 ($p < 0.001$). For AnxA2, a significant difference was also found between grade I and grade II. For AnxA5, the difference between grade II and grade IV was less significant ($p < 0.05$). All three proteins were expressed at a very low level in the normal human brain (*left bars*). A representative core is shown *above* each quantification box plot. The *black bar* in the box indicates the median sample value; the *whiskers* indicate the *upper* and *lower* quartile, respectively. *Open circles* represent observations, which lie more than 1.5 times the interquartile range from the first and third quartile. *ctr br.*, control brain; *Gr*, grade.

that Calnexin, AnxA2, and AnxA5 may be further exploited as biomarkers for malignant glioma.

Kristiansen *et al.*[73] applied [18]O labeling based quantitative proteomics to identify proteins that are highly expressed in cholangiocarcinomas. Golgi membrane protein 1, annexin IV, and epidermal growth factor receptor pathway substrate 8 are identified and verified as candidate biomarkers for cholangiocarcinomas. Among them, Golgi membrane protein 1 was observed to be overexpressed in 89% of 36 cholangiocarcinoma cases analyzed by staining tissue microarrays and has the potential to be biomarker for early detection of cholangiocarcinomas.

4.2. *Cancer cell lines*

Lund *et al.*[74] performed SILAC-based quantitative proteomic analysis to the plasma membrane proteome of two human breast carcinoma cell lines, NM-2C5 and M-4A4, which were found to be equally tumorigenic but have different metastasis ability. They identified 16 cell surface proteins as potential markers of the ability of breast cancer cells to form distant metastases. Liang *et al.*[75] identified approximately 70 differentially expressed proteins among 830 membrane or membrane-associated proteins in the normal and malignant breast cancer cells by SILAC methods. Leth-Larsen and co-workers applied SILAC method to quantify 1919 identified proteins in metastasis-related plasma membrane protein fraction of human breast cancer cell. Thirteen membrane proteins were highly expressed and three were underexpressed in the metastatic cell line compared with the nonmetastasis cell line.[76] The study demonstrated a quantitative proteomic strategy to identify clinically relevant key molecules in the early events of metastasis, some of which may prove to be potential targets for cancer therapy. Stockwin *et al.*[77] performed quantitative proteomic analysis of plasma membrane from hypoxia-adapted murine B16F10 melanoma using differential ^{16}O/^{18}O stable isotopic labeling and multidimensional liquid chromatography-tandem mass spectrometry. The analysis resulted in the quantification of 2,433 proteins and 39.8% of the 1620 annotated proteins were identified as membrane proteins and 10.2% as extracellular proteins. For a subset of plasma membrane and secreted proteins, consistent increases at the proteomic and transcriptomic levels were observed for aminopeptidase N (CD13), carbonic anhydrase IX, potassium-transporting ATPase, matrix metalloproteinase 9, and stromal cell derived factor I (SDF-1) (Fig. 8). Antibody-based analysis of a panel of human melanoma

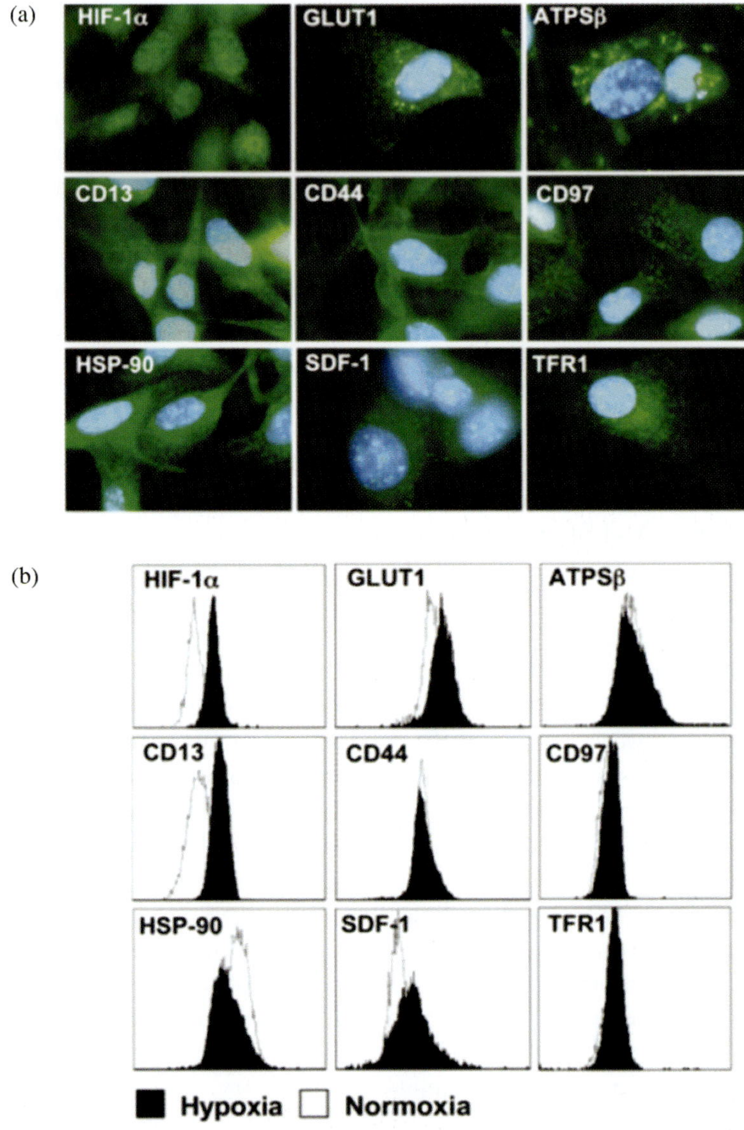

Fig. 8. Flow cytometric validation of hypoxia-induced changes in protein expression. (a) Hypoxic B16F10 cells were fixed, permeabilized, and stained with nine different antibodies to confirm both expression and subcellular localization. Control slides stained with isotype-matched monoclonal antibody were included and showed negligible fluorescence. (b) Hypoxic and normoxic cells were then processed in parallel using the identical protocol and then analyzed by flow cytometry. Results shown are fold changes in FL1 fluorescence (FL1-H/FL1-N) normalized to isotype control fluorescence values. (adapt from Ref. 77).

cell lines confirmed that CD13 and SDF-1 were consistently upregulated during hypoxia.

5. Future Perspective

Membrane proteins are attractive targets for cancer diagnosis, prognosis, monitoring, treatment prediction, and may serve as drug target for therapy. However, membrane proteins are highly hydrophobic and low abundant which require more sophisticated proteomics methods for their purification, digestion, separation, and identification. Although most proteomics studies of membrane proteins are based on cancer cell lines, new strategies for identification of membrane protein markers in tissues and cancer cells obtained by using laser capture microdissection from patients are being developed. Other than the discovery of biomarker candidates by membrane proteomics, the validation and functional assay of these candidates are crucial for the future daily clinical practice. The collection of a large cohort of proper patient materials, production of antibodies targeting these membrane proteins, assay development, and testing of large numbers of clinical samples remains the most difficult mission and need close collaborations between translational researchers, pathologists, and clinicians.

References

1. Zheng YZ, Foster LJ. (2009) Biochemical and proteomic approaches for the study of membrane microdomains. *J Proteomics* **72**(1): 12–22.
2. Uetrecht AC, Bear JE. (2006) Coronins: The return of the crown. *Trends Cell Biol* **16**(8): 421–426.
3. Josic D, Clifton JG, Kovac S, Hixson DC. (2008) Membrane proteins as diagnostic biomarkers and targets for new therapies. *Curr Opin Mol Ther* **10**(2): 116–123.
4. Hopkins AL, Groom CR. (2002) The druggable genome. *Nat Rev Drug Discov* **1**(9): 727–730.
5. Yildirim MA, Goh KI, Cusick ME, Barabasi AL, Vidal M. (2007) Drug-target network. *Nat Biotechnol* **25**(10): 1119–1126.
6. Srinivas PR, Kramer BS, Srivastava S. (2001) Trends in biomarker research for cancer detection. *Lancet Oncol* **2**(11): 698–704.
7. Ludwig JA, Weinstein JN. (2005) Biomarkers in cancer staging, prognosis and treatment selection. *Nat Rev Cancer* **5**(11): 845–856.
8. Leth-Larsen R, Lund RR, Ditzel HJ. (2010) Plasma membrane proteomics and its application in clinical cancer biomarker discovery. *Mol Cell Proteomics* **9**(7): 1369–1382.

9. Perkins GL, Slater ED, Sanders GK, Prichard JG. (2003) Serum tumor markers. *Am Fam Physician* **68**(6): 1075–1082.

10. Shariat SF, Karam JA, Margulis V, Karakiewicz PI. (2008) New blood-based biomarkers for the diagnosis, staging and prognosis of prostate cancer. *BJU Int* **101**(6): 675–683.

11. Kufe DW. (2009) Mucins in cancer: Function, prognosis and therapy. *Nat Rev Cancer* **9**(12): 874–885.

12. Hayes DF, Sekine H, Ohno T, Abe M, Keefe K, Kufe DW. (1985) Use of a murine monoclonal antibody for detection of circulating plasma DF3 antigen levels in breast cancer patients. *J Clin Invest* **75**(5): 1671–1678.

13. Bast RC, Jr., Hennessy B, Mills GB. (2009) The biology of ovarian cancer: New opportunities for translation. *Nat Rev Cancer* **9**(6): 415–428.

14. Steinberg W. (1990) The clinical utility of the CA 19-9 tumor-associated antigen. *Am J Gastroenterol* **85**(4): 350–355.

15. Wishart DS, Knox C, Guo AC, Shrivastava S, Hassanali M, Stothard P, Chang Z, Woolsey J. (2006) DrugBank: A comprehensive resource for in silico drug discovery and exploration. *Nucleic Acids Res* **34**(Database issue): D668–D672.

16. Yonesaka K, Zejnullahu K, Lindeman N, Homes AJ, Jackman DM, Zhao F, Rogers AM, Johnson BE, Janne PA. (2008) Autocrine production of amphiregulin predicts sensitivity to both gefitinib and cetuximab in EGFR wild-type cancers. *Clin Cancer Res* **14**(21): 6963–6973.

17. Normanno N, Bianco C, Strizzi L, Mancino M, Maiello MR, De Luca A, Caponigro F, Salomon DS. (2005) The ErbB receptors and their ligands in cancer: An overview. *Curr Drug Targets* **6**(3): 243–257.

18. Slamon DJ, Clark GM, Wong SG, Levin WJ, Ullrich A, McGuire WL. (1987) Human breast cancer: Correlation of relapse and survival with amplification of the HER-2/neu oncogene. *Science* **235**(4785): 177–182.

19. Bange J, Zwick E, Ullrich A. (2001) Molecular targets for breast cancer therapy and prevention. *Nat Med* **7**(5): 548–552.

20. Ross JS, Fletcher JA, Linette GP, Stec J, Clark E, Ayers M, Symmans WF, Pusztai L, Bloom KJ. (2003) The Her-2/neu gene and protein in breast cancer 2003: Biomarker and target of therapy. *Oncologist* **8**(4): 307–325.

21. Pegram MD, Finn RS, Arzoo K, Beryt M, Pietras RJ, Slamon DJ. (1997) The effect of HER-2/neu overexpression on chemotherapeutic drug sensitivity in human breast and ovarian cancer cells. *Oncogene* **15**(5): 537–547.

22. Carlomagno C, Perrone F, Gallo C, De Laurentiis M, Lauria R, Morabito A, Pettinato G, Panico L, D'Antonio A, Bianco AR, De Placido S. (1996) c-erb B2 overexpression decreases the benefit of adjuvant tamoxifen in early-stage breast cancer without axillary lymph node metastases. *J Clin Oncol* **14**(10): 2702–2708.

23. Lu B, McClatchy DB, Kim JY, Yates JR, 3rd. (2008) Strategies for shotgun identification of integral membrane proteins by tandem mass spectrometry. *Proteomics* **8**(19): 3947–3955.

24. Fenselau C. (2007) A review of quantitative methods for proteomic studies. *J Chromatogr B Analyt Technol Biomed Life Sci* **855**(1): 14–20.

25. Krogh A, Larsson B, von Heijne G, Sonnhammer EL. (2001) Predicting transmembrane protein topology with a hidden Markov model: Application to complete genomes. *J Mol Biol* **305**(3): 567–580.

26. Goshe MB, Blonder J, Smith RD. (2003) Affinity labeling of highly hydrophobic integral membrane proteins for proteome-wide analysis. *J Proteome Res* **2**(2): 153–161.

27. Speers AE, Wu CC. (2007) Proteomics of integral membrane proteins — theory and application. *Chem Rev* **107**(8): 3687–3714.

28. Tan S, Tan HT, Chung MC. (2008) Membrane proteins and membrane proteomics. *Proteomics* **8**(19): 3924–3932.

29. Schindler J, Lewandrowski U, Sickmann A, Friauf E, Nothwang HG. (2006) Proteomic analysis of brain plasma membranes isolated by affinity two-phase partitioning. *Mol Cell Proteomics* **5**(2): 390–400.

30. Schindler J, Nothwang HG. (2006) Aqueous polymer two-phase systems: Effective tools for plasma membrane proteomics. *Proteomics* **6**(20): 5409–5417.

31. Zhao Y, Zhang W, Kho Y. (2004) Proteomic analysis of integral plasma membrane proteins. *Anal Chem* **76**(7): 1817–1823.

32. McDonald CA, Yang JY, Marathe V, Yen TY, Macher BA. (2009) Combining results from lectin affinity chromatography and glycocapture approaches substantially improves the coverage of the glycoproteome. *Mol Cell Proteomics* **8**(2): 287–301.

33. Lee A, Kolarich D, Haynes PA, Jensen PH, Baker MS, Packer NH. (2009) Rat liver membrane glycoproteome: Enrichment by phase partitioning and glycoprotein capture. *J Proteome Res* **8**(2): 770–781.

34. Ghosh D, Krokhin O, Antonovici M, Ens W, Standing KG, Beavis RC, Wilkins JA. (2004) Lectin affinity as an approach to the proteomic analysis of membrane glycoproteins. *J Proteome Res* **3**(4): 841–850.

35. Blonder J, Goshe MB, Moore RJ, Pasa-Tolic L, Masselon CD, Lipton MS, Smith RD. (2002) Enrichment of integral membrane proteins for proteomic analysis using liquid chromatography-tandem mass spectrometry. *J Proteome Res* **1**(4): 351–360.

36. Han J, Schey KL. (2004) Proteolysis and mass spectrometric analysis of an integral membrane: Aquaporin 0. *J Proteome Res* **3**(4): 807–812.

37. Yu YQ, Gilar M, Gebler JC. (2004) A complete peptide mapping of membrane proteins: A novel surfactant aiding the enzymatic digestion of bacteriorhodopsin. *Rapid Commun Mass Spectrom* **18**(6): 711–715.

38. Norris JL, Porter NA, Caprioli RM. (2003) Mass spectrometry of intracellular and membrane proteins using cleavable detergents. *Anal Chem* **75**(23): 6642–6647.

39. Washburn MP, Wolters D, Yates JR, 3rd. (2001) Large-scale analysis of the yeast proteome by multidimensional protein identification technology. *Nat Biotechnol* **19**(3): 242–247.

40. Russell WK, Park ZY, Russell DH. (2001) Proteolysis in mixed organic-aqueous solvent systems: Applications for peptide mass mapping using mass spectrometry. *Anal Chem* **73**(11): 2682–2685.

41. Wu CC, MacCoss MJ, Howell KE, Yates JR, 3rd. (2003) A method for the comprehensive proteomic analysis of membrane proteins. *Nat Biotechnol* **21**(5): 532–538.

42. Nielsen PA, Olsen JV, Podtelejnikov AV, Andersen JR, Mann M, Wisniewski JR. (2005) Proteomic mapping of brain plasma membrane proteins. *Mol Cell Proteomics* **4**(4): 402–408.

43. Rodriguez-Ortega MJ, Norais N, Bensi G, Liberatori S, Capo S, Mora M, Scarselli M, Doro F, Ferrari G, Garaguso I, Maggi T, Neumann A, Covre A, Telford JL, Grandi G. (2006) Characterization and identification of vaccine candidate proteins through analysis of the group A Streptococcus surface proteome. *Nat Biotechnol* **24**(2): 191–197.

44. Karsan A, Blonder J, Law J, Yaquian E, Lucas DA, Conrads TP, Veenstra T. (2005) Proteomic analysis of lipid microdomains from lipopolysaccharide-activated human endothelial cells. *J Proteome Res* **4**(2): 349–357.

45. Blackler AR, Speers AE, Ladinsky MS, Wu CC. (2008) A shotgun proteomic method for the identification of membrane-embedded proteins and peptides. *J Proteome Res* **7**(7): 3028–3034.

46. Lu X, Zhu H. (2005) Tube-gel digestion: A novel proteomic approach for high throughput analysis of membrane proteins. *Mol Cell Proteomics* **4**(12): 1948–1958.

47. Han CL, Chien CW, Chen WC, Chen YR, Wu CP, Li H, Chen YJ. (2008) A multiplexed quantitative strategy for membrane proteomics: Opportunities for mining therapeutic targets for autosomal dominant polycystic kidney disease. *Mol Cell Proteomics* **7**(10): 1983–1997.

48. Phadke ND, Molloy MP, Steinhoff SA, Ulintz PJ, Andrews PC, Maddock JR. (2001) Analysis of the outer membrane proteome of Caulobacter crescentus by two-dimensional electrophoresis and mass spectrometry. *Proteomics* **1**(5): 705–720.

49. Bunai K, Nozaki M, Kakeshita H, Nemoto T, Yamane K. (2005) Quantitation of de novo localized (15)N-labeled lipoproteins and membrane proteins having one and two transmembrane segments in a Bacillus subtilis secA temperature-sensitive mutant using 2D-PAGE and MALDI-TOF MS. *J Proteome Res* **4**(3): 826–836.

50. Helling S, Schmitt E, Joppich C, Schulenborg T, Mullner S, Felske-Muller S, Wiebringhaus T, Becker G, Linsenmann G, Sitek B, Lutter P, Meyer HE, Marcus K. (2006) 2-D differential membrane proteome analysis of scarce protein samples. *Proteomics* **6**(16): 4506–4513.

51. Park YM, Kim JY, Kwon KH, Lee SK, Kim YH, Kim SY, Park GW, Lee JH, Lee B, Yoo JS. (2006) Profiling human brain proteome by multi-dimensional separations coupled with MS. *Proteomics* **6**(18): 4978–4986.

52. Shi R, Kumar C, Zougman A, Zhang Y, Podtelejnikov A, Cox J, Wisniewski JR, Mann M. (2007) Analysis of the mouse liver proteome using advanced mass spectrometry. *J Proteome Res* **6**(8): 2963–2972.

53. Gorg A, Postel W, Gunther S. (1988) The current state of two-dimensional electrophoresis with immobilized pH gradients. *Electrophoresis* **9**(9): 531–546.

54. Hubner NC, Ren S, Mann M. (2008) Peptide separation with immobilized pI strips is an attractive alternative to in-gel protein digestion for proteome analysis. *Proteomics* **8**(23-24): 4862–4872.

55. McQuade LR, Schmidt U, Pascovici D, Stojanov T, Baker MS. (2009) Improved membrane proteomics coverage of human embryonic stem cells by peptide IPG-IEF. *J Proteome Res* **8**(12): 5642–5649.

56. Link AJ, Eng J, Schieltz DM, Carmack E, Mize GJ, Morris DR, Garvik BM, Yates JR, 3rd. (1999) Direct analysis of protein complexes using mass spectrometry. *Nat Biotechnol* **17**(7): 676–682.

57. Durr E, Yu J, Krasinska KM, Carver LA, Yates JR, Testa JE, Oh P, Schnitzer JE. (2004) Direct proteomic mapping of the lung microvascular endothelial cell surface *in vivo* and in cell culture. *Nat Biotechnol* **22**(8): 985–992.

58. Blonder J, Rodriguez-Galan MC, Chan KC, Lucas DA, Yu LR, Conrads TP, Issaq HJ, Young HA, Veenstra TD. (2004) Analysis of murine natural killer cell microsomal proteins using two-dimensional liquid chromatography coupled to tandem electrospray ionization mass spectrometry. *J Proteome Res* **3**(4): 862–870.

59. Manadas B, English JA, Wynne KJ, Cotter DR, Dunn MJ. (2009) Comparative analysis of OFFGel, strong cation exchange with pH gradient, and RP at high pH for first-dimensional separation of peptides from a membrane-enriched protein fraction. *Proteomics* **9**(22): 5194–5198.

60. Michel PE, Reymond F, Arnaud IL, Josserand J, Girault HH, Rossier JS. (2003) Protein fractionation in a multicompartment device using Off-Gel isoelectric focusing. *Electrophoresis* **24**(1-2): 3–11.

61. Elschenbroich S, Ignatchenko V, Sharma P, Schmitt-Ulms G, Gramolini AO, Kislinger T. (2009) Peptide separations by on-line MudPIT compared to isoelectric focusing in an off-gel format: Application to a membrane-enriched fraction from C2C12 mouse skeletal muscle cells. *J Proteome Res* **8**(10): 4860–4869.

62. Van den Bergh G, Arckens L. (2004) Fluorescent two-dimensional difference gel electrophoresis unveils the potential of gel-based proteomics. *Curr Opin Biotechnol* **15**(1): 38–43.

63. Ong SE, Blagoev B, Kratchmarova I, Kristensen DB, Steen H, Pandey A, Mann M. (2002) Stable isotope labeling by amino acids in cell culture, SILAC, as a simple and accurate approach to expression proteomics. *Mol Cell Proteomics* **1**(5): 376–386.

64. Du YC, Gu S, Zhou J, Wang T, Cai H, Macinnes MA, Bradbury EM, Chen X. (2006) The dynamic alterations of H2AX complex during DNA repair detected by a proteomic approach reveal the critical roles of Ca(2+)/calmodulin in the ionizing radiation-induced cell cycle arrest. *Mol Cell Proteomics* **5**(6): 1033–1044.

65. Beynon RJ, Pratt JM. (2005) Metabolic labeling of proteins for proteomics. *Mol Cell Proteomics* **4**(7): 857–872.

66. Gygi SP, Rist B, Gerber SA, Turecek F, Gelb MH, Aebersold R. (1999) Quantitative analysis of complex protein mixtures using isotope-coded affinity tags. *Nat Biotechnol* **17**(10): 994–999.

67. Ross PL, Huang YN, Marchese JN, Williamson B, Parker K, Hattan S, Khainovski N, Pillai S, Dey S, Daniels S, Purkayastha S, Juhasz P, Martin S, Bartlet-Jones M, He F, Jacobson A, Pappin DJ. (2004) Multiplexed protein quantitation in Saccharomyces cerevisiae using amine-reactive isobaric tagging reagents. *Mol Cell Proteomics* **3**(12): 1154–1169.

68. Ramus C, Gonzalez de Peredo A, Dahout C, Gallagher M, Garin J. (2006) An optimized strategy for ICAT quantification of membrane proteins. *Mol Cell Proteomics* **5**(1): 68–78.

69. Yao X, Freas A, Ramirez J, Demirev PA, Fenselau C. (2001) Proteolytic 18O labeling for comparative proteomics: Model studies with two serotypes of adenovirus. *Anal Chem* **73**(13): 2836–2842.

70. Alfonso P, Canamero M, Fernandez-Carbonie F, Nunez A, Casal JI. (2008) Proteome analysis of membrane fractions in colorectal carcinomas by using 2D-DIGE saturation labeling. *J Proteome Res* **7**(10): 4247–4255.

71. Pawlik TM, Hawke DH, Liu Y, Krishnamurthy S, Fritsche H, Hunt KK, Kuerer HM. (2006) Proteomic analysis of nipple aspirate fluid from women with early-stage breast cancer using isotope-coded affinity tags and tandem mass spectrometry reveals differential expression of vitamin D binding protein. *BMC Cancer* **6**: 68.

72. Rajcevic U, Petersen K, Knol JC, Loos M, Bougnaud S, Klychnikov O, Li KW, Pham TV, Wang J, Miletic H, Peng Z, Bjerkvig R, Jimenez CR, Niclou SP. (2009) iTRAQ-based proteomics profiling reveals increased metabolic activity and cellular cross-talk in angiogenic compared with invasive glioblastoma phenotype. *Mol Cell Proteomics* **8**(11): 2595–2612.

73. Kristiansen TZ, Harsha HC, Gronborg M, Maitra A, Pandey A. (2008) Differential membrane proteomics using 18O-labeling to identify biomarkers for cholangiocarcinoma. *J Proteome Res* **7**(11): 4670–4677.

74. Lund R, Leth-Larsen R, Jensen ON, Ditzel HJ. (2009) Efficient isolation and quantitative proteomic analysis of cancer cell plasma membrane proteins for identification of metastasis-associated cell surface markers. *J Proteome Res* **8**(6): 3078–3090.

75. Liang X, Zhao J, Hajivandi M, Wu R, Tao J, Amshey JW, Pope RM. (2006) Quantification of membrane and membrane-bound proteins in normal and malignant breast cancer cells isolated from the same patient with primary breast carcinoma. *J Proteome Res* **5**(10): 2632–2641.

76. Leth-Larsen R, Lund R, Hansen HV, Laenkholm AV, Tarin D, Jensen ON, Ditzel HJ. (2009) Metastasis-related plasma membrane proteins of human breast cancer cells identified by comparative quantitative mass spectrometry. *Mol Cell Proteomics* **8**(6): 1436–1449.

77. Stockwin LH, Blonder J, Bumke MA, Lucas DA, Chan KC, Conrads TP, Issaq HJ, Veenstra TD, Newton DL, Rybak SM. (2006) Proteomic analysis of plasma membrane from hypoxia-adapted malignant melanoma. *J Proteome Res* **5**(11): 2996–3007.

Chapter 14

Structure-Based Systems Biology and Drug Design in Cancer Research

Chun-Hua Hsu*

1. Introduction

Structural biology now has an increasingly prominent role in cancer drug discovery, not only in understanding the function and mode of action of cancer targets by elucidating their tertiary structures but also by informing and accelerating the iterative process of drug design. Systems biology is also an emerging area which integrates information from diverse biological components and data into models of the system as a whole. As such, systems biology is associated with genome-wide studies, having little to do with structural biology. Structural biology and systems biology look like two different universes; however, it may be considered that structural biology could play a significant role in systems biology.[1,2] Structures of macromolecules, especially molecular machines or protein complexes, could provide quantitative parameters, help to elucidate functional networks, or enable rational designed perturbation experiments for drug development. Thus, structure-based systems biology, the determination and prediction of atomic resolution 3D structures of proteins on a specific pathway for better understanding their structure–function relationships, has now provided a new rationale for structural and systems biology and has become a major initiative in biotechnology. In the next stage, structures of macromolecular complexes discovered from systems biology could be solved

*Department of Agricultural Chemistry and Program of Genome and Systems Biology, National Taiwan University, Taipei 106, Taiwan. andyhsu@ntu.edu.tw

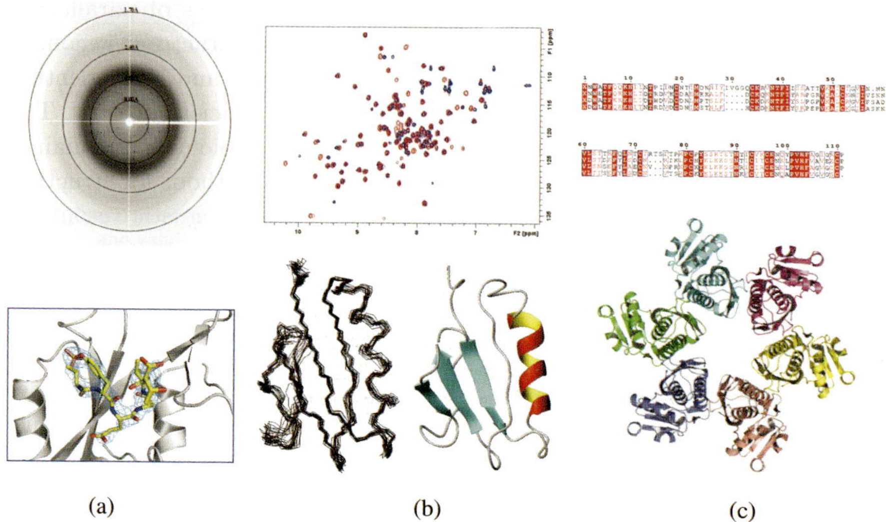

(a)   (b)   (c)

Fig. 2. Methods for the atomic level structural determination of macromolecular. (a) Electron diffraction map and 3D structure of Grb2-SH2/FAK complex determined by X-ray crystallography. X-ray crystallography integrates the diffraction patterns collected after bombarding a crystallized protein or complex with X-rays to construct its 3D structure. In principle, there is no size limit on the structures studied using this technique, although it is often difficult to obtain sufficient material for crystallization. (b) 3D structure and HSQC spectrum of C35 protein solved by NMR spectroscopy. NMR spectroscopy extracts distances between atoms by measuring transitions between different nuclear spin states within a magnetic field. These distances are then used as restraints to build 3D structure. NMR spectroscopy also provides atomic-resolution structures, but is generally limited to protein of about 400 residues. It plays an increasingly important role in studying interaction interfaces between structures determined independently. (c) Molecular modeling and the predicated 3D structure. Molecular modeling encompasses all theoretical methods and computational techniques used to model or mimic the behavior of molecules. Based on the protein sequence, 3D structure of target protein could be constructed according to their homologous protein structure or the prediction of the sequence property.

is limited to proteins less than 45,000 Daltons molecular weight. X-ray crystallography is not limited in these respects but does require the formation of a single, high-order, 3D crystal. Unfortunately, most proteins have been selected during evolution on the basis of properties other than their ability to crystallize, and little is known about the mechanism of protein crystallization. A strictly empirical approach is generally taken, searching as systematically as possible the many parameters that affect crystal formation. Current automatic robot systems in protein crystallization takes advantage of commercially available liquid-handling systems initially developed for other application such as high-throughput screening, as well as developments of image analysis for monitoring crystal growth.[6]

2.3. *NMR techniques*

The general topology of the polypeptide chain in solution can be determined by NMR spectroscopy. The structure obtained in this way is not as detailed and accurate as that obtained crystallographically, but NMR has the advantage of using a protein in solution rather than in a crystal lattice. NMR spectra are generated by placing a sample in a magnetic field and applying different radio-frequency pulses for various experimental designs. NMR spectroscopy allows the determination of atomic structures of ever-larger subunits and even their complexes (Fig. 2). Increasingly, it is used to identify residues involved in protein interactions. General speaking, structure determination by X-ray crystallography is more effective and faster than NMR. However protein crystallization is still an experiment by trial and error, the information from NMR then could provide crystallographer to do the "meaningful trial." For the purpose of drug screening, NMR spectroscopy was the first technique to be exploited for use in fragment screening.[7] In "SAR by NMR," perturbations to the NMR spectra of a protein are used to indicate that ligand binding is taking place and to give some indication of the location of the binding site.[8] Recent reports have emphasized the complementarity of NMR screening methods and crystallography in applications to inhibitor design.

2.4. *Molecular modeling*

Molecular modeling is the science of representing molecular structures numerically and simulating their behavior with the equations of quantum and classical physics (Fig. 2). The common feature of molecular modeling techniques is the atomic level description of the molecular systems; the lowest level of information is individual atoms (or a small group of atoms). Since computing technology is rapidly developing, molecular modeling methods are now routinely used to investigate the molecular structure and dynamics. The types of biological activity that have been investigated using molecular modeling include protein folding, enzyme catalysis, protein stability, and conformational changes associated with biomolecular function. Also, studies of molecular recognition of protein and ligand could be applied to molecular docking and virtual screening for drug discovery.[9]

3. Structure-Based Systems Biology and Drug Design

It has long been recognized that knowledge of the 3D structures of proteins has the potential to accelerate drug discovery. Combined with structure-based systems biology, many new protein targets would be identified from pathway analysis and then studied by either X-ray analysis and NMR spectroscopy or molecular modeling. Detailed structural knowledge of the interactions between macromolecules, such as that provided by X-ray crystallography, NMR, and other biophysical approaches, and made publicly available through the Protein Data Bank (PDB), can also contribute to the understanding of cellular networks.

Structure-based systems biology and drug design could be illustrated by taking example of an important kinase, focal adhesion kinase (FAK). Cellular interactions with extracellular matrix (ECM) play essential roles in many aspects of cancer initiation, progression, and metastasis.[10−12] Integrin family cell adhesion receptors are the major mediators of cell adhesion to ECM, which link ECM to actin cytoskeleton at the cellular structure called focal adhesion (or focal contacts).[13,14] In addition to the clustered integrins themselves, multiple structural and signaling molecules have been localized to focal adhesions, which highlight the importance of focal adhesions in the regulation of cellular structure and functions.[15] Chief among these proteins is FAK, which is the earliest identified and one of the most prominent signaling molecules in focal adhesion.[16−18]

FAK is overexpressed in many tumors including those of the brain, breast, ovary, liver, colon, prostate, and thyroid.[17,19−25] From systems biological views, FAK overexpression is highly correlated with an invasive phenotype in these tumors. Inhibition of FAK signaling by overexpression of dominant negative fragments of FAK reduces invasion of ovarian cancer cells[26] and glioblastomas.[27] FAK therefore represents a significant molecular target for the development of anti-cancer drugs. Various small molecule inhibitors for use as potential cancer therapies have been developed.[28−31] However, there still some problems in the development of drugs that obstruct FAK function. Most of the small molecules are targeting inhibition of kinase domain, thereby preventing the activation of downstream signaling cascades. However, FAK activity may not be absolutely essential for its signaling functions. For example, reports showed that FAK FERM–mediated nuclear localization of FAK promotes enhanced cell survival through the inhibition of tumor suppressor p53 independent of its kinase activity.[32] Another problem is the specificity of the kinase inhibitor, since kinase domains of a range of different proteins show a high degree of amino acid conservation in the catalytic domains.[33] The other way for drug design is to block the adaptor function of FAK, such as preventing the binding of proteins to one or multiple tyrosines, to the proline-rich domains or by preventing localization of FAK to the focal adhesions. The small molecules or peptide-analogs could be developed to disrupt such FAK-domain/molecule interactions. Both targeting FAK kinase domain and its adaptor function could provide a prospective approach for cancer therapy.

FAK is a nonreceptor tyrosine kinase with a molecular mass of 125 kDa and encoded by the *fak* gene located on human chromosome 8q24.[34] Although transcript variants encoding two slightly different protein isoforms have been found for *fak* gene (isoform α has 1,052 a.a. and β has additional 22 a.a. to the N-terminus of α), FAK shares more than 95% amino acid identity across different species.[35] Structurally, FAK is composed of an N-terminal FERM (band 4.1, ezrin, radixin, moesin homology) domain, followed by an $\sim$40 residue linker region, a central kinase domain, an $\sim$220 residue proline-rich low-complexity region, and a C-terminal focal adhesion targeting (FAT) domain (Fig. 3).

The FERM domain of FAK, containing an autophosphorylation site (Tyr-397), is a three-lobed (F1–F3) architecture characteristic (PDB code:2AL6).[36] and the homologues are found in a number of cytoskeletal proteins.[37,38] FERM domains have been indicated to mediate both protein–protein interactions and protein–membrane interactions.[39,40] Some studies showed that truncation of the FERM domain of FAK resulted in an increase in phosphorylation of FAK and its kinase activity, suggesting a negative regulatory role in FAK activation.[41,42] In addition, the FERM domain of FAK was shown to interact with the kinase domain of FAK *in vitro* and *in vivo*, proposing a direct autoinhibitory working model of FAK regulation.[39] The crystal structure of FAK residue 31-686 containing the FERM and kinase domains is in accord with this model (pdb code: 2J0J and 2J0L).[43] and speculates that FAK activation is initiated by breaking an intramolecular autoinhibitory interaction between the N-terminal FERM and kinase domain. The C-terminal FAT domain of FAK is critical

Fig. 3. Schematic showing the domain organization of FAK. FAK is composed of an N-terminal FERM domain, followed by an $\sim$40 residue linker region, a central kinase domain, an $\sim$220 residue proline-rich low-complexity region, and a C-terminal FAT domain.

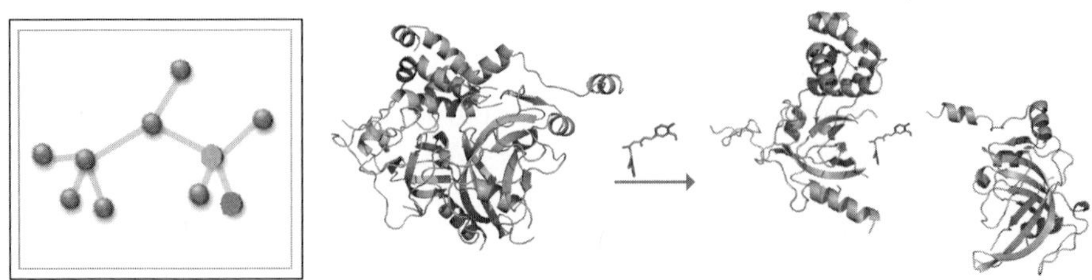

Fig. 4. From nodes and edges to atomic detail for drug design. We should be able to use large-scale protein interaction data to obtain meaningful insight regarding cellular functions, which could provide significant drug targets. The drug could be specifically disrupting interaction between two molecules.

for targeting FAK to focal adhesions. X-ray crystallography and NMR analysis showed that the FAT domain consists of a four-helix bundle (pdb code:1K40 and 1KTM).[44,45] Both the N-terminal and C-terminal domains have been shown to mediate FAK interaction with a variety of other proteins or molecules critical for activation of FAK by integrins or other cell surface receptors as well as FAK regulation of different cellular functions. Understanding molecular interaction from nodes and edges of systems biology approaches, we can determine the structures of interacting complexes to atomic detail for drug design. We should be able to use large-scale protein interaction data to obtain meaningful insight regarding cellular functions, which could provide significant drug targets (Fig. 4). The drug could be specifically disrupting interaction between two molecules. On the other hand, the disrupting drugs could be evaluated and examined how they influence cellular signaling by systems biology approaches.

Generally speaking, structure-based drug design relies on molecular-level structural knowledge of a disease-associated receptor or ligand. Such targets are usually proteins, allowing the living organism to take advantage of the cascade effect inherent in many enzymatic processes. After the protein target is identified and its structure determined, a chemical is specifically designed to bind this target irreversibly and thus negate its activity. Such a structure-based drug design minimizes the extensive screening typically used to identify an appropriate drug. With techniques such as protein–ligand X-ray crystallography, NMR spectroscopy, and molecular modeling providing valuable insight into protein–ligand interactions, structure-based systems biology and drug design is promising the reliable approaches for novel anticancer drug discovery.

References

1. Aloy P, Russell RB. (2006) Structural systems biology: Modelling protein interactions. *Nat Rev Mol Cell Biol* **7**: 188–197.
2. Aloy P, Russell RB. (2005) Structure-based systems biology: A zoom lens for the cell. *FEBS Lett* **579**: 1854–1858.
3. Burley SK. (2000) An overview of structural genomics. *Nat Struct Biol* **7**(Suppl): 932–934.
4. Lundstrom K. (2007) Structural genomics and drug discovery. *J Cell Mol Med* **11**: 224–238.
5. Shih YP *et al.* (2002) High-throughput screening of soluble recombinant proteins. *Protein Sci* **11**: 1714–1719.
6. Manjasetty BA, Turnbull AP, Panjikar S, Bussow K, Chance MR. (2008) Automated technologies and novel techniques to accelerate protein crystallography for structural genomics. *Proteomics* **8**: 612–625.
7. Yee A, Gutmanas A, Arrowsmith CH. (2006) Solution NMR in structural genomics. *Curr Opin Struct Biol* **16**: 611–617.
8. Shuker SB, Hajduk PJ, Meadows RP, Fesik SW. (1996) Discovering high-affinity ligands for proteins: SAR by NMR. *Science* **274**: 1531–1534.
9. Grant MA. (2009) Protein structure prediction in structure-based ligand design and virtual screening. *Comb Chem High Throughput Screen* **12**: 940–960.
10. Provenzano PP, Keely PJ. (2009) The role of focal adhesion kinase in tumor initiation and progression. *Cell Adh Migr* **3**: 347–350.
11. Lomberk G. (2010) The extracellular matrix and cell migration. *Pancreatology* **10**: 4–5.
12. Hynes RO. (2009) The extracellular matrix: Not just pretty fibrils. *Science* **326**: 1216–1219.
13. Hynes RO. (2002) Integrins: Bidirectional, allosteric signaling machines. *Cell* **110**: 673–687.
14. Arnaout MA, Mahalingam B, Xiong JP. (2005) Integrin structure, allostery, and bidirectional signaling. *Annu Rev Cell Dev Biol* **21**: 381–410.
15. Dubash AD *et al.* (2009) Chapter 1. Focal adhesions: New angles on an old structure. *Int Rev Cell Mol Biol* **277**: 1–65.
16. Schaller MD. (2010) Cellular functions of FAK kinases: Insight into molecular mechanisms and novel functions. *J Cell Sci* **123**: 1007–1013.
17. Golubovskaya VM, Kweh FA, Cance WG. (2009) Focal adhesion kinase and cancer. *Histol Histopathol* **24**: 503–510.
18. Schwock J, Dhani N, Hedley DW. (2010) Targeting focal adhesion kinase signaling in tumor growth and metastasis. *Expert Opin Ther Targets* **14**: 77–94.
19. Bonome T *et al.* (2005) Expression profiling of serous low malignant potential, low-grade, and high-grade tumors of the ovary. *Cancer Res* **65**: 10602–10612.
20. Cance WG *et al.* (2000) Immunohistochemical analyses of focal adhesion kinase expression in benign and malignant human breast and colon tissues: Correlation with preinvasive and invasive phenotypes. *Clin Cancer Res* **6**: 2417–2423.

21. Golubovskaya VM *et al.* (2009) FAK overexpression and p53 mutations are highly correlated in human breast cancer. *Int J Cancer* **125**: 1735–1738.

22. Natarajan M, Hecker TP, Gladson CL. (2003) FAK signaling in anaplastic astrocytoma and glioblastoma tumors. *Cancer J* **9**: 126–133.

23. Han NM, Fleming RY, Curley SA, Gallick GE. (1997) Overexpression of focal adhesion kinase (p125FAK) in human colorectal carcinoma liver metastases: Independence from c-src or c-yes activation. *Ann Surg Oncol* **4**: 264–268.

24. Owens LV *et al.* (1996) Focal adhesion kinase as a marker of invasive potential in differentiated human thyroid cancer. *Ann Surg Oncol* **3**: 100–105.

25. Tremblay L *et al.* (1996) Focal adhesion kinase (pp125FAK) expression, activation and association with paxillin and p50CSK in human metastatic prostate carcinoma. *Int J Cancer* **68**: 164–171.

26. Sood AK *et al.* (2004) Biological significance of focal adhesion kinase in ovarian cancer: Role in migration and invasion. *Am J Pathol* **165**: 1087–1095.

27. Jones G, Machado J Jr, Tolnay M, Merlo A. (2001) PTEN-independent induction of caspase-mediated cell death and reduced invasion by the focal adhesion targeting domain (FAT) in human astrocytic brain tumors which highly express focal adhesion kinase (FAK). *Cancer Res* **61**: 5688–5691.

28. Slack-Davis JK *et al.* (2007) Cellular characterization of a novel focal adhesion kinase inhibitor. *J Biol Chem* **282**: 14845–14852.

29. Jones ML, Shawe-Taylor AJ, Williams CM, Poole AW. (2009) Characterization of a novel focal adhesion kinase inhibitor in human platelets. *Biochem Biophys Res Commun* **389**: 198–203.

30. Shi Q *et al.* (2007) A novel low-molecular weight inhibitor of focal adhesion kinase, TAE226, inhibits glioma growth. *Mol Carcinog* **46**: 488–496.

31. Lietha D, Eck MJ. (2008) Crystal structures of the FAK kinase in complex with TAE226 and related bis-anilino pyrimidine inhibitors reveal a helical DFG conformation. *PLoS One* **3**: e3800.

32. Lim ST *et al.* (2008) Nuclear FAK promotes cell proliferation and survival through FERM-enhanced p53 degradation. *Mol Cell* **29**: 9–22.

33. van Nimwegen MJ, van de Water B. (2007) Focal adhesion kinase: A potential target in cancer therapy. *Biochem Pharmacol* **73**: 597–609.

34. Fiedorek FT Jr, Kay ES. (1995) Mapping of the focal adhesion kinase (FADK) gene to mouse chromosome 15 and human chromosome 8. *Mamm Genome* **6**: 123–126.

35. Whitney GS *et al.* (1993) Human T and B lymphocytes express a structurally conserved focal adhesion kinase, pp125FAK. *DNA Cell Biol* **12**: 823–830.

36. Ceccarelli DF, Song HK, Poy F, Schaller MD, Eck MJ. (2006) Crystal structure of the FERM domain of focal adhesion kinase. *J Biol Chem* **281**: 252–259.

37. Chishti AH *et al.* (1998) The FERM domain: A unique module involved in the linkage of cytoplasmic proteins to the membrane. *Trends Biochem Sci* **23**: 281–282.

38. Meurice N *et al.* (2010) Structural conservation in band 4.1, ezrin, radixin, moesin (FERM) domains as a guide to identify inhibitors of the proline-rich tyrosine kinase 2. *J Med Chem* **53**: 669–677.

39. Cooper LA, Shen TL, Guan JL. (2003) Regulation of focal adhesion kinase by its amino-terminal domain through an autoinhibitory interaction. *Mol Cell Biol* **23**: 8030–8041.

40. Cohen LA, Guan JL. (2005) Residues within the first subdomain of the FERM-like domain in focal adhesion kinase are important in its regulation. *J Biol Chem* **280**: 8197–8207.

41. Chan PY, Kanner SB, Whitney G, Aruffo A. (1994) A transmembrane-anchored chimeric focal adhesion kinase is constitutively activated and phosphorylated at tyrosine residues identical to pp125FAK. *J Biol Chem* **269**: 20567–20574.

42. Schlaepfer DD, Hunter T. (1996) Evidence for *in vivo* phosphorylation of the Grb2 SH2-domain binding site on focal adhesion kinase by Src-family protein-tyrosine kinases. *Mol Cell Biol* **16**: 5623–5633.

43. Lietha D *et al.* (2007) Structural basis for the autoinhibition of focal adhesion kinase. *Cell* **129**: 1177–1187.

44. Hayashi I, Vuori K, Liddington RC. (2002) The focal adhesion targeting (FAT) region of focal adhesion kinase is a four-helix bundle that binds paxillin. *Nat Struct Biol* **9**: 101–106.

45. Liu G, Guibao CD, Zheng J. (2002) Structural insight into the mechanisms of targeting and signaling of focal adhesion kinase. *Mol Cell Biol* **22**: 2751–2760.

Chapter 15

Discovering Drug Targets
for Cancer Therapy

Tsui-Chin Huang[*], Hsin-Yi Chang[*] and Hsueh-Fen Juan[*,†]

1. Introduction

Cancer is a class of life-threatening diseases that constitutes one of the leading causes of death worldwide. Its presence in human beings has been known for centuries; the first evidence of cancer was found on ancient Egyptian papyrus that can be dated back to 1500 BC. Cancer results from the mutation of genes within cells, which leads to the rapid proliferation of abnormal cells. In addition to out-of-control growth, cancer cells can also invade the surrounding normal tissue or migrate to other organs through the circulatory or lymphatic systems by a process known as metastasis.[1,2] The method for treating a cancer depends upon the location of the neoplasm, the tumor grade, physical spread, type of malignancy, and the patient's physical condition, and can include surgical resection, chemotherapy, radiotherapy, and monoclonal antibody therapy, amongst other methods.

In humans, acquired immunity can be divided into humoral (in which the protective effect is derived from B cell antibody production) and cellular immunity. Both Cytotoxic T cells and natural killer cells (NK cells) play important roles in cellular immunity.[3] Cytotoxic T cells can recognize abnormal cells and destroy them within 4 to 7 days, sometimes with the assistance of helper T cells.[4] NK cells are derived from bone marrow lymphoid stem cells and exist mainly in the peripheral blood and spleen. They are the first line of defense against abnormal cells once they

[*]Department of Life Science, Institute of Molecular and Cellular Biology, and Graduate Institute of Biomedical Electronics and Bioinformatics, National Taiwan University, Taipei 106, Taiwan.

[†]yukijuan@ntu.edu.tw

have transformed and bear specific tumor cell surface molecules.[5,6] Both cytotoxic T cells and NK cells can induce cell death by the same mechanism, using degranulation, a process that releases perforin and granzymes by exocytosis to trigger cell apoptosis. Studies have demonstrated that depressed NK cell activity is shown to be linked to increased rates of cancer recurrence and metastasis and reduced patient survival.[7]

Conventional cancer treatments mainly focus on inhibiting the rapid proliferation of cancer cells and the induction of apoptosis. Apoptosis is a basic physiological phenomenon in multicellular organisms responsible for the removal of unwanted or abnormal cells during development.[8] When undergoing apoptosis, cells shrink and their DNA is degraded by an endonuclease into 180 to 200 base pair (bp) fragments. The cells finally split into several apoptotic bodies and are engulfed by neighboring phagocytic cells, such as macrophages.[9,10]

Cancer cells are mostly heterogeneous. High-grade tumors can often survive after chemotherapy and radiotherapy or escape detection by cytotoxic T cells and NK cells, leading to cancer recurrence. As described by H. Kitano (2003; 2004), cancer cells are robust due to their characteristics of heterogeneity, redundancy, and feedback.[11,12] When heterogeneity occurs by mutation and aneuploidy, it results in cancer cells having multiple redundant pathways for survival and growth, giving them the dynamic ability to escape from the regular control of the cell cycle and resist anticancer drug treatments, thus becoming harder to eliminate.[11,12] Take for instance the MDM2 oncogene, which is transactivated by p53. The overexpression of MDM2 in cancer cells negatively regulates the stability of p53 by its E3 ligase activity; this forms a feedback loop, which removes cancer cells from the surveillance of the apoptosis mechanism.[12]

With the establishment of gene regulatory networks and protein–protein interaction networks in cancer cells, the development of cancer therapy has been moved into the era of a systems biology–based strategy, which provides a global perspective for understanding dynamic cancer behavior. In this chapter, we will introduce current cancer treatments and discuss the drug targets (Fig. 1) involving tumorigenesis from a systems biology point of view.

1.1. *Cancer therapy*

Current best practice in cancer treatment involves combined multidisciplinary treatment (also known as multi-modality treatment). This treatment model may consist of surgical resection, radiation therapy

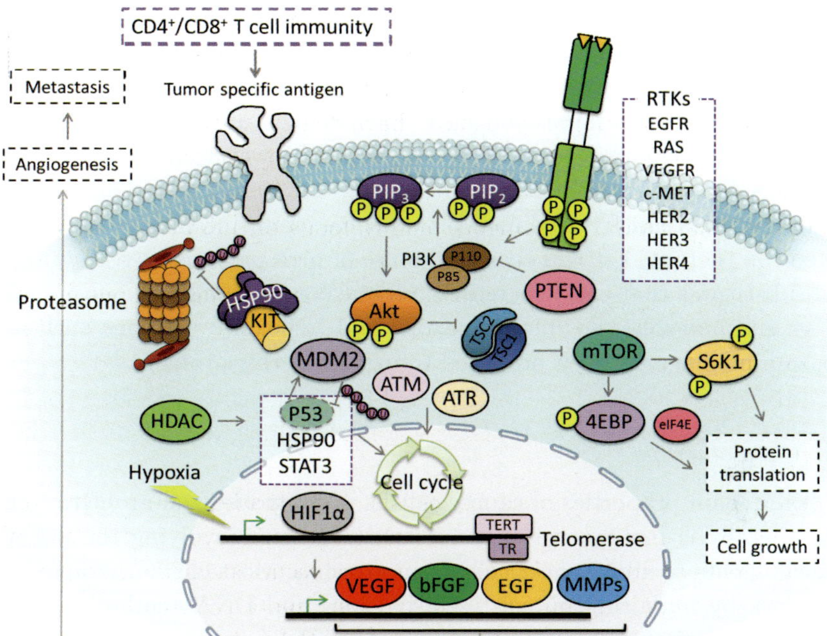

Fig. 1. Intracellular molecular drug targets for the cancer therapies discussed in this chapter.

with systemic chemotherapy, targeted therapy, or other various adjuvant medications. Combinations of these modalities can maximize therapeutic effects, improve patient survival, and give patients a higher quality of life.

1.1.1. *Surgery*

Surgery is the oldest method of removing tumor and remains the most effective treatment for low-grade solid cancers.[13–16] Significant improvements in surgical outcomes in recent decades have come about because of developments in surgical tools, concepts, and techniques. Recent studies describe a number of new breakthrough techniques, including radical surgery,[17] minimally invasive surgery,[18] and endoscopic surgery.[19] Combined with radiotherapy, chemotherapy, hormone therapy, and immunotherapy, surgical resection of primary tumors usually provides a good prognosis.

1.1.2. *Radiation therapy*

Radiation therapy works by damaging DNA during cell division. It has two distinct methods of preventing DNA replication; directly, through ionizing DNA to cause strand breakage, or indirectly, by ionizing water and oxygen to form free radicals that attack DNA.[20] It would seem promising to use radiation therapy to treat most cancers because cancer cells are generally highly proliferative and have a diminished ability to repair sublethal damage compared to most healthy cells. However, the cells in the middle of solid tumors often stop reproducing due to hypoxia. This lack of oxygen, the source of free radicals, makes some cancer cells more resistant to radiation therapy. For this reason, the use of high pressure oxygen tanks, blood substitutes that carry increased oxygen, hypoxic cell radiosensitizers such as misonidazole and metronidazole, and hypoxic cytotoxins such as tirapazamine, have all been proposed to overcome this problem.[21]

1.1.3. *Chemotherapy*

One of the main properties of cancer cells is uncontrolled cell proliferation. Chemotherapy is the general term for any treatment involving the use of chemical agents to stop cancer cells from growing. Most chemotherapeutic drugs work by impairing mitosis (cell division) and DNA synthesis; some drugs can also cause cells to undergo apoptosis.[22] Chemotherapy regimens utilize a diverse combination of chemical agents for different cancers. Recent reports indicate that the most commonly used chemotherapeutic agents are: etoposide, VM26, m-AMSA, dexamethasone, vincristine, cis-platinum, cyclophosphamide, paclitaxel, 5′-fluoro-deoxyuridine, 5′-fluorouracil, and adriamycin.[23–25]

Chemotherapeutic agents can eliminate cancer cells not only at the original cancer site but also at remote sites. For this reason, chemotherapy is considered a systemic treatment. The major task of chemotherapeutic agents is to destroy rapidly dividing cells. However, most chemotherapeutic agents cannot differentiate between cancer cells and normal cells. As a result, some continually proliferating somatic cells, such as hair and blood cells, will often also be killed through chemotherapy.

The latest concept in chemotherapy is targeted therapy, which is designed to overcome toxicity to normal cells and aims to induce fewer unwanted side effects. This concept is discussed further below.

1.1.4. *Targeted therapy*

In the late 1990s to early 2000s, oncologists used many more new drugs than had been used in the past three decades. Drugs such as paclitaxel, docetaxel, gemcitabine, irinotecan, oxaliplatin, and pemetrexed have all led to considerable progress in the treatment of many cancers. Although these new-generation chemotherapeutic agents are effective in breast, ovarian, lung, stomach, esophagus, pancreas, bladder, and colon cancer, many large-scale clinical trials have demonstrated that the overall survival of cancer patients is still poor. The fact remains that chemotherapy causes systemic toxic effects, and drug resistance continues to be a problem.[26,27]

In order to further improve treatment efficiency, the scientific community and pharmaceutical companies have turned to the development of new treatments with less toxicity.[28] Scientists with an in-depth understanding of molecular cancer biology emphasize the importance of finding new drug targets that are expressed specifically in cancer cells. These drugs could target cancer cells with (1) specific antigens,[29] (2) specific growth factor receptors,[30] (3) angiogenesis-related factors,[31] (4) cancer-specific signaling pathways,[32] (5) molecules involved in uncontrolled cell cycle progression,[33] and (6) molecules that can prevent cells from programmed cell death.[34] These drugs often work by introducing mutations into the genes of cancer cells, sometimes with the aid of viruses. These abnormal molecules thereby affect tumor survival, proliferation, local invasion, and tumor ability to metastasize to distant sites. The development of new drugs to bind to these targets can reduce the toxicity caused in chemotherapy and improve cancer treatment outcomes. The concept of targeting specific cancer cell molecules is known as targeted therapy.[35]

2. Characterization of Drug Targets

Cancer cells result from the genetic modification and transformation of normal cells. Currently, there are two kinds of known cancer genes that are associated with the formation of cancer: oncogenes and tumor suppressor genes. Cancer formation is often initiated by the activation of several oncogenes or the deletion of tumor suppressor genes.[36]

The establishment of genetics, molecular biology, and human genome sequencing has provided us with a new understanding of cancer cell transformation and has continued to help scientists discover advanced treatments. Table 1 lists the FDA-approved drugs that are currently in their clinical trial phase.

In this section, we will focus on characterization of cancer cells and the basis of cancer cell physiology to draw out which might be drug targets for cancer therapies.

Table 1. Anticancer drugs undergoing clinical trials, listed according to information public from NCI.

Cancer drug	Target	Disease indication
Alemtuzumab (Campath®)	CD52	BCLL
Alitretinoin (Panretin®)	Retinoic acid receptor; retinoid X receptor	AIDS-related Kaposi sarcoma
Anastrozole (Arimidex®)	Aromatase	Breast cancer
Bevacizumab (Avastin®)	VEGF	NSCLC; metastasic breast cancer; metastatic colorectal cancer
Bexarotene (Targretin®)	Retinoic acid receptor; retinoid X receptor	CTCL
Bortezomib (Velcade®)	Proteasome	Multiple myeloma
Cetuximab (Erbitux®)	EGFR	Colorectal cancer.
Dasatinib (Sprycel®)	Tyrosine kinases	CML; acute lymphoblastic leukemia
Denileukin diftitox (Ontak®)	Cell surface IL-2 receptors	CTCL
Erlotinib hydrochloride (Tarceva®)	Tyrosine kinase activity of EGFR	NSCLC; pancreatic cancer
Everolimus (Afinitor®)	mTOR kinase	Kidney cancer
Exemestane (Aromasin®)	Aromatase	Breast cancer
Fulvestrant (Faslodex®)	ER	Breast cancer
Gefitinib (Iressa®)	EGFR	NSCLC
Ibritumomab tiuxetan (Zevalin®)	CD20	B-cell non-Hodgkin lymphoma
Imatinib mesylate (Gleevec®)	Tyrosine kinases	Gastrointestinal stromal tumor; leukemia
Lapatinib ditosylate (Tykerb®)	Tyrosine kinases	Advanced or metastatic breast cancer
Letrozole (Femara®)	Aromatase	Breast cancer
Nilotinib (Tasigna®)	Tyrosine kinases	CML
Ofatumumab (Arzerra®)	CD20	CLL
Panitumumab (Vectibix®)	EGFR	Metastatic colon cancer
Pazopanib hydrochloride (Votrient®)	Tyrosine kinases	Advanced renal cell carcinoma
Pralatrexate (Folotyn®)	Dihdrofolate reductase	PTCL
Rituximab (Rituxan®)	CD20	B-cell non-Hodgkin lymphoma
Romidepsin (Istodax®)	HDACs	CTCL
Sorafenib tosylate (Nexavar®)	Tyrosine kinases	Advanced renal cell carcinoma; hepatocellular carcinoma

(Continued)

Table 1. (*Continued*)

Cancer drug	Target	Disease indication
Sunitinib malate (Sutent®)	Tyrosine kinases	Metastatic renal cell carcinoma; gastrointestinal stromal tumor
Tamoxifen	ER	Breast cancer
Temsirolimus (Torisel®)	mTOR kinase	Renal cell carcinoma
Toremifene (Fareston®)	ER	Breast cancer
Tositumomab and 131I-tositumomab (Bexxar®)	CD20	B-cell non-Hodgkin lymphoma
Trastuzumab (Herceptin®)	HER-2	Breast cancer
Tretinoin (Vesanoid®)	Retinoic acid receptors	Acute promyelocytic leukemia.
Vorinostat (Zolinza®)	HDACs	CTCL

Abbreviation: BCLL, B-cell chronic lymphocytic leukemia; NSCLC, nonsmall cell lung cancer; CTCL, cutaneous T-cell lymphoma; CML, Chronic myelogenous leukemia; CLL, chronic lymphocytic leukemia; PTCL, peripheral T-cell lymphoma.

2.1. *Molecules*

2.1.1. *Receptor tyrosine kinases*

Tyrosine kinase receptors (TKRs) are located on the plasma membrane and play an important role in signal transduction leading to cell growth. TKRs such as EGFR, HER2, HER3, HER4, VEGRF, c-MET, as well as downstream effecter RAS, are often overexpressed or unregulated due to genetic mutations in several cancers (Fig. 2). As a result, the use of tyrosine kinase receptor inhibitors can dramatically inhibit growth signals of cancer cells.[37,38] Some tyrosine kinase receptor inhibitors that are currently in clinical use include gefitinib, erlotinib, lapatinib, imatinib, sorafenib, and sunitini.[39,40] Gefitinib and erlotinib demonstrate good effects for patients with poor prognosis nonsmall cell lung cancer after chemotherapy.[41] Lapatinib is used to treat advanced or metastatic breast cancer, particularly breast cancer patients with brain metastases.[42,43] Imatinib is used in the treatment of chronic myelogenous leukemia and gastrointestinal stromal tumors.[44] Sorafenib and sunitinib are two multi-target tyrosine kinase receptor inhibitors. In addition to blocking the proliferation of cancer cells, they also reduce angiogenesis and metastasis by blocking the uptake of oxygen and other extracellular nutrition.[45,46] Another antibody

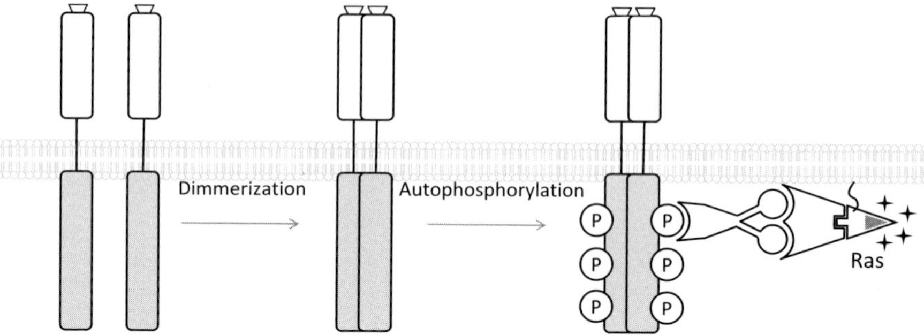

Fig. 2.   Tyrosine kinases receptors can be activated by dimerization and autophosphorylation. As a consequence, TKRs transmit signals through Ras by the hydrolysis of GTP.

inhibitor, Trastuzumab, can directly target the HER2/Neu receptor to block PI3K/AKT/mTOR signaling. It can also inactivate the TK nonreceptor, Src, which can inhibit tumor suppressor PTEN. Activated PTEN in return inhibits the PI3K pathway.[47]

2.1.2. *PI3K/AKT/mTOR*

The mammalian target of rapamycin kinase (mTOR) is a key regulatory kinase responsible for the transmission of signals regulating cell proliferation. When mTOR is activated by PI3K signaling, it activates downstream signals to promote cell differentiation or proliferation. When this regulatory mechanism is out of control, abnormal cell proliferation and incomplete differentiation results, leading to tumor formation. Consequently, mTOR has been a new anti-cancer drug target (Fig. 3).[48–50]

Temsirolimus specifically inhibits mTOR kinase, causing the cell cycle to be arrested at the G1 phase (pre-cell synthesis) and inhibits cell division and proliferation. Temsirolimus may also regulate the translation of hypoxia-inducible factor-1a (HIF-1a), which in turn induces the expression of vascular endothelial growth factor (VEGF). Temsirolimus thus inhibits cancer cell growth and angiogenesis, reducing metastasis. It is approved for the treatment of advanced renal cell carcinoma.[51]

The inhibition of mTOR has also been found to increase sensitivity to the anticancer drug, etoposide. Etoposide is derived from podophyllotoxin, found in the American Mayapple, and is currently used to treat several cancers, including small cell lung cancer, Kaposi's sarcoma, germ cell

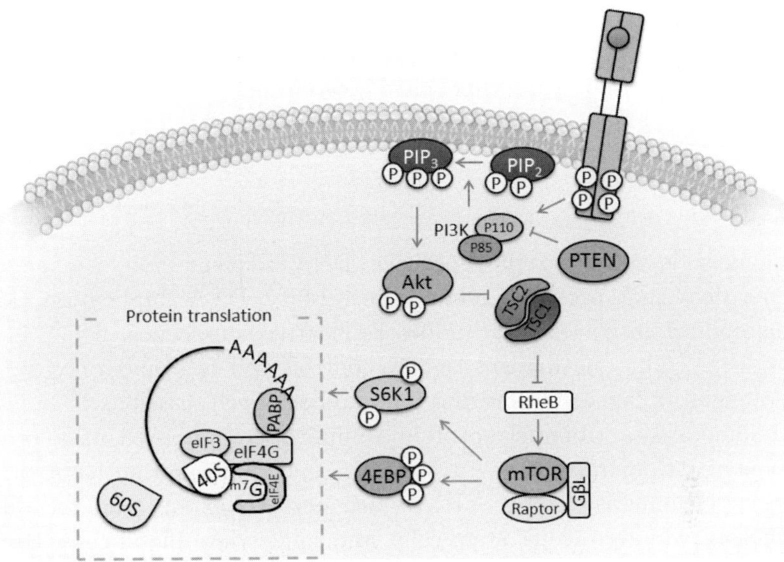

Fig. 3. Activation of the PTEN/Akt/mTOR pathway results in promotion of protein translation.

cancer, acute leukemia, and lymphoma. It inhibits DNA topoisomerase II, which destroys the double helix structure of DNA and thus inhibits cancer cell proliferation.[52]

2.1.3. *Histone deacetylases*

Chromosome remodeling plays an important role in the regulation of gene transcription. Histone acetylase (HATs) and histone deacetylase (HDACs) regulate the balance of acetylation on lysines of core histones. Hypoacetylation of histones is responsible for chromatin condensation, while acetylated histones can open the chromatin structures for transcriptional regulators and therefore increase transcription. In addition to histones, other acetylated proteins, including p53, STAT3, and HSP90, have also been shown to be substrates for the HDACs; these proteins are also highly correlated with cancer formation.[53]

The use of HDAC inhibitors in lymphoma cells can re-establish normal gene expression. HDAC inhibitors can increase the expression of pro-apoptotic proteins, such as BAX and death receptors, as well as regulating the activation of a number of proteins such as p53,

NFκB, HSP-90, and β-catenin. There are several HDAC inhibitors under clinical trials for cutaneous T-cell lymphoma. Vorinostat and romidepsin are two HDAC inhibitors that have proven to be effective for cancer therapy.[54,55]

2.1.4. Telomerase

A telomere is a region of repetitive DNA sequences at the ends of chromosomes that prevents genomic instability. DNA duplication causes telomere shortening; this can induce replicative senescence, which blocks cell division. This mechanism tightly controls cell fate and prevents the development of cancer by limiting the number of cell divisions.[56]

Telomerase, a ribonucleoprotein complex, is composed of a protein component (telomerase reverse transcriptase, TERT) and an RNA primer sequence (telomerase RNA, TR), which acts to extend the termini of chromosomes in germ cells, stem cells, and some white blood cells. Human somatic cells lacking telomerase divide in a restricted manner and eventually become senescent. Malignant cells, however, bypass this restriction and become immortalized due to telomere extension by the activation of telomerase.[57]

Many investigators believe that telomerase is an ideal target for cancer therapeutics. GV1001 is a peptide vaccine that can be processed and presented as MHC-I epitopes after immunization, and can induce CD4$^+$ and CD8$^+$ T-cell immunity specific for hTERT. GV1001 is currently being investigated in phase III trials for pancreatic cancer, and phase II trials for hepatoma and nonsmall cell lung cancer (NSCLC).[58–60] GRN163L is an oligonucleotide that targets the active site of telomerase (the telomerase RNA) and is currently being investigated in phase-II clinical trials for the treatment of leukemia, chronic lymphocytic (CLL), breast cancer, NSCLC, and multiple myeloma patients.[61]

2.1.5. HSP90 and Proteasome

The proteasome–ubiquitin system is an important cellular protein metabolic pathway. In response to a changing physiological environment, cells utilize this system to execute rapid protein metabolism for adaptation without altering their cell structures. Proteins that are poly-ubiquitinized would be decomposed by proteasomes. These proteins include cell cycle–related CDKs, oncogene kinases, tumor suppressor genes (such as p53),

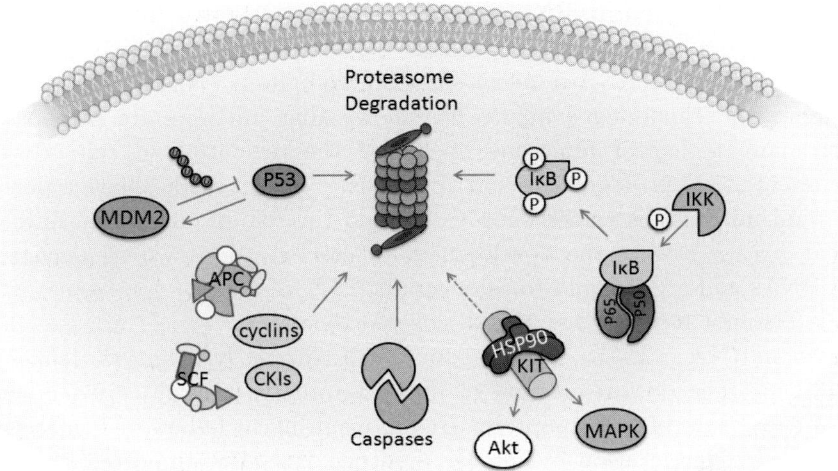

Fig. 4. The ubiquitination of P53, cell cycle regulators, apoptosis regulators, oncogenes, and IκB play an important role in tumorigenesis.

apoptosis factors, antiapoptosis factors, inflammatory cytokines, and angiogenesis factors.[62]

An example of this is NF-κB, an important transcription factor that may facilitate cancer cell survival and is found in multiple myeloma cells. NF-κB binding with I-κB is restricted to the cytoplasm. When I-κB is degraded by the proteasome system, it releases NF-κB, which can enter the nucleus and function as a transcription factor.[63] Bortezomib can inhibit the proteasome function and reduce the decomposition of I-κB, thereby inhibiting NF-κB.[64] Heat shock protein 90 (referred to as HSP90) is an auxiliary folding protein (chaperon protein). It can bind to various proteins and protect them from proteasome–ubiquitin system degradation.[65]

Studies have found that oncogenes expressed by cancer cells, such as KIT, are protected by HSP90. The use of HSP90 inhibitors, such as 17-AAG, can inhibit the chaperone activity of HSP90. As a result, these oncogenes will then be ubiquitinized and degraded by proteasomes.[66]

2.1.6. *miRNAs*

microRNA (miRNA) is an endogenous noncoding small molecule RNA with a length of roughly 18 to 25 nucleotides. miRNA is highly conserved in evolution and has a function in post-transcriptional gene regulation. It can be transcribed by RNA polymerase II or generated from introns by alternative splicing. These small molecules can target mRNA 3′UTR with the assistance of RNA-induced silencing complexes (RISCs) and inhibit translation. Emerging evidence has shown that miRNAs are responsible for many biological functions, including the regulation of cell growth, differentiation, proliferation, and apoptosis.[67] Recent studies have reported several miRNAs as tumor suppressors and the role of oncogenes. miRNA that play a role in the development of cancer are known as oncogenic miRNAs and are referred to as oncomiR.[68]

The first miRNA that was discovered to suppress cancer was the clustered miRNA miR-15a/miR-16-1 in B-cell chronic lymphocytic leukemia (CLL).[69] The downregulation of miR-15a/miR-16-1 negatively regulates the expression of anti-apoptotic *BCL2*, resulting in failure of hematopoietic cell differentiation.[70] Further evidence has also supported the role of miRNA in tumorigenesis, such as miR-145 in colon cancer, miR-21 in glioblastoma, and let-7 in lung, breast, urothelial, and cervical cancers.[71] Emerging studies have indicated that miRNA is not only a key regulator in development but might also be an ideal target for cancer therapy.[72]

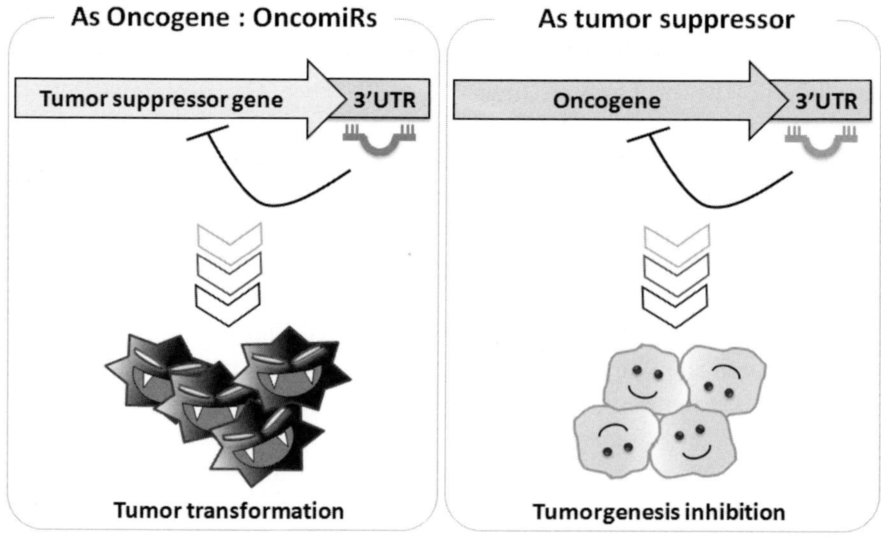

Fig. 5.   MiRNAs play a distinguished role in tumor development.

2.2. Pathways

2.2.1. Cell cycle regulators

The cell cycle consists of a series of events that take place within a cell and lead to its division and duplication. During development, the cell cycle is a vital process for growing a single fertilized cell into a mature organism. The process is tightly controlled by cell cycle regulators in the transitions between each phase, thus preventing functionless cells from duplicating.[73] The 2001 Nobel Prize in Physiology or Medicine was awarded for discoveries that showed that cell cycle progression is directly regulated by different cyclins and cyclin-dependent kinases (CDKs).[74,75]

From DNA replication to cell division, each transition period in the cell cycle is under strict surveillance, which ensures the completeness and accuracy of every step. This regulatory mechanism, known as the cell cycle checkpoint, consists of a series of proteins.[75] Two main checkpoints exist: the G1/S checkpoint and the G2/M checkpoint. Checkpoint sensor proteins (e.g. ATM and ATR) are thought to bind directly to damaged DNA, activating signal transducer kinases, which then activate downstream effectors (such as Chk2, and p53) that regulate the ultimate cellular response, which can be cell-cycle arrest for DNA repair or apoptosis.[76–79]

In addition to cyclins and CDKs, the cell cycle can be regulated by cell-cycle inhibitors that prevent cycle progression. There are two main families of inhibitors: cip/kip and INK4a/ARF.[80,81] Because these genes are involved in the prevention of tumor formation, they are known as tumor suppressor genes.

Cell-cycle dysregulation is a critical determinant of cancer progression. The permanent proliferative ability of tumor cells is obviously the result of out-of-control cell cycle progression.[82] Understanding the regulatory mechanisms of cell division will help with exploring the causes of cancer and the development of new anticancer drugs.

2.2.2. Apoptosis pathways

Apoptosis plays an important role in the homeostasis of multicellular organisms. It is also known as programmed cell death and is a normal process in healthy multicellular individuals.[83] The 2002 Nobel Prize in Physiology or Medicine was awarded for the study of apoptosis and its involvement in development, growth, and the cell suicide mechanism. This research clearly demonstrated that most cancer cell development results from a loss of apoptotic function.[84–87]

Apoptosis can be induced by both intrinsic and extrinsic pathways and is regulated by multiple molecules, including the caspase cascade, the Bcl-2 family, and mitogen-activated protein kinase (MAPK).[88–90] There are many anticancer drugs in clinical trials that target apoptosis regulators, including Genasense[91] and GX01[92] for Bcl-2; the E1A-lipid complex for E1A[93]; TRAIL for ApoII[94]; and INGN201,[95] ONYX-015,[96] and SCH58500[97] for p53.

2.2.3. *Angiogenesis*

Because they grow rapidly, cancer cells require much more nutrition than do normal cells. Angiogenesis is necessary to provide an adequate environment for the development of cancer cells[98] and is a process that involves the formation of new blood vessels from existing ones. In the early stage of tumor development, cancer cells can easily acquire nutrition by diffusion. As tumor size increases, diffusion is no longer sufficient to serve the need for proliferation, especially for the central region of the tumor.[99] The hypoxic tumor microenvironment that results may lead to upregulation of chemokines and other factors such as vascular endothelial growth factor (VEGF), basic fibroblast growth factor (bFGF), and epidermal growth factor (EGF). These factors then facilitate the proliferation and migration of endothelial cells to form new blood vessels around the tumor, as well as migrating and invading into adjacent tissues.[100]

Drugs that prevent the growth of new blood vessels provide a new strategy to indirectly restrict the growth of tumors. In recent years, some new treatments, such as Endostatin,[101] Angiostatin,[102] and Avastin,[103] as well as the 40-year-old sedative Thalidomide,[104] have shown positive results in clinical trials, particularly for liver and metastasic cancer.

2.2.4. *Metastasis*

Cancer metastasis is a major cause of death in cancer patients. Metastasis involves a series of processes in which cancer cells remodel their cytoskeleton, secreting matrix metalloproteinases (MMPs) to degrade all kinds of extracellular matrix that constitute the basement membrane of the carcinoma's border.[105,106] Cancer cells begin to migrate and invade the surrounding normal tissue, before entering the circulation and lymphatic system and spreading to the rest of the body to form secondary tumors (metastatic tumors). These metastatic tumors become the most difficult part of cancer treatment. One of the most critical conditions necessary for cancer metastasis is the formation of a new network of blood vessels; i.e. tumor angiogenesis, as discussed above. For this reason, angiogenesis inhibitors can be used to prevent the growth of metastases.[107–109]

2.2.5. *Protein–protein interaction network*

The high-throughput data of gene co-expression and protein–protein interactions (PPIs) (derived from large scale techniques such as DNA microarrays, yeast two hybrids, and mass spectrometry pull down experiments), have allowed scientists to build tumor pathways and compare them with normal cells to find cancer specific nodes or modules that could be targets for cancer therapy.[110–112] These data can be obtained from public databases like BioGrid,[113] the Database of Interacting Proteins,[114] the Human Protein Reference Database (HPRD),[115] IntACT,[116] and MINT[117] for PPI data; and the Stanford Microarray Database,[118] and the Gene Expression Omnibus[119] (GEO) for expression data sets. One recent study integrated the existing data sets and built a pathway-based analysis system that can be applied to several types of cancers, including glioblastoma, breast, colon, and pancreatic cancers. It identified several modules that are highly activated in cancers. Specifically, the study found that the focal adhesion pathway and transcription, cell cycle, and DNA repair stages could be tumor-specific modules and thus may be ideal targets for cancer therapy.[120]

2.2.6. *Synthetic lethality*

Synthetic lethality refers to a situation in which mutations in two or more genes leads to cell death, whereas a single mutation in one of these genes does not. This idea has recently been applied in cancer therapy to reduce side effects on normal cells.[121,122]

In 2007, Whitehurst *et al.* performed a genome-wide screening using RNAi to silence several candidate genes, thus increasing the effect of anticancer drugs.[123] The development of the new agent olaparib has also demonstrated the possibility that blocking PARP in BRCA-deficient patients can significantly increase anticancer efficiency on the basis of synthetic lethality.[124] PARP and BRCA are proteins involved in DNA repair, and cells require both proteins to be functional in order to accomplish DNA repair. Blocking PARP in tumors bearing a BRCA mutation means that cells are unable to overcome the stress, resulting in cell death. However, disrupting the DNA repair mechanism is more difficult to achieve in cells with normal BRCA genes. Applying this new point of view into cancer therapy thus can gradually reduce the systematic toxicity to the normal tissues in patients.

3. Obstacles in Cancer Therapy

Recently, evidence has served to reinforce our understanding of how difficult cancer is to treat. Emerging evidence has indicated that cancer cells are heterogeneous and has suggested that the ability of a tumor to grow is dependent on a small subset of cells within a tumor, termed cancer stem cells (CSCs). CSCs have been identified in several cancers, including leukemia and breast, brain, lung, liver, ovarian, prostate, colon, and oral cancers. CSCs and normal stem cells have many similar features, including self-renewal and the capacity to differentiate.[125]

The proliferation of CSCs is extremely slow, or even paused, and the expression of ABC transport proteins is often found. These properties mean that it is hard to kill CSCs with chemotherapy or radiation therapy (which target rapidly-growing cells), leading to cancer recurrence and metastasis.[126,127]

ABC transporters are a large group of transport proteins that use ATP hydrolysis for energy in active transport. They were first discovered in the inner membrane of *Escherichia coli* and have since been noted in more than 100 species, including humans. P-glycoprotein, an ABC transporter, is responsible for multidrug resistance in tumor cells; the ATP hydrolysis activity of P-glycoprotein means that anti cancer drugs can be pumped out of cells. Patients with P-glycoprotein expression generally have a poor prognosis.[128] Another feature of CSCs is their high capacity to initiate tumorigenesis. It only takes several hundred CSCs to form tumors in immune-deficient mice, while it takes more than 1 million cancer cells in a xenograft cancer model. The excellent tumorigenesis ability of CSCs thus becomes an important topic in the development of a cancer therapy strategy.[129–133]

4. Conclusion

In this chapter, we have attempted to link important molecular discoveries to contemporary strategies in cancer therapy. With the introduction of a theoretical basis and the practical application of molecular targeted therapies and a systems biology–based approach to cancer, cancer therapy is moving into the next generation of techniques, with high efficiency, fewer side-effects, and lower rates of recurrence. There have been overwhelmingly powerful drug targets discovered in recent decades, and still others are under investigation at the present time. It will be worthwhile to undertake more clinical trials to verify the feasibility of various treatment combinations. Since pharmacologic strategies now exist that effectively prevent cancer development in several types of tumors, a concerted effort to eliminate all threatening cancers in humans will become a real possibility in the near future.

Acknowledgments

This work was supported by National Research Program for Genomic Medicine, Department of Health (DOH), National Science Council of Taiwan, and the National Taiwan, University Frontier and Innovative Research Program.

References

1. Diamandopoulos GT. (1996) Cancer: An historical perspective. *Anticancer Res* **16**(4A): 1595–1602.
2. Kardinal CG, Yarbro JW. (1979) A conceptual history of cancer. *Semin Oncol* **6**(4): 396–408.
3. Hayakawa Y, Smyth MJ. (2006) Innate immune recognition and suppression of tumors. *Adv Cancer Res* **95**: 293–322.
4. Barry M, Bleackley RC. (2002) Cytotoxic T lymphocytes: All roads lead to death. *Nat Rev Immunol* **2**(6): 401–409.
5. Biron CA, Nguyen KB, Pien GC *et al.* (1999) Natural killer cells in antiviral defense: Function and regulation by innate cytokines. *Annu Rev Immunol* **17**: 189–220.
6. Carnaud C, Lee D, Donnars O *et al.* (1999) Cutting edge: Cross-talk between cells of the innate immune system: NKT cells rapidly activate NK cells. *J Immunol* **163**(9): 4647–4650.
7. Rosenberg SA. (1984) Adoptive immunotherapy of cancer: Accomplishments and prospects. *Cancer Treat Rep* **68**(1): 233–255.
8. Kerr JF, Wyllie AH, Currie AR. (1972) Apoptosis: A basic biological phenomenon with wide-ranging implications in tissue kinetics. *Br J Cancer* **26**(4): 239–257.
9. Susin SA, Daugas E, Ravagnan L *et al.* (2000) Two distinct pathways leading to nuclear apoptosis. *J Exp Med* **192**(4): 571–580.
10. Nagata S. (2000) Apoptotic DNA fragmentation. *Exp Cell Res* **256**(1): 12–18.

11. Kitano H. (2003) Cancer robustness: Tumor tactics. *Nature* **426**(6963): 125.

12. Kitano H. (2004) Cancer as a robust system: Implications for anticancer therapy. *Nat Rev Cancer* **4**(3): 227–235.

13. Cady B. (1997) Basic principles in surgical oncology. *Arch Surg* **132**(4): 338–346.

14. Soderstrom MJ, Gilson SD. (1995) Principles of surgical oncology. *Vet Clin North Am Small Anim Pract* **25**(1): 97–110.

15. Hohenberger W, Meyer T. (2000) Principles of surgical oncology. I. *Zentralbl Chir* **125**(5): W31–W38.

16. Hohenberger W, Meyer T. (2000) Basic principles in surgical oncology (2). *Zentralbl Chir* **125**(6): W39–W48.

17. Amdal CD, Jacobsen AB, Tausjo JE *et al.* (2010) Radical treatment for oesophageal cancer patients unfit for surgery and chemotherapy. A 10-year experience from the Norwegian Radium Hospital. *Acta Oncol* **49**(2): 209–218.

18. Schmitto JD, Mokashi SA, Cohn LH. (2010) Minimally-invasive valve surgery. *J Am Coll Cardiol* **56**(6): 455–462.

19. Matsuwaki Y, Moriyama H. (2010) Progress in endoscopic sinus surgery. *Nippon Rinsho* **68**(7): 1360–1365.

20. Hazard L, Yang G, McAleer MF *et al.* (2008) Principles and techniques of radiation therapy for esophageal and gastroesophageal junction cancers. *J Natl Compr Canc Netw* **6**(9): 870–878.

21. Harrison LB, Chadha M, Hill RJ *et al.* (2002) Impact of tumor hypoxia and anemia on radiation therapy outcomes. *Oncologist* **7**(6): 492–508.

22. Hirsch J. (2006) An anniversary for cancer chemotherapy. *Jama* **296**(12): 1518–1520.

23. Hannun YA. (1997) Apoptosis and the dilemma of cancer chemotherapy. *Blood* **89**(6): 1845–1853.

24. Kaufmann SH. (1989) Induction of endonucleolytic DNA cleavage in human acute myelogenous leukemia cells by etoposide, camptothecin, and other cytotoxic anticancer drugs: A cautionary note. *Cancer Res* **49**(21): 5870–5878.

25. Walker PR, Smith C, Youdale T *et al.* (1991) Topoisomerase II-reactive chemotherapeutic drugs induce apoptosis in thymocytes. *Cancer Res* **51**(4): 1078–1085.

26. Zhukov NV, Tjulandin SA. (2008) Targeted therapy in the treatment of solid tumors: Practice contradicts theory. *Biochemistry (Mosc)* **73**(5): 605–618.

27. Green MR. (2004) Targeting targeted therapy. *N Engl J Med* **350**(21): 2191–2193.

28. Melisi D, Troiani T, Damiano V *et al.* (2004) Therapeutic integration of signal transduction targeting agents and conventional anti-cancer treatments. *Endocr Relat Cancer* **11**(1): 51–68.

29. Huang T, Chang H, Hsu C *et al.* (2008) Targeting therapy for breast carcinoma by ATP synthase inhibitor aurovertin B. *J Proteome Res* **7**(4): 1433–1444.

30. Soker S, Takashima S, Miao H *et al.* (1998) Neuropilin-1 is expressed by endothelial and tumor cells as an isoform-specific receptor for vascular endothelial growth factor. *Cell* **92**(6): 735–745.
31. Yoshida S, Ono M, Shono T *et al.* (1997) Involvement of interleukin-8, vascular endothelial growth factor, and basic fibroblast growth factor in tumor necrosis factor alpha-dependent angiogenesis. *Mol Cell Biol* **17**(7): 4015.
32. Sordella R, Bell D, Haber D *et al.* (2004) Gefitinib-sensitizing EGFR mutations in lung cancer activate anti-apoptotic pathways. *Science* **305**(5687): 1163.
33. Chang F, Lee J, Navolanic P *et al.* (2003) Involvement of PI3K/Akt pathway in cell cycle progression, apoptosis, and neoplastic transformation: A target for cancer chemotherapy. *Leukemia* **17**(3): 590–603.
34. Reed J. (1994) Bcl-2 and the regulation of programmed cell death. *J Cell Biol* **124**(1): 1.
35. Tsuruo T, Naito M, Tomida A *et al.* (2003) Molecular targeting therapy of cancer: Drug resistance, apoptosis and survival signal. *Cancer Sci* **94**(1): 15–21.
36. Merlo LM, Pepper JW, Reid BJ *et al.* (2006) Cancer as an evolutionary and ecological process. *Nat Rev Cancer* **6**(12): 924–935.
37. Blume-Jensen P, Hunter T. (2001) Oncogenic kinase signalling. *Nature* **411**(6835): 355–365.
38. Gschwind A, Fischer OM, Ullrich A. (2004) The discovery of receptor tyrosine kinases: Targets for cancer therapy. *Nat Rev Cancer* **4**(5): 361–370.
39. Arora A, Scholar EM. (2005) Role of tyrosine kinase inhibitors in cancer therapy. *J Pharmacol Exp Ther* **315**(3): 971–979.
40. Ellis PM, Morzycki W, Melosky B *et al.* (2009) The role of the epidermal growth factor receptor tyrosine kinase inhibitors as therapy for advanced, metastatic, and recurrent non-small-cell lung cancer: A Canadian national consensus statement. *Curr Oncol* **16**(1): 27–48.
41. Shao YY, Lin CC, Yang CH. (2010) Gefitinib or erlotinib in the treatment of advanced non-small cell lung cancer. *Discov Med* **9**(49): 538–545.
42. Burris HA, 3rd (2004) Dual kinase inhibition in the treatment of breast cancer: Initial experience with the EGFR/ErbB-2 inhibitor lapatinib. *Oncologist* **9**(Suppl 3): 10–15.
43. Higa GM, Abraham J. (2007) Lapatinib in the treatment of breast cancer. *Expert Rev Anticancer Ther* **7**(9): 1183–1192.
44. Droogendijk HJ, Kluin-Nelemans HJ, van Doormaal JJ *et al.* (2006) Imatinib mesylate in the treatment of systemic mastocytosis: A phase II trial. *Cancer* **107**(2): 345–351.
45. Wilhelm SM, Adnane L, Newell P *et al.* (2008) Preclinical overview of sorafenib, a multikinase inhibitor that targets both RAF and VEGF and PDGF receptor tyrosine kinase signaling. *Mol Cancer Ther* **7**(10): 3129–3140.
46. Hiles JJ, Kolesar JM. (2008) Role of sunitinib and sorafenib in the treatment of metastatic renal cell carcinoma. *Am J Health Syst Pharm* **65**(2): 123–131.

47. Hudis CA. (2007) Trastuzumab — mechanism of action and use in clinical practice. *N Engl J Med* **357**(1): 39–51.

48. LoPiccolo J, Blumenthal GM, Bernstein WB *et al.* (2008) Targeting the PI3K/Akt/mTOR pathway: Effective combinations and clinical considerations. *Drug Resist Updat* **11**(1–2): 32–50.

49. Dancey JE. (2006) Therapeutic targets: MTOR and related pathways. *Cancer Biol Ther* **5**(9): 1065–1073.

50. Rubio-Viqueira B, Hidalgo M. (2006) Targeting mTOR for cancer treatment. *Curr Opin Investig Drugs* **7**(6): 501–512.

51. Hudes G, Carducci M, Tomczak P *et al.* (2007) Temsirolimus, interferon alfa, or both for advanced renal-cell carcinoma. *N Engl J Med* **356**(22): 2271–2281.

52. Xu Q, Thompson JE, Carroll M. (2005) mTOR regulates cell survival after etoposide treatment in primary AML cells. *Blood* **106**(13): 4261–4268.

53. Yang XJ, Seto E. (2007) HATs and HDACs: From structure, function and regulation to novel strategies for therapy and prevention. *Oncogene* **26**(37): 5310–5318.

54. Lane AA, Chabner BA. (2009) Histone deacetylase inhibitors in cancer therapy. *J Clin Oncol* **27**(32): 5459–5468.

55. Gloghini A, Buglio D, Khaskhely NM *et al.* (2009) Expression of histone deacetylases in lymphoma: Implication for the development of selective inhibitors. *Br J Haematol* **147**(4): 515–525.

56. Blackburn EH, Greider CW, Szostak JW. (2006) Telomeres and telomerase: The path from maize, Tetrahymena and yeast to human cancer and aging. *Nat Med* **12**(10): 1133–1138.

57. Artandi SE, DePinho RA. (2010) Telomeres and telomerase in cancer. *Carcinogenesis* **31**(1): 9–18.

58. Kyte JA. (2009) Cancer vaccination with telomerase peptide GV1001. *Expert Opin Investig Drugs* **18**(5): 687–694.

59. Brunsvig PF, Aamdal S, Gjertsen MK *et al.* (2006) Telomerase peptide vaccination: A phase I/II study in patients with non-small cell lung cancer. *Cancer Immunol Immunother* **55**(12): 1553–1564.

60. Bernhardt SL, Gjertsen MK, Trachsel S *et al.* (2006) Telomerase peptide vaccination of patients with non-resectable pancreatic cancer: A dose escalating phase I/II study. *Br J Cancer* **95**(11): 1474–1482.

61. Harley CB. (2008) Telomerase and cancer therapeutics. *Nat Rev Cancer* **8**(3): 167–179.

62. Glickman MH, Ciechanover A. (2002) The ubiquitin-proteasome proteolytic pathway: Destruction for the sake of construction. *Physiol Rev* **82**(2): 373–428.

63. Di Napoli M, McLaughlin B. (2005) The ubiquitin-proteasome system as a drug target in cerebrovascular disease: Therapeutic potential of proteasome inhibitors. *Curr Opin Investig Drugs* **6**(7): 686–699.

64. Voorhees PM, Dees EC, O'Neil B *et al.* (2003) The proteasome as a target for cancer therapy. *Clin Cancer Res* **9**(17): 6316–6325.

65. Csermely P, Schnaider T, Soti C *et al.* (1998) The 90-kDa molecular chaperone family: Structure, function, and clinical applications. A comprehensive review. *Pharmacol Ther* **79**(2): 129–168.
66. Waza M, Adachi H, Katsuno M *et al.* (2005) 17-AAG, an Hsp90 inhibitor, ameliorates polyglutamine-mediated motor neuron degeneration. *Nat Med* **11**(10): 1088–1095.
67. Bartel DP. (2004) MicroRNAs: Genomics, biogenesis, mechanism, and function. *Cell* **116**(2): 281–297.
68. Hammond SM. (2006) RNAi, microRNAs, and human disease. *Cancer Chemother Pharmacol* **58**(Suppl 1): s63–s68.
69. Calin GA, Dumitru CD, Shimizu M *et al.* (2002) Frequent deletions and down-regulation of micro-RNA genes miR15 and miR16 at 13q14 in chronic lymphocytic leukemia. *Proc Natl Acad Sci USA* **99**(24): 15524–15529.
70. Cimmino A, Calin GA, Fabbri M *et al.* (2005) miR-15 and miR-16 induce apoptosis by targeting BCL2. *Proc Natl Acad Sci USA* **102**(39): 13944–13949.
71. Esquela-Kerscher A, Slack FJ. (2006) Oncomirs — microRNAs with a role in cancer. *Nat Rev Cancer* **6**(4): 259–269.
72. Blenkiron C, Miska EA. (2007) miRNAs in cancer: Approaches, aetiology, diagnostics and therapy. *Hum Mol Genet* **16**(Spec No 1): R106–R113.
73. Spellman P, Sherlock G, Zhang M *et al.* (1998) Comprehensive identification of cell cycle-regulated genes of the yeast *Saccharomyces cerevisiae* by microarray hybridization. *Mole Biol Cell* **9**(12): 3273.
74. Hartwell LH, Kastan MB. (1994) Cell cycle control and cancer. *Science* **266**(5192): 1821–1828.
75. Hartwell L, Weinert T. (1989) Checkpoints: Controls that ensure the order of cell cycle events. *Science (Washington)* **246**(4930): 629–629.
76. Kuerbitz S, Plunkett B, Walsh W *et al.* (1992) Wild-type p53 is a cell cycle checkpoint determinant following irradiation. *Proc Nat Acad Sci* **89**(16): 7491.
77. Matsuoka S, Huang M, Elledge S. (1998) Linkage of ATM to cell cycle regulation by the Chk2 protein kinase. *Science* **282**(5395): 1893.
78. Taylor W, Stark G. (2001) Regulation of the G2/M transition by p53. *Oncogene* **20**(15): 1803–1815.
79. Kastan M, Bartek J. (2004) Cell-cycle checkpoints and cancer. *Nature* **432**(7015): 316–323.
80. Sherr C. (2001) The INK4a/ARF network in tumor suppression. *Nature Rev Mole Cell Biol* **2**(10): 731–737.
81. Sherr C. (2000) The Pezcoller lecture: Cancer cell cycles revisited. *Cancer Res* **60**(14): 3689.
82. Kamb A, Gruis N, Weaver-Feldhaus J *et al.* (1994) A cell cycle regulator potentially involved in genesis of many tumor types. *Science* **264**(5157): 436.
83. Danial N, Korsmeyer S. (2004) Cell Death: Critical Control Points. *Cell* **116**(2): 205–219.

84. Ambrosini G, Adida C, Altieri D. (1997) A novel anti-apoptosis gene, survivin, expressed in cancer and lymphoma. *Nature Med* **3**(8): 917–921.

85. Yuan J, Shaham S, Ledoux S *et al.* (1993) The C. elegans cell death gene ced-3 encodes a protein similar to mammalian interleukin-1 [beta]-converting enzyme. *Cell* **75**(4): 641–652.

86. Ellis H, Horvitz H. (1986) Genetic control of programmed cell death in the nematode C. elegans. *Cell* **44**(6): 817–829.

87. Sulston J, Schierenberg E, White J *et al.* (1983) The embryonic cell lineage of the nematode Caenorhabditis elegans. *Dev Biol* **100**(1): 64–119.

88. Fulda S, Debatin K. (2006) Extrinsic versus intrinsic apoptosis pathways in anticancer chemotherapy. *Oncogene* **25**(34): 4798–4811.

89. Wolf B, Green D. (1999) Suicidal tendencies: Apoptotic cell death by caspase family proteinases. *J Biol Chem* **274**(29): 20049.

90. Yuan J, Yankner B. (2000) Apoptosis in the nervous system. *Nature* **407**(6805): 802–809.

91. Chi K, Gleave M, Klasa R *et al.* (2001) A phase I dose-finding study of combined treatment with an antisense Bcl-2 oligonucleotide (Genasense) and mitoxantrone in patients with metastatic hormone-refractory prostate cancer. *Clin Cancer Res* **7**(12): 3920.

92. Letai A. (2005) Pharmacological manipulation of Bcl-2 family members to control cell death. *J Clin Invest* **115**(10): 2648–2655.

93. Yoo G, Hung M, Lopez-Berestein G *et al.* (2001) Phase I trial of intratumoral liposome E1A gene therapy in patients with recurrent breast and head and neck cancer. *Clin Cancer Res* **7**(5): 1237.

94. Nicholson D. (2000) From bench to clinic with apoptosis-based therapeutic agents. *Nature* **407**(6805): 810–816.

95. Merritt J, Roth J, Logothetis C. (2001) In: *Clinical Evaluation of Adenoviral-Mediated p53 Gene Transfer: Review of INGN 201 Studies*, pp. 105–114. Elsevier.

96. Heise C, Sampson-Johannes A, Williams A *et al.* (1997) ONYX-015, an E1B gene-attenuated adenovirus, causes tumor-specific cytolysis and antitumoral efficacy that can be augmented by standard chemotherapeutic agents. *Nature Med* **3**(6): 639–645.

97. Nielsen L, Shi B, Hajian G *et al.* (1999) Combination therapy with the farnesyl protein transferase inhibitor SCH66336 and SCH58500 (p53 adenovirus) in preclinical cancer models. *Cancer Res* **59**(23): 5896.

98. Hanahan D, Folkman J. (1996) Patterns and emerging mechanisms of the angiogenic switch during tumorigenesis. *Cell* **86**(3): 353–364.

99. Carmeliet P, Jain R. (2000) Angiogenesis in cancer and other diseases. *Nature* **407**(6801): 249–257.

100. Relf M, LeJeune S, Scott P *et al.* (1997) Expression of the angiogenic factors vascular endothelial cell growth factor, acidic and basic fibroblast growth factor, tumor growth factor {beta}-1, platelet-derived endothelial cell growth factor, placenta growth factor, and pleiotrophin in human primary breast cancer and its relation to angiogenesis. *Cancer Res* **57**(5): 963.

101. O'Reilly M, Boehm T, Shing Y *et al.* (1997) Endostatin: An endogenous inhibitor of angiogenesis and tumor growth. *Cell* **88**(2): 277–285.
102. O'Reilly M, Holmgren L, Shing Y *et al.* (1994) Angiostatin: A novel angiogenesis inhibitor that mediates the suppression of metastases by a Lewis lung carcinoma. *Cell* **79**(2): 315–328.
103. Avery R, Pieramici D, Rabena M *et al.* (2006) Intravitreal bevacizumab (Avastin) for neovascular age-related macular degeneration. *Ophthalmology* **113**(3): 363–372.
104. D'Amato R, Loughnan M, Flynn E *et al.* (1994) Thalidomide is an inhibitor of angiogenesis. *Proc Nat Acad Sci USA* **91**(9): 4082.
105. Liotta L, Steeg P, Stetler-Stevenson W. (1991) Cancer metastasis and angiogenesis: An imbalance of positive and negative regulation. *Cell* **64**(2): 327–336.
106. Ramaswamy S, Ross K, Lander E *et al.* (2002) A molecular signature of metastasis in primary solid tumors. *Nat Genet* **33**(1): 49–54.
107. Zetter BR. (1998) Angiogenesis and tumor metastasis. *Ann Rev Med* **49**: 407–424.
108. Keshet E, Ben-Sasson S. (1999) Anticancer drug targets: Approaching angiogenesis. *J Clin Invest* **104**(11): 1497–1501.
109. Rak J, St Croix B, Kerbel R. (1995) Consequences of angiogenesis for tumor progression, metastasis and cancer therapy. *Anti-Cancer Drugs* **6**(1): 3.
110. Jones R, Gordus A, Krall J *et al.* (2005) A quantitative protein interaction network for the ErbB receptors using protein microarrays. *Nature* **439**(7073): 168–174.
111. Rual J, Venkatesan K, Hao T *et al.* (2005) Towards a proteome-scale map of the human protein–protein interaction network. *Nature* **437**(7062): 1173–1178.
112. Olayioye M, Neve R, Lane H *et al.* (2000) NEW EMBO MEMBERS'REVIEW: The ErbB signaling network: Receptor heterodimerization in development and cancer. *Science's STKE* **19**(13): 3159.
113. Stark C, Breitkreutz B, Reguly T *et al.* (2006) BioGRID: A general repository for interaction datasets. *Nucleic Acids Res* **34**(Database Issue): D535.
114. Xenarios I, Salwinski L, Duan X *et al.* (2002) DIP, the Database of Interacting Proteins: A research tool for studying cellular networks of protein interactions. *Nucleic Acids Res* **30**(1): 303.
115. Peri S, Navarro J, Amanchy R *et al.* (2003) Development of human protein reference database as an initial platform for approaching systems biology in humans. *Genome Res* **13**(10): 2363.
116. Hermjakob H, Montecchi-Palazzi L, Lewington C *et al.* (2004) IntAct: An open source molecular interaction database. *Nucleic Acids Res* **32**(Database Issue): D452.
117. Zanzoni A, Montecchi-Palazzi L, Quondam M *et al.* (2002) MINT: A Molecular INTeraction database. *FEBS Lett* **513**(1): 135–140.
118. Sherlock G, Hernandez-Boussard T, Kasarskis A *et al.* (2001) The Stanford microarray database. *Nucleic Acids Res* **29**(1): 152.

119. Edgar R, Domrachev M, Lash A. (2002) Gene Expression Omnibus: NCBI gene expression and hybridization array data repository. *Nucleic Acids Res* **30**(1): 207.
120. Wu G, Feng X, Stein L. (2010) A human functional protein interaction network and its application to cancer data analysis. *Genome Biol* **11**(5): R53.
121. Kaelin W. (2005) The concept of synthetic lethality in the context of anticancer therapy. *Nature Rev Cancer* **5**(9): 689–698.
122. Iglehart J, Silver D. (2009) Synthetic lethality — a new direction in cancer-drug development. *N Eng J Med* **361**(2): 189.
123. Whitehurst A, Bodemann B, Cardenas J *et al.* (2007) Synthetic lethal screen identification of chemosensitizer loci in cancer cells. *Nature* **446**(7137): 815–819.
124. Fong P, Boss D, Yap T *et al.* (2009) Inhibition of poly (ADP-ribose) polymerase in tumors from BRCA mutation carriers. *N Eng J Med* **361**(2): 123–134.
125. Reya T, Morrison S, Clarke M *et al.* (2001) Stem cells, cancer, and cancer stem cells. *Nature* **414**(6859): 105–111.
126. Gottesman M, Fojo T, Bates S. (2002) Multidrug resistance in cancer: Role of ATP-dependent transporters. *Nature Rev Cancer* **2**(1): 48–58.
127. Zhou S, Schuetz J, Bunting K *et al.* (2001) The ABC transporter Bcrp1/ABCG2 is expressed in a wide variety of stem cells and is a molecular determinant of the side-population phenotype. *Nature Med* **7**(9): 1028–1034.
128. Smit J, Schinkel A, Elferink R *et al.* (1993) Homozygous disruption of the murine mdr2 P-glycoprotein gene leads to a complete absence of phospholipid from bile and to liver disease. *Cell* **75**(3): 451–462.
129. Collins A, Berry P, Hyde C *et al.* (2005) Prospective identification of tumorigenic prostate cancer stem cells. *Cancer Res* **65**(23): 10946.
130. O'Brien C, Pollett A, Gallinger S *et al.* (2006) A human colon cancer cell capable of initiating tumor growth in immunodeficient mice. *Nature* **445**(7123): 106–110.
131. Prince M, Sivanandan R, Kaczorowski A *et al.* (2007) Identification of a subpopulation of cells with cancer stem cell properties in head and neck squamous cell carcinoma. *Proc Natl Acad Sci* **104**(3): 973.
132. Dick J. (2003) Breast cancer stem cells revealed. *Proc Nat Acad Sci* **100**(7): 3547.
133. Chiba T, Kita K, Zheng Y *et al.* (2006) Side population purified from hepatocellular carcinoma cells harbors cancer stem cell-like properties. *Hepatology* **44**(1): 240–251.

Index